L. Guglielmo A. Ianora (Eds.)

Atlas of Marine Zooplankton Straits of Magellan

COPEPODS

M.G. Mazzocchi, G. Zagami, A. Ianora,
L. Guglielmo, N. Crescenti, J. Hure

With 183 Figures

Springer Verlag
Berlin Heidelberg New York London Paris
Tokyo Hong Kong Barcelona Budapest

Professor Dr. Letterio Guglielmo
Università degli Studi di Messina
Dipartimento di Biologia Animale e Ecologia Marina
98166 S. Agata
Italia

Dr. Adrianna Ianora
Stazione Zoologica A. Dohrn
80121 Napoli
Italia

Guglielmo/Ianora (Eds.):
Atlas of Marine Zooplankton / Straits of Magellan / Copepods (1995)
ISBN-13 : 978-3-642-79141-3

Atlas of Marine Zooplankton / Straits of Magellan
Amphipods, Mysids, Euphausiids, Ostracods, Chaetognaths (1996)
ISBN-13 : 978-3-642-79141-3

ISBN-13 : 978-3-642-79141-3 e-ISBN-13 : 978-3-642-79139-0
DOI: 10.1007/978-3-642-79139-0

Atlas of marine zooplankton, Straits of Magellan / L. Guglielmo, A. Ianora (eds.). p. cm. Includes bibliographical references (v. 1, p.). Contents: [1] Copepods. ISBN-13 : 978-3-642-79141-3 - 1. Marine zooplankton-Magellan, Strait of (Chile and Argentina) – sAtlases. I. Guglielmo, L. Letterio), 1945-. II. Ianora, A. (Adrianna), 1953–QL 123.A85 1995 592.09294–dc20 94-39998

Softcover reprint of the hardcover 1st edition 1995

Cover design: E. Kirchner, Heidelberg
Typesetting: Best-Set Typesetters Ltd. Hong Kong
SPIN 10468242 39/3130-5 4 3 2 1 0 - Printed on acid-free paper

**In memory
of Bruno Scotto di Carlo,
mentor and friend**

Italian Ministry of the University and Scientific and Technological Research
National Research Program for Antarctica (PNRA)

Foreword

The Antarctic represents the last of the world's still unexplored continents. Since 1985, Italy has sent nine expeditions to this region, making it one of the most scientifically active nations in the Antarctic. Italy has built a permanent base there and has begun a systematic collection of data and samples which will contribute to our understanding of this peculiar and extremely fascinating continent.

Within this framework, Italy has also conducted research in a sub-Antarctic region of great scientific interest, the Straits of Magellan. This region, with its characteristic oceanography and biota, represents a transition point between the faunal assemblages of temperate and antarctic seas. This atlas on the zooplankton of the region is an important contribution to the study of the faunal assemblages of the Straits and will hopefully be followed by others. The initiative is part of a vast program launched by the Italian Ministry of Universities and Scientific and Technological Research to promote the exploration of Antarctic and sub-Antarctic seas, the aim of which is to preserve one of the Earth's largest and most undisturbed natural laboratories.

National Scientific Co-ordinator
Ecology and Biogeochemistry of the Southern Ocean
Italian Programme for Antarctic Research

FRANCESCO FARANDA

The current need to study and understand the biodiversity of the world's oceans makes the Magellan Straits a key region for taxonomic and ecological studies. Located at the southern tip of South America, this is the longest of the Chilean fjords, connecting the Atlantic and Pacific Oceans. The Straits used to be, before the construction of the Panama Canal, the most important navigation route between these oceans. Nevertheless, its oceanography and biogeography are relatively unknown.

The Straits constitute a highly heterogeneous area and are a major link between the pelagic faunas of temperate and Antarctic regions. In this regard, this volume represents a valuable contribution to the understanding of its copepod fauna. Marine copepods are, in fact, the most abundant component of the zooplankton and play a key role in the carbon flux dynamics of the area. Also, the copepod fauna is an interesting group to study for taxonomic and evolutionary purposes. While some species, such as *Calanoides patagoniensis* are endemic to the area, others, such as *Paraeuchaeta antarctica*, have an Antarctic origin.

The close collaboration between Chilean and Italian scientists in recent years has been auspicious and will hopefully continue to contribute to the greater understanding of the biogeography of this region.

Coordinador Programa Antarctico
Universidad de Chile

VICTOR M. MARIN

Regional faunal studies have proved their usefulness time and time again. They summarize and provide easy access to the great wealth of scattered information on local biodiversity. The lack of such studies often hinders the work of researchers interested in important, but remote and inaccessible regions of the world. Charles Darwin, in his Journal of Researches, written on the voyage of H.M.S. Beagle, noted the comparative richness of the marine animal life in the waters of the Straits of Magellan and around Tierra del Fuego. However, in more than a century and a half since Darwin's visit, few detailed studies have been made of the marine animals inhabiting those cold sub-Antarctic waters. The atlas of Zooplankton of the Straits of Magellan will help to fill this gap.

The first volume of the atlas deals with the copepods, a dominant and often diverse group in the pelagic community. Knowledge of the composition and productivity of the copepod fraction of a zooplankton community is essential for the development of a predictive understanding of fishery dynamics in any area. This volume should greatly facilitate such work. It enables the identification of the most common copepod species of the region and also provides information on their distributions and abundances, as well as background data on the physical and biological environment. Its emphasis is strongly on illustrations, with only brief text descriptions, and the use of top quality scanning electron micrographs to supplement the traditional line drawings is unique; thus adding considerably to the information content and aiding interpretation of the drawings by non-specialists. This volume will serve as a sound platform for future copepod research in the region and I congratulate the authors on their innovative presentation.

Museum of Natural History
London, England

GEOFFREY A. BOXSHALL

Preface

An Atlas on the Zooplankton Assemblages of the Straits of Magellan represents one of the few attempts in recent years to describe the pelagic fauna of a sub-Antarctic region. Its primary aim is to provide a rapid and practical handbook for the easy identification of the plankton species living in the area. The idea of preparing such an Atlas also derives from the fact that much of the faunistic information for sub-Antarctic waters is very old and usually contained in rare monographs and reports. Also, since most recent taxonomic records are generally focused on genera or families, a comprehensive treatise will hopefully facilitate future studies on plankton communities in these regions.

The Atlas is intended for both experts and beginners in plankton ecology. In preparing it, we have tried to amalgamate older studies with newer ones that incorporate modern techniques. Our purpose was to preserve the descriptive format of illustrative volumes published at the turn of the century while using advanced instrumentation such as the electron microscope and computer-arranged line drawings. The format of our catalogue is extremely succinct, with brief taxonomic descriptions of females and males, when possible. More attention is given to illustrations than to descriptions since most of the species sampled within the Straits are exhaustively described elsewhere.

The first volume of the Atlas describes the copepod assemblages, based on the material collected during the course of an Italian expedition in the Straits of Magellan in 1991. Forty-five copepod species are described, 18 of which are new for the region. Detailed information is also given on their distribution and abundances, together with a series of environmental parameters sampled at the time of collection.

Acknowledgements

It is a pleasure to record our gratitude to all those who contributed to the preparation of this Atlas. These include Giorgio Dafnis for his competence and skill in the preparation of material for Scanning Electron Microscopy and for his technical assistance at the SEM microscope. Gennaro Iamunno assisted in the preparation of SEM material. Giuseppe Gargiulo printed all the SEM micrographs. Gisella Princivalli typed the manuscript. Thanks are due also to Frank Kurbjeweit for sending us samples of *Microcalanus* from the Antarctic for comparative purposes. In addition, we also wish to record our appreciation to our colleagues and to those of the crew of the R/V *Cariboo* who participated in the 1991 cruise to the Magellan Straits. Of these, special thanks are due to Pino Arena for his invaluable work at sea and help in generating the copepod distribution maps and to Vincenzo Bonanzinga for his technical assistance in data analysis. Thanks are also due to the National Antarctic Research Program (PNRA) for their financial support.

Contents

Chapter 1 Study Area

1.1 Introduction

The Straits of Magellan are a natural confluence of water masses from the Pacific and Atlantic Oceans and are closely linked to the Antarctic ecosystem (Fig. 1.1). The distance from the Pacific (Capo Deseado) to Atlantic (Punta Dungeness) entrances is about 570 km with a distance from northern to southern shores that ranges from 4–30 km. Along the main axis, different sills and shelfs separate specific basins (Fig. 1.2) that play a crucial role in driving water mass circulation and shaping current patterns. According to Antezana et al. (1992), the Straits are divided as follows:

A. EASTERN SOUTH PACIFIC: located 90 km off the entrance to the Straits (Evangelistas Island) with a depth range from 4000 m rising to 2000 m.

B. MAGELLAN SLOPE: located 50–90 km off Evangelistas Island with a depth range from 2000 m rising to 200 m. The slope is very narrow (20 km wide), bordered by a well-defined sill, 20–50 m deep, separating Pacific waters from those in the Straits. The sill is located 36 km SE of Evangelistas Island, immediately beyond Pilar Cape.

C. DEEP WESTERN ENTRANCE: located between Pilar Cape to NE of San Carlos III Islands at Paso Tortuoso, with a depth range from 50–1200 m. This basin is limited by two shallow sills and can in turn be divided, from west to east, into three sub-basins:

1. Pacific Mouth, from Pilar Cape to Tamar Island (50–500 m depth, 63 km long, 18 km wide);
2. Paso del Mar, extending from Tamar Island to Havannah Point (500–1200 m depth, 40 km long, 10 km wide);
3. Paso Largo, extending from Havannah Point to San Carlos III Island (50–700 m depth, 75 km long, 45 km wide).

D. INTERMEDIATE PASSAGE: extending from San Carlos Island to Pelicano Point (N of Punta Arenas), 50–500 m deep (mean depth of 300 m) and 200 km long. This basin is widely connected to the west via Bahia Inutil and Canal Whitehead-Seno Almirantazgo. Three sub-basins can be identified:

1. Paso Froward, extending from San Carlos Island to Cape Froward (50–400 m deep on a SE downward slope, 72 km long, 9.9 km wide);
2. Seno Magdalena, a transition area on the southern junction of eastern and western passages;
3. Paso Ancho, extending from San Isidro Cape to Pelicano Point (100–350 m deep, 90 km long on the northward slope, 27 km wide).

E. SHALLOW EASTERN ENTRANCE: extending from Pelicano Point to Dungenes Point at the Atlantic entrance to the Straits, 30–50 m deep, 200 km long. Three sub-basins are bordered by the Segunda Angostura and the Primera Angostura; this is the easternmost (Atlantic Mouth) as well as the longest and widest of the channel basins.

F. WESTERN SOUTH ATLANTIC SHELF: extending gradually beyond the entrance at Dungenes Point into the Atlantic shelf with a depth range from 50–60 m. There is no clear

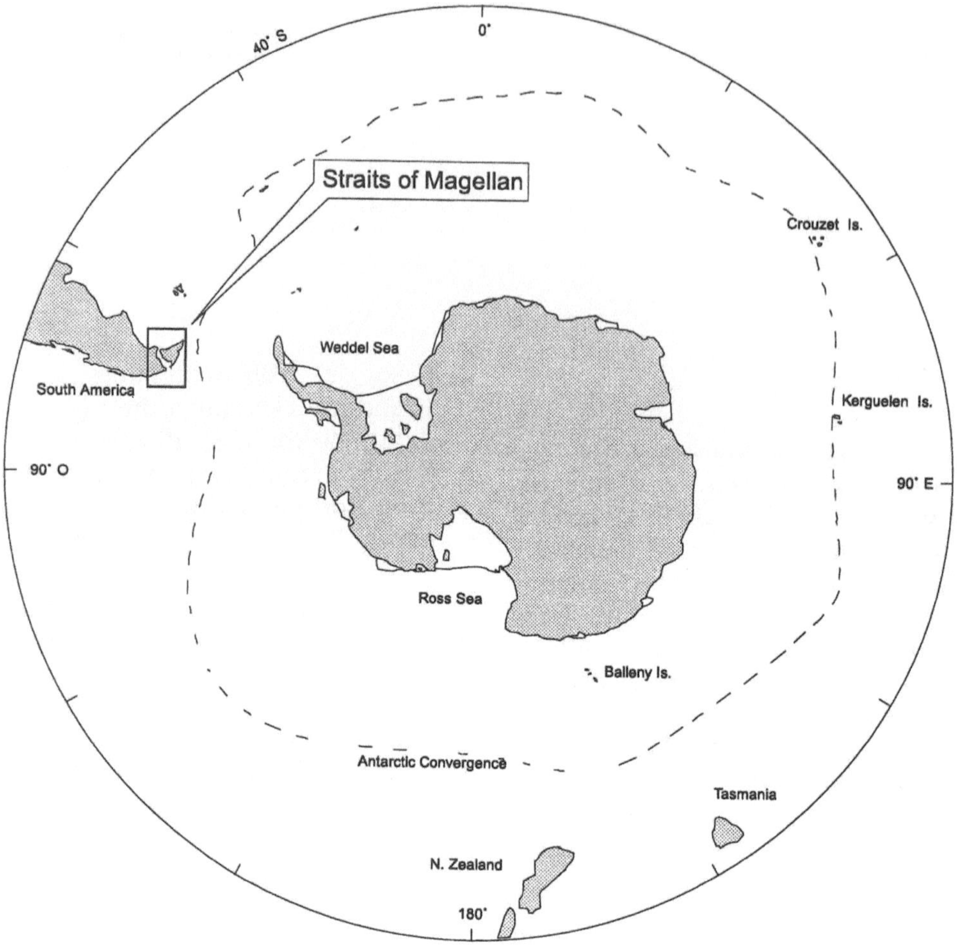

Fig. 1.1. Antarctic continent and surrounding seas

orographic or bathygraphic distinction towards the mouth of the Passage.

Below, we briefly report on the main hydrographic and biological results from the two Italian expeditions in the Straits of Magellan in the austral spring (November) 1989 and austral summer (February–March) 1991.

1.2 Hydrology

In the austral spring 1989 (Panella et al. 1991b), at the eastern entrance of the Straits, the water column was homogeneous with thermohaline characters comparable to those found in coastal waters of the corresponding Atlantic shelf (Figs. 1.3, 1.4, 1.5). This water is thought to originate from the mixing of the top layer of Intermediate Atlantic Waters coming from the south, with more diluted waters of continental origin, flowing from the Straits of Magellan and spreading over a large area of the shelf (Boltovskoy 1970; Lusquinos and Valdez 1971; Glorioso 1987). In the stretch between the Atlantic sill and Ancho Canal, water masses are vertically homogeneous as a result of the significant mixing by strong tidal currents present in this area.

Moving towards Canal Ancho, temperature and salinity ranges decrease gradually as the water mixes with inner waters of the Straits and with freshwater from the continental channel network. The dilution and

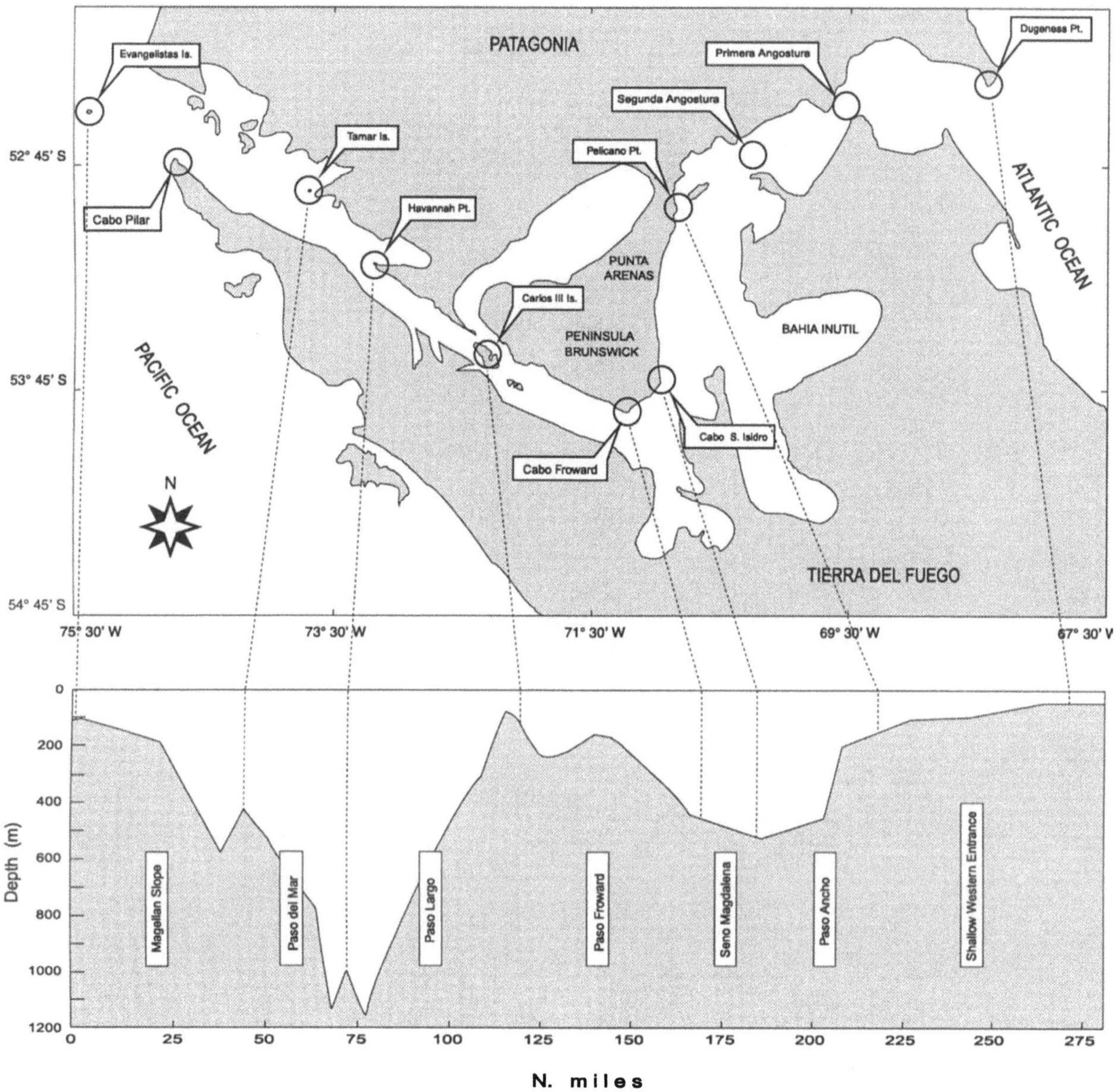

Fig. 1.2. Subdivisions of microbasins and depth profiles along the main axis within the Straits of Magellan

cooling processes become more marked in the southern section of Canal Ancho, reflecting a further input of continental water. In addition, this area is also influenced by water from the Whiteside Canal and Seno Almirantazgo which receive waters from Rio Azopardo and fluvial-glacial water from glaciers located on the northern side of Cordillera Darwin.

Near Capo Froward, where the central sub-basin of the Straits turns northwest toward the Carlos III Island sill, temperature and salinity increase slightly, particularly in the surface layer and down to the 100–200 m depth. On the whole, the vertical thermohaline structure of the water masses making up most of the central sub-basin are very similar. Intense water mass exchange takes place in the area of the sill. Near Carlo III Island, a frontal system can be observed due to strong tidal mixing resulting from the marked narrowing of the Straits. The vertical

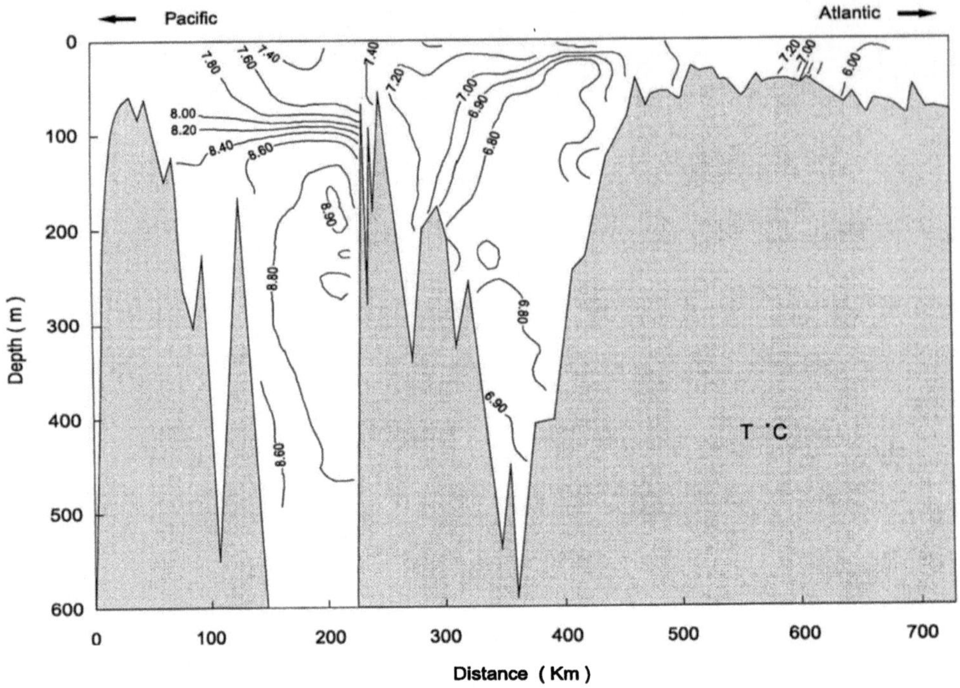

Fig. 1.3. Isotherms along the main axis of the Straits of Magellan during the austral spring 1989. (After Panella et al. 1991b)

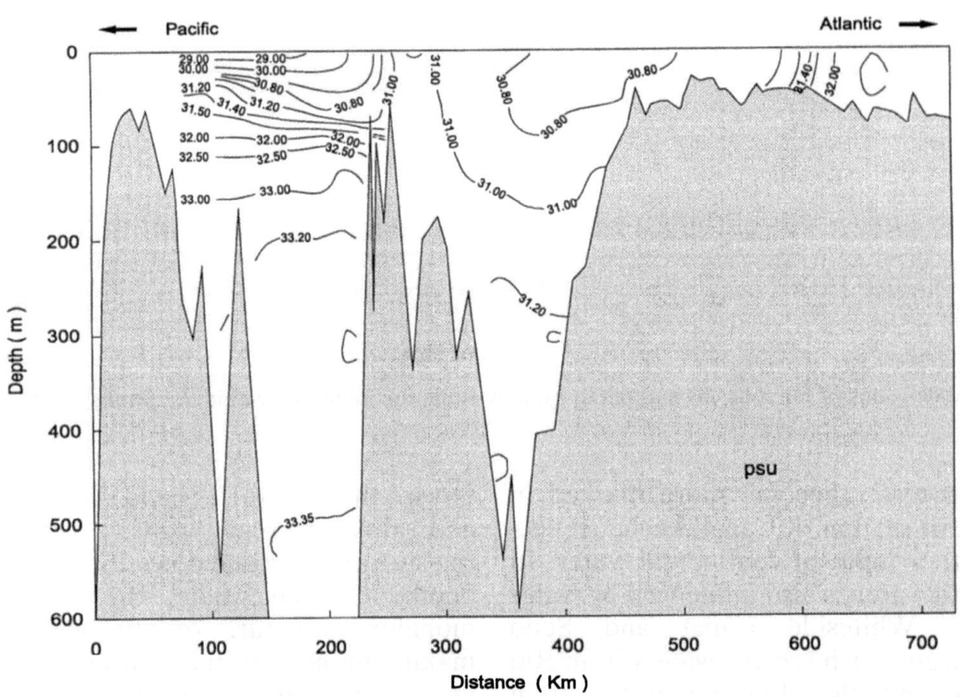

Fig. 1.4. Isohalines along the main axis of the Straits of Magellan during the austral spring 1989. (After Panella et al. 1991b)

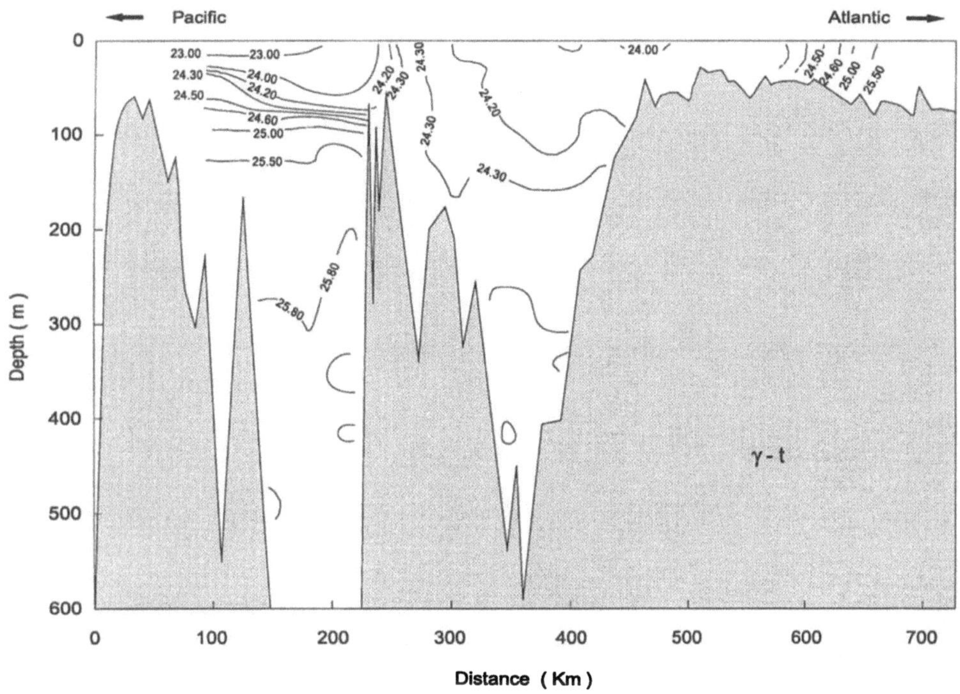

Fig. 1.5. Isopycnals along the main axis of the Straits of Magellan during the austral spring 1989. (After Panella et al. 1991b)

structure of the water masses occupying the western sub-basin beyond the sill is completely different from that of the central sector of the Straits. Colder and more diluted water masses are found in the underlying layer down to 500 m. The moderate thermocline and the well-defined halocline characterizing the top layer give rise to a stratified and stable structure.

The warmer and saltier water masses occupying the underlying layer indicate a strong Pacific influence due to the inflow of sub-Antarctic water into the sub-basin. This water, which originates north of the Polar Front, is known to travel towards the Chilean coast via the West Drift Current. Then it moves southward at 43°S latitude, carried by the southern branch of the West Drift Current. This produces the Cape Horn Current which flows towards the Strait of Drake. Near the coast, sub-Antarctic water undergoes significant surface dilution due to continental water runoff. Surface temperatures of 9–10 °C and salinities below 33.6 psu were recorded in summer in the ocean area near the Pacific entrance of the Straits (Silva and Neshyba 1977).

1.3 Tidal Currents

The most striking hydrodynamic feature of the Straits are the strong tidal currents originating from semi-diurnal tidal waves in the Atlantic Ocean (Medeiros and Kierfve 1988). The tidal regime of the Straits is macrotidal, deriving from the amplification of waves along the wide continental shelf of the Atlantic where tidal amplitude can reach a maximum of 9 m during the spring tide. Tidal amplitude within the Straits is much lower, varying from 7.8 m at the channel entrance to 1.2 m at a distance of 150 km (Fig. 1.6). A rapid diminution in the height of the tides is observed west of the Primera and Segunda Angostura. The mean and spring tide ranges drop, respectively, from 7.8 and 9.5 m on the eastern side of the Primera Angostura to 4.0 and 7.8 m immediately west of this basin. By contrast, mean and spring

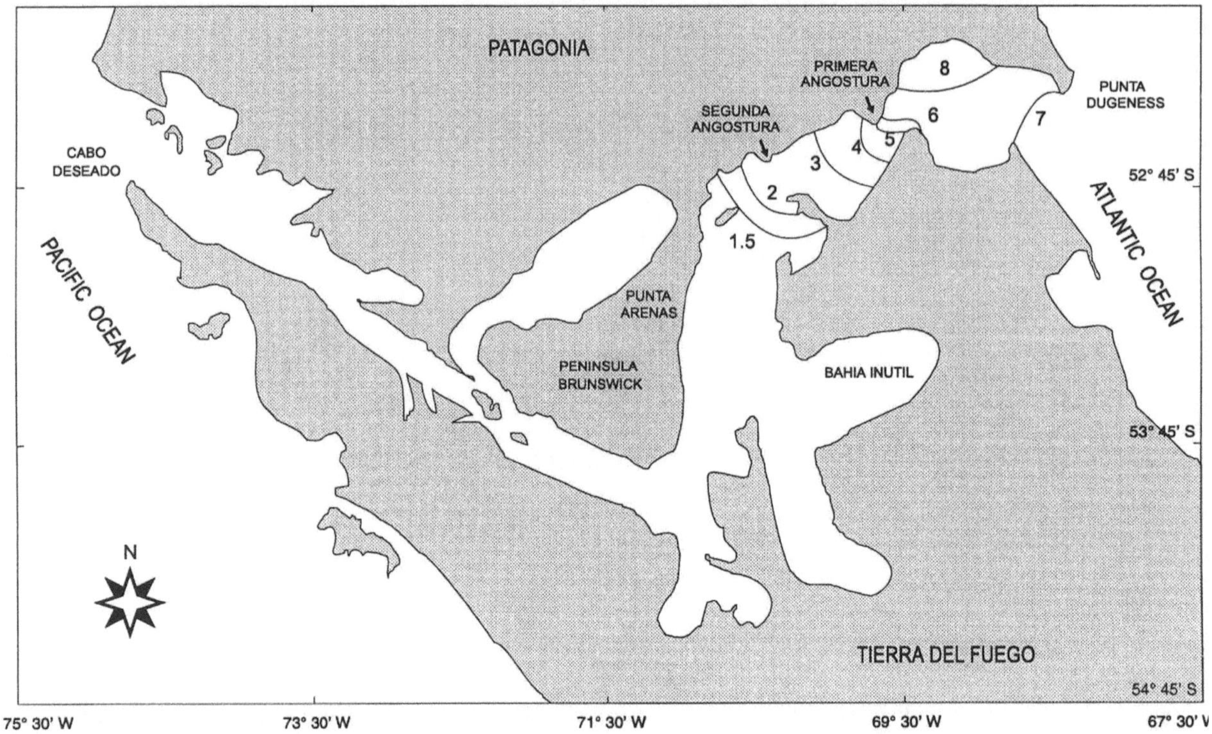

Fig. 1.6. Mean values of tidal amplitude (m) in the Straits of Magellan. (After Medeiros and Kjerfve 1988)

tide ranges are 1.2 and 1.8 m for the Segunda Angostura.

Tidal currents can reach velocities greater than 4.5 m/s in narrow passages such as the Primera Angostura and 3 m/s in the Segunda Angostura. Lower speeds are recorded at the Atlantic entrance (2.5 m/s) and in the Ancho Canal (1.9 m/s). At the Atlantic opening at a depth of 40 m, speeds of 80–100 cm/s have been measured in an east-west direction (Michelato et al. 1991). At the same time, bottom speeds are less than half these values due to friction. Maximum speeds of up to 1.25–1.30 m/s have been recorded at the 10–20 m depth between the two Angosturas. On the Pacific side of the Straits, the tidal regime is mixed, mainly semidiurnal, with speeds of 1.1–1.2 m/s.

1.4 Chemical Characteristics

The western part of the Straits, which is deeper and narrower, differs from the eastern sector due to the strong vertical gradient of chemical properties at a depth of about 100 m as reported by Catalano and Goffart (1991) for the austral summer 1991. The water column in this region differs with respect to the rest of the Straits in the generally smaller content of oxygen (Fig. 1.7) and the greater presence of nitrite plus nitrate (Fig. 1.8). The Straits therefore appear to be sub-divided by the sill of Carlos III Island into at least two large zones, each one of which is further divided into smaller zones owing to particular local phenomena that seem to be mainly related to the morphology of the area.

1.5 Photosynthetic Pigments and Primary Production

Phytoplankton pigment distributions along the entire channel are given by Panella et al. (1991a) for the austral spring 1989 when low levels (<1 μg/l) of total Chl *a* were observed in the Pacific area. The richest area, in terms

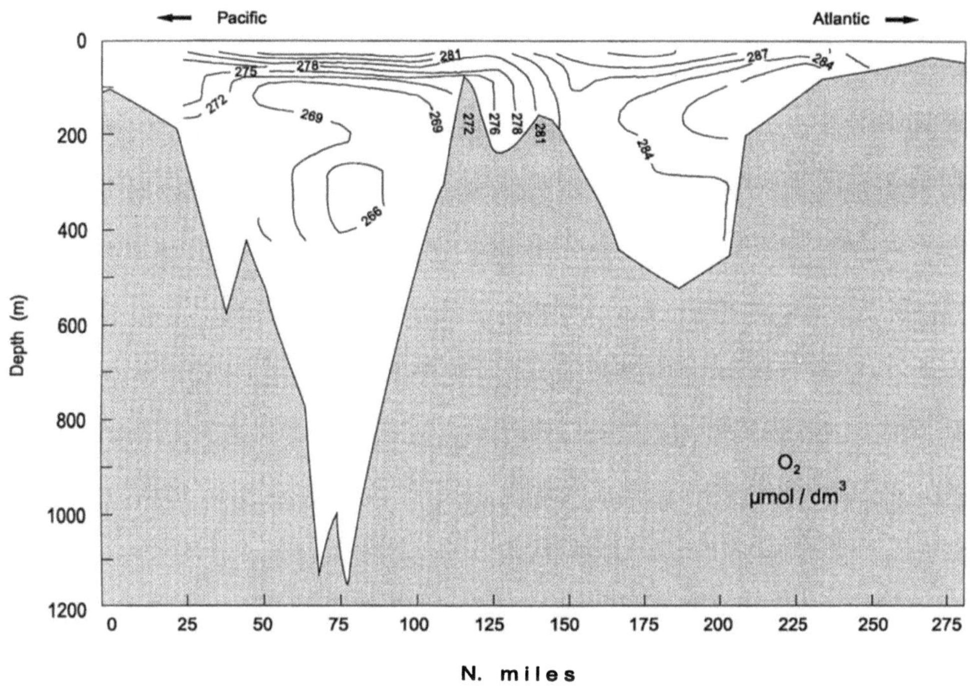

Fig. 1.7. Dissolved oxygen in the Straits of Magellan during the austral summer 1991. (After Catalano and Goffart 1991)

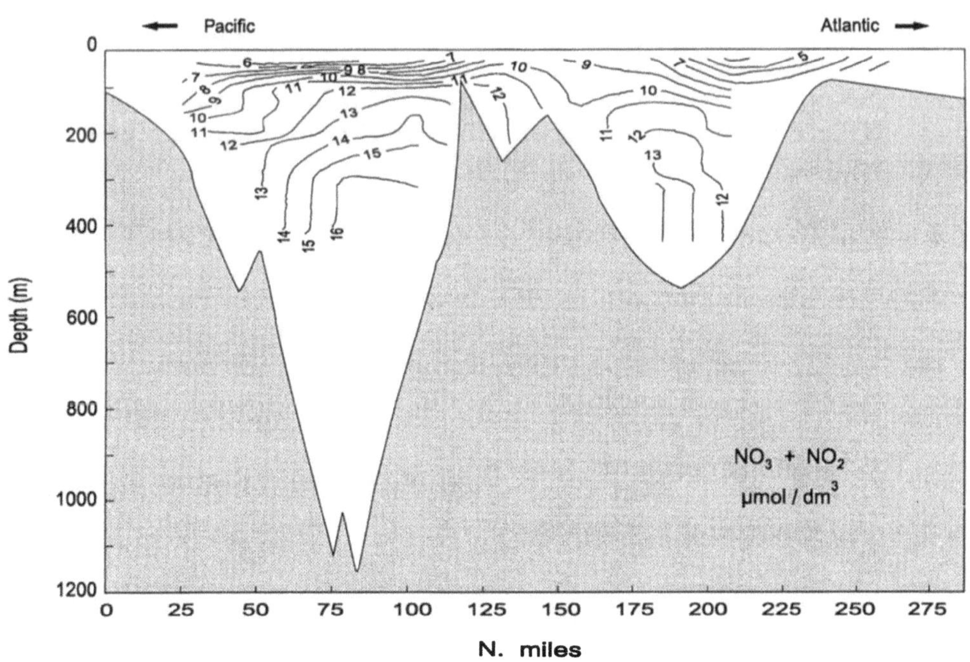

Fig. 1.8. Nitrate plus nitrite in the Straits of Magellan during the austral summer 1991. (After Catalano and Goffart 1991)

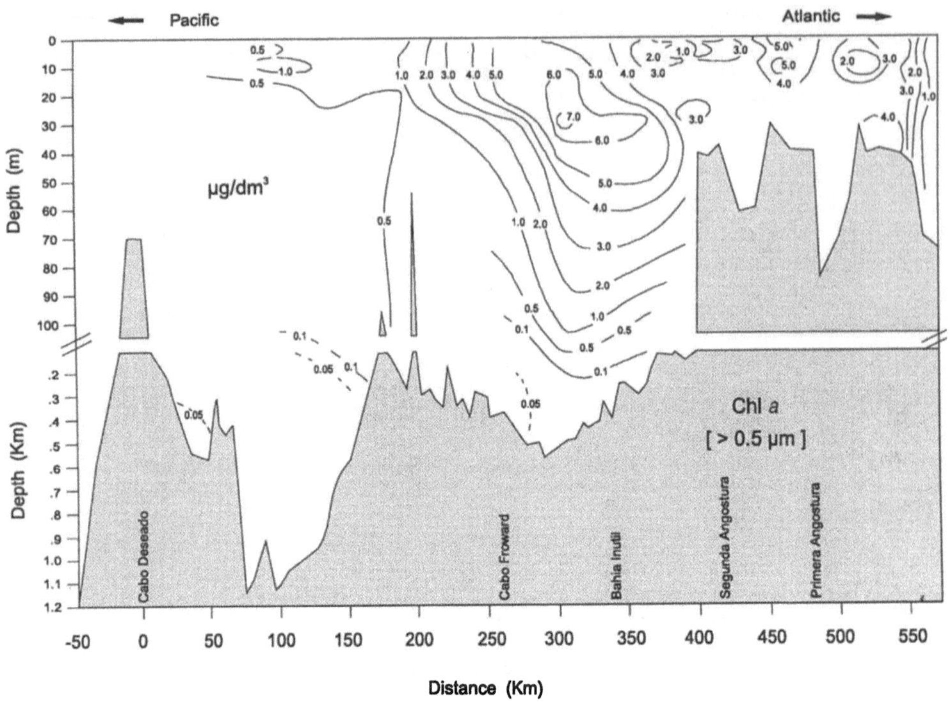

Fig. 1.9. Distribution of total chlorophyll *a* along the main axis of the Straits of Magellan during the austral spring 1989. (After Panella et al. 1991a)

of total Chl *a*, was in the Central region, from Carlos III Island to Paso Ancho. Maximum values of total Chl *a* (>7 µg/l) occurred east of Cape Froward (Fig. 1.9). Along the axis of the Straits, the highest percentage of Chl *a* was found in the micro- and nanophytoplankton fractions (>2 µm). The picophytoplankton fraction (<2 µm) showed a similar pattern to the other fractions but with much lower concentrations (<0.6 µg/l) (Fig. 1.10).

During the austral summer 1991, the richest area was that of Paso Ancho with values in excess of 2 µg/l (Fig. 1.11) (Saggiomo et al. 1994). The microphytoplankton fraction was mainly confined to the Pacific entrance and to the embayment between the two Narrows while in the rest of the Straits it was always less than 5% of the total assemblage. The nano- and picoplankton fractions occurred throughout the entire lengths of the Straits, comprising a mean of 33 and 63% of the entire phytoplankton assemblage, respectively. The integrated values of Chl *a* during the same cruise are given by Magazzù et al. (1991) (Fig. 1.12) who also report values for primary production (Fig. 1.13). In the Andean area lowest values of primary production were observed (282 mg C/m^2/d) in an extremely mixed water column (Artegiani et al. 1991) with high nutrient concentrations (Catalano and Goffart 1991). By contrast, the highest values of primary production (above 1000 mg C/m^2/d) were observed in Paso Ancho and Bahia Inutil, in a thermally stratified water column characterized by low nutrient concentrations.

1.6 Particulate Organic Matter

In the austral summer 1991, the Straits were characterized by low concentrations and large variations in quali-quantitative composition of the seston. The surface values of total suspended matter (TSM) ranged from to 0.5 mg/dm^3 in Pacific waters to 2 mg/dm^3 in the area between the two Angosturas (Fig. 1.14). High surface TSM values occurred near Paso Ancho with average values of

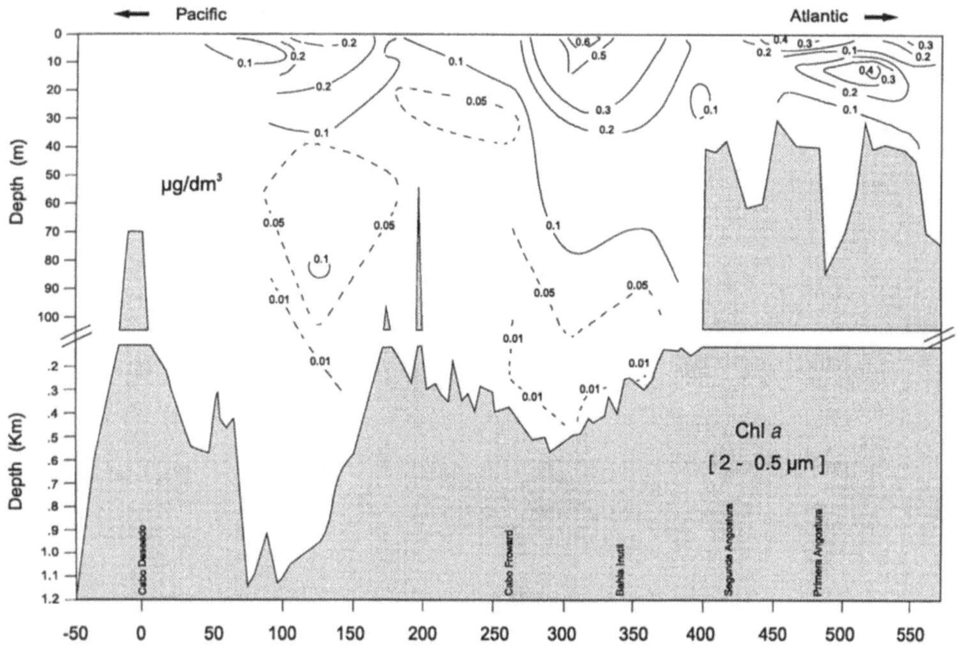

Fig. 1.10. Distribution of picoplanktonic chlorophyll *a* along the main axis of the Straits of Magellan during the austral spring 1989. (After Panella et al. 1991a)

0.81 mg/dm³. TSM concentrations below the 50 m depth were usually lower than 0.4 mg/dm³ (Povero et al. 1991).

The distribution of particulate organic carbon (POC) (Fig. 1.15) and nitrogen (PON) were similar to those of TSM. POC and PON concentrations in the stations between the Primera and Segunda Angostura (average 120.4 and 16.9 µg/dm³ and 137.1 and 19.4 µg/dm³, respectively) were less than expected based on TSM values. Inside the Straits, POC and PON surface concentrations ranged between 38–178 µg/dm³ and 6–30 µg/dm³, respectively, with maximum values in Paso Ancho. Below the 50 m depth, values ranged between 23–150 µg/dm³ for POC and 3–20 µg/dm³ for PON. Particulate organic matter (POM) distribution was similar to that of POC (Fig. 1.16).

In brief, the area close to the Pacific Ocean was characterised by low quantities of seston, mainly composed of organic matter of detrital-heterotrophic origin. The Paso Froward area was characterised by the scarcity of POM whereas in the Punta Arenas Basin large amounts of POM were related to high primary production values. Finally, large amount of seston mostly composed of inorganic matter probably due to resuspension processes were found in the area close to the Atlantic Ocean.

1.7 Phytoplankton

Information on the abundance and composition of phytoplankton populations in the Straits of Magellan are scarce and fragmentary and are mainly associated with red tide blooms (Campodonico et al. 1975; Guzman et al. 1975; Lembeye et al. 1975, 1978; Lembeye 1981; Uribe 1988a,b; Iriarte et al. 1993). A more extensive study over the entire area of the Straits was conducted in the austral spring 1989 (Cabrini and Fonda Umani 1991; Uribe 1991). At this time, phytoplankton populations consisted mainly of diatoms, microflagellates and dinoflagellates. Of these, diatoms were the main organisms both at the surface and deeper

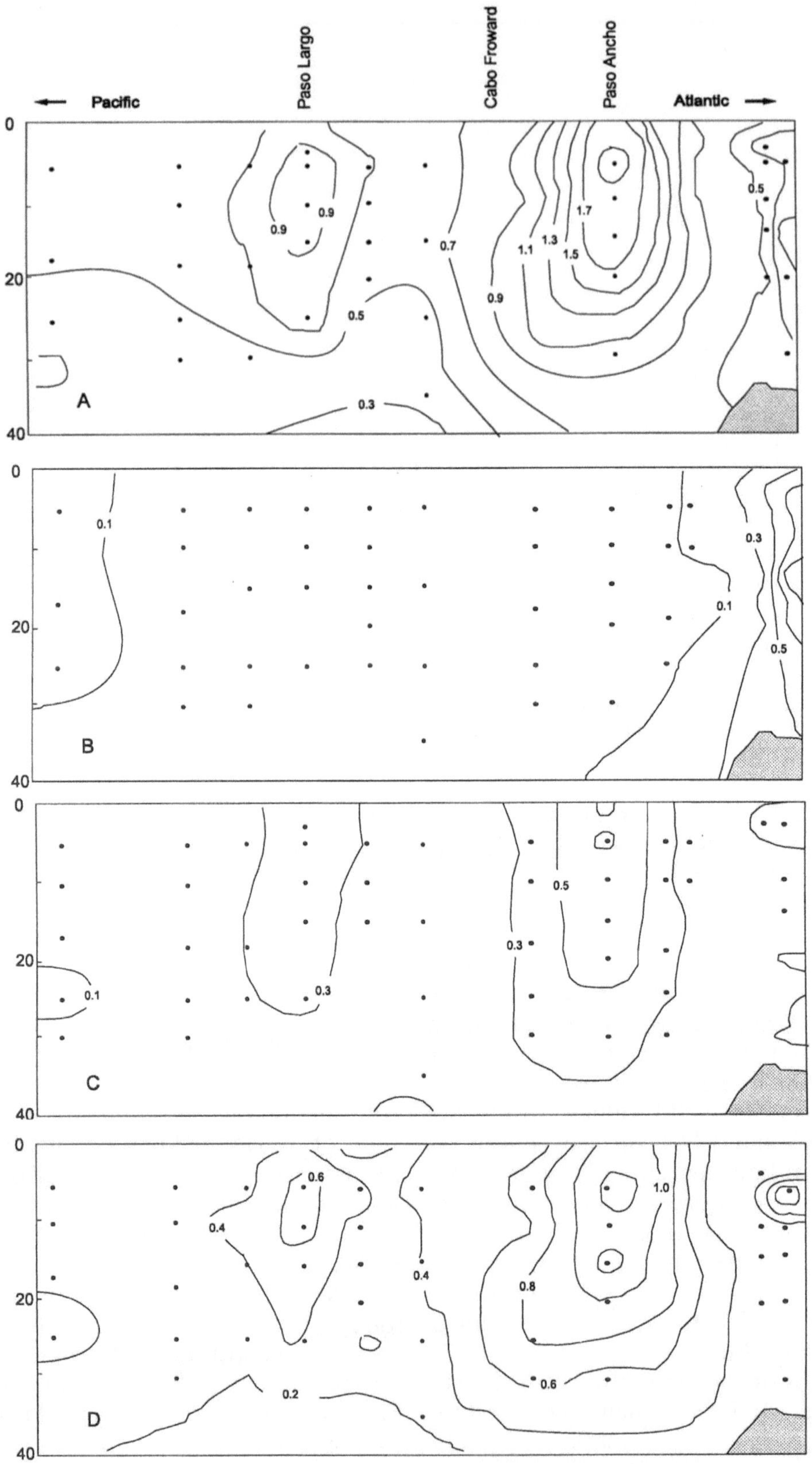

Fig. 1.11A–D. Distribution of chorophyll *a* along the main axis of the Straits of Magellan during the austral summer 1991. **A** Total phytoplankton (>0.5 μm). **B** Microphytoplankton (>10 μm). **C** Nanophytoplankton (10–2 μm). **D** Picoplankton (2–0.2 μm). (After Saggiomo et al. 1994)

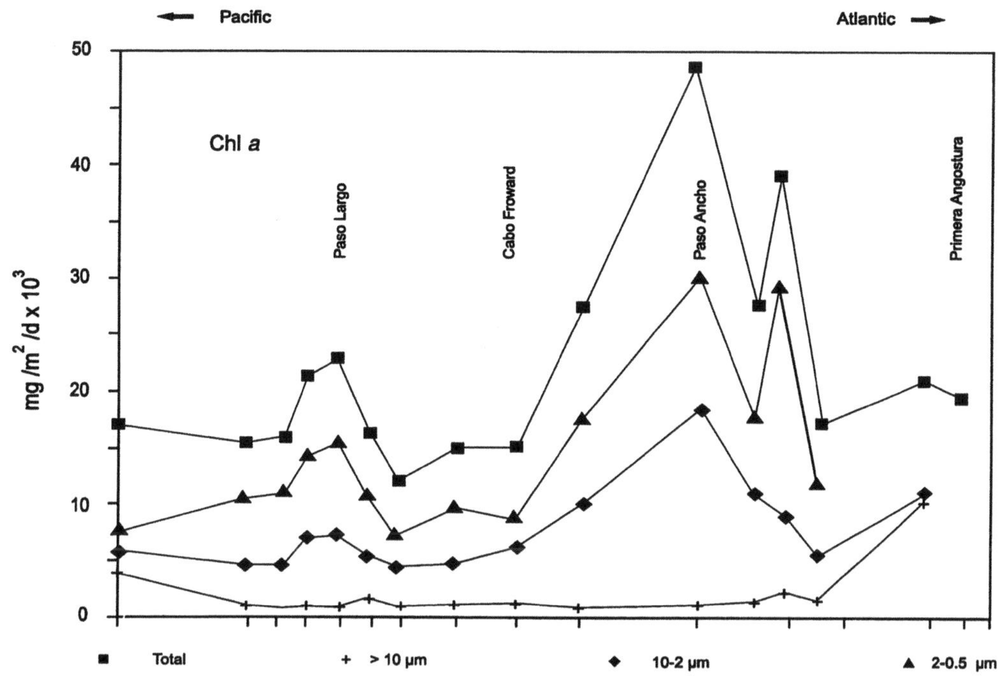

Fig. 1.12. Integrated values of total and size-fractionated chlorophyll *a* along the main axis of the Straits of Magellan during the austral summer 1991. (After Magazzù et al. 1991)

depths with abundances ranging from 7.2×10^5 cells/dm^3 to 4.5×10^6 cells/dm^3. Diatoms were particularly abundant in the Punta Arenas Basin while they were quite rare at stations at the Atlantic outlet and on the west side of the Straits. The most abundant species was *Asterionella japonica* (syn. *A. glacialis*) while the most widespread were *Thalassionema nitzschioides*, *Thalassiosira aestivalis*, *T. decipiens* and *T. rotula*. Dinoflagellates were found in most of the Straits with values ranging from 4×10^3 cells/dm^3 to 10^5 cells/dm^3. Of these, only the genera *Protoperidinium* and *Scrippsiella* were present in significant numbers.

On the basis of net-phytoplankton data, Uribe (1991) divides the Straits into three topographic basins: an eastern area from the Primera Angostura to the Atlantic Ocean; a largest, middle area from the Primera Angostura to Carlos III Island; and a third area from Carlos III Island to the Pacific Ocean. The abundance of benthic species in the first sector may be explained by the turbulence produced by tides in a very shallow basin. The occurrence of dinoflagellates in high numbers is believed to reflect different hydrographic conditions from Isla Carlos III to the Pacific Ocean.

During the austral summer 1991 (Marino et al. 1991), cell numbers ranged from 17.2×10^4 cells/l in the Atlantic Ocean to 12.5×10^6 cells/l in the Punta Arenas Basin. Four main areas were identified on the basis of phytoplankton abundances, and biomass and species composition (Fig. 1.17). At the Pacific entrance, where low cell concentrations were recorded, small coccoid cells, coccolithophorids, and small flagellates constituted the bulk of the population. In the area between Cabo Deseado and Cabo Froward, notably higher cell concentrations were recorded and small coccoid cells markedly dominated the population. In Punta Arenas Basin and Bahia Inutil, where cell numbers reached maximum values, small coccoid cells were still dominant in term of cell numbers and small-sized diatoms reached high concentrations. *Lennoxia faveolata*, a recently described centric diatom

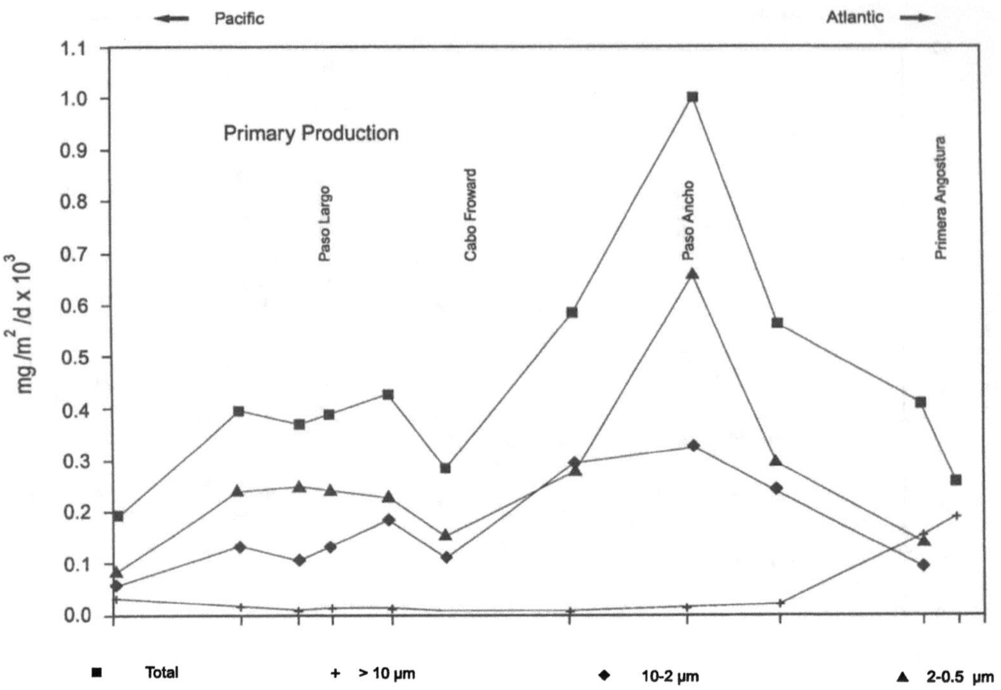

Fig. 1.13. Integrated values of primary production along the main axis of the Straits of Magellan during the austral summer 1991. (After Magazzù et al. 1991)

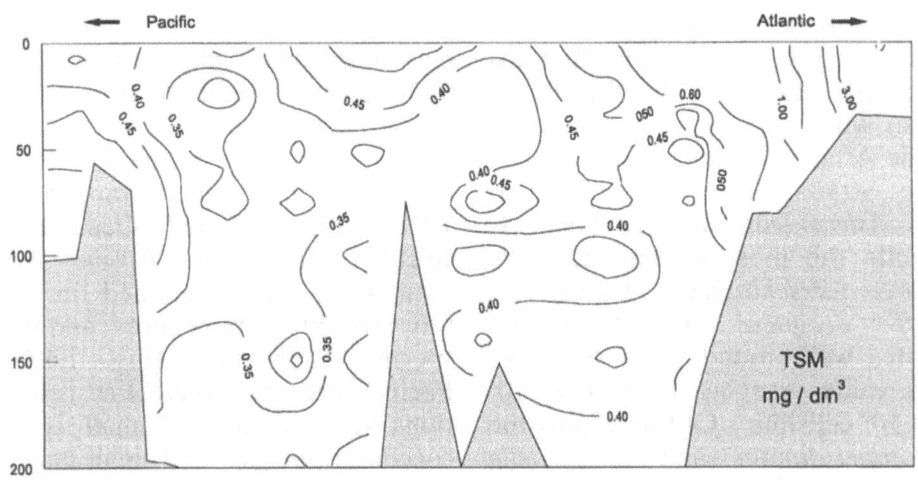

Fig. 1.14. Distribution of seston (*TSM*) in the Straits of Magellan during the austral summer 1991. (After Povero et al. 1991)

(Thomsen et al. 1993), was particularly abundant in this area. A fourth subregion was identified in the area beyond the Segunda Angostura, influenced by coastal Atlantic waters and exposed to a strong macrotidal regime. In this area, cell concentrations dropped to values below 10^6 cells/l, and coccolithophorids and flagellates replaced small coccoid cells in abundance. These results indicate that in the late summer of 1991, the Straits of Magellan, from the Pacific entrance to the Segunda Angostura,

Fig. 1.15. Distribution of the organic fraction of seston (*POC*) in the Straits of Magellan in the austral summer 1991. (After Povero et al. 1991)

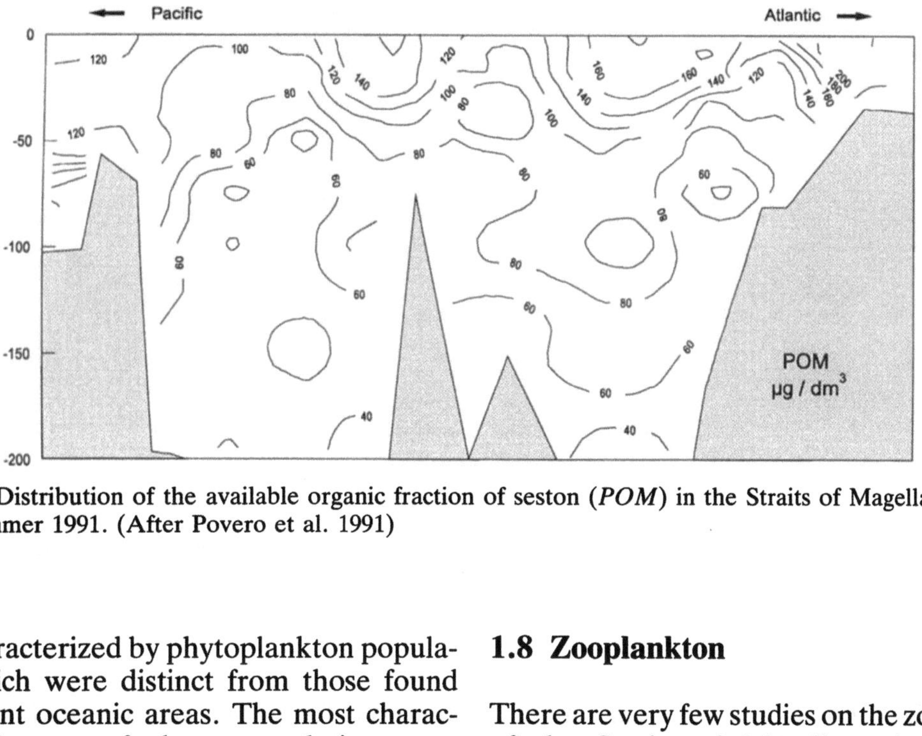

Fig. 1.16. Distribution of the available organic fraction of seston (*POM*) in the Straits of Magellan during the austral summer 1991. (After Povero et al. 1991)

were characterized by phytoplankton populations which were distinct from those found in adjacent oceanic areas. The most characteristic feature of these populations was the numerical dominance and relative importance, in terms of biomass, of very small (<5 µm) eukaryotic algae. Among these, an important role was played by non-flagellate species, such as very small coccoid cells and diatoms.

1.8 Zooplankton

There are very few studies on the zooplankton of the Straits of Magellan. Arcos (1974) identified the Calanoid copepods of the Magellan region along the southern coast of Chile but not within the Straits proper. Mazzocchi and Ianora (1991) were the first to describe the copepod assemblages of the Straits of Magellan. They identified a total of 26 species during the austral spring 1989. Lowest values were recorded in the Atlantic

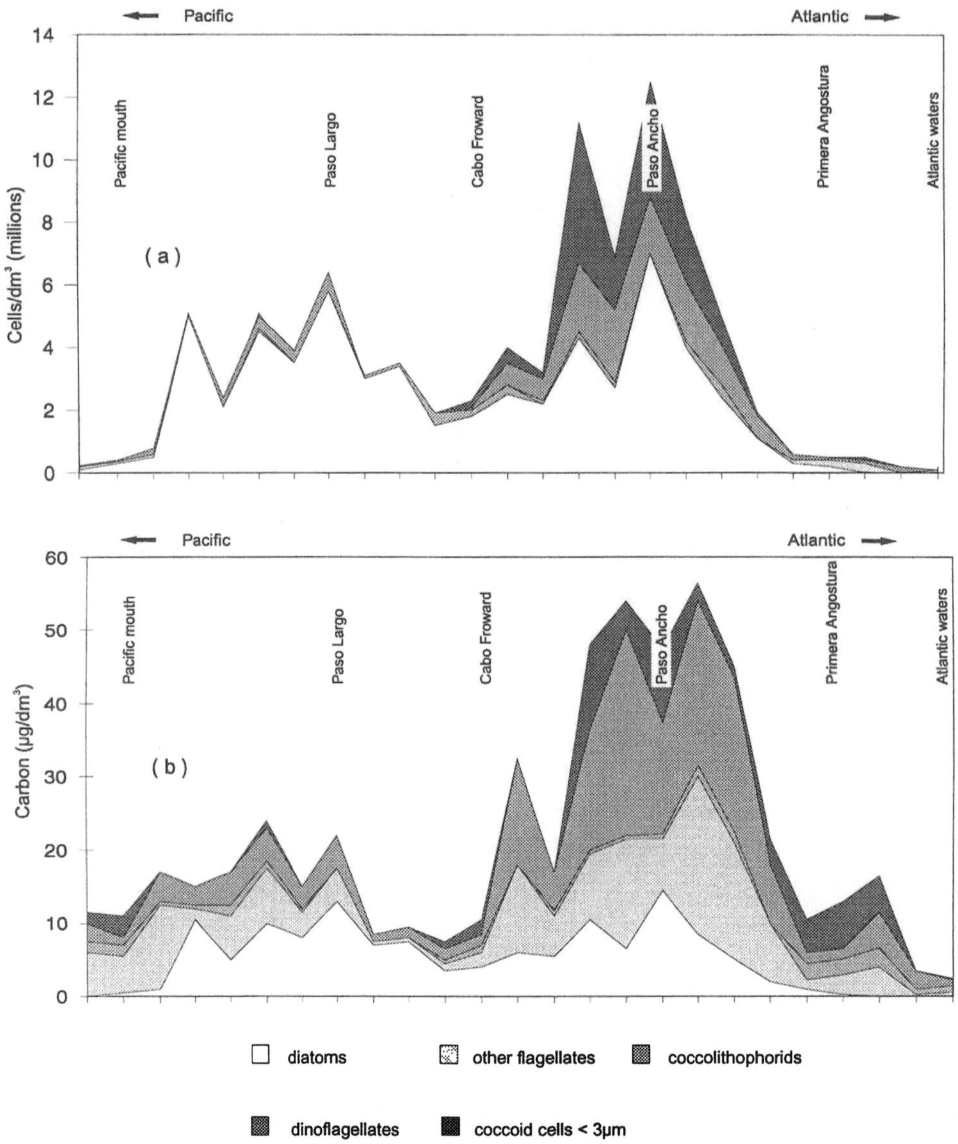

Fig. 1.17a,b. Phytoplankton abundances (**a**) and biomass values (**b**) at 25 surface stations in the Straits of Magellan during the austral summer 1991. (After Marino et al. 1991)

sector (124.4 ind. m^{-3}) whereas highest values occurred in the central and Pacific areas (5995 ind. m^{-3}). The species identified for the Straits included those that are typical of sub-Antarctic areas (e.g. *Calanus simillimus*, *Ctenocalanus citer*, *Drepanopus forcipatus*, *Clausocalanus brevipes*, *Candacia cheirura*) and those with a more cosmopolitan distribution (e.g. *Oithona similis*, *Paracalanus parvus*, *Acartia tonsa*) that are widely reported for neritic regions of the Southern and Northern hemispheres.

A limited number of species constituted the bulk of the copepod assemblage at all stations sampled (Fig. 1.18). *O. similis*, *C. citer* and *D. forcipatus* were the dominant species, together comprising about 85% of total copepod numbers. Of these, *O. similis* was the most abundant, comprising mean values of 40.2% of all copepods sampled.

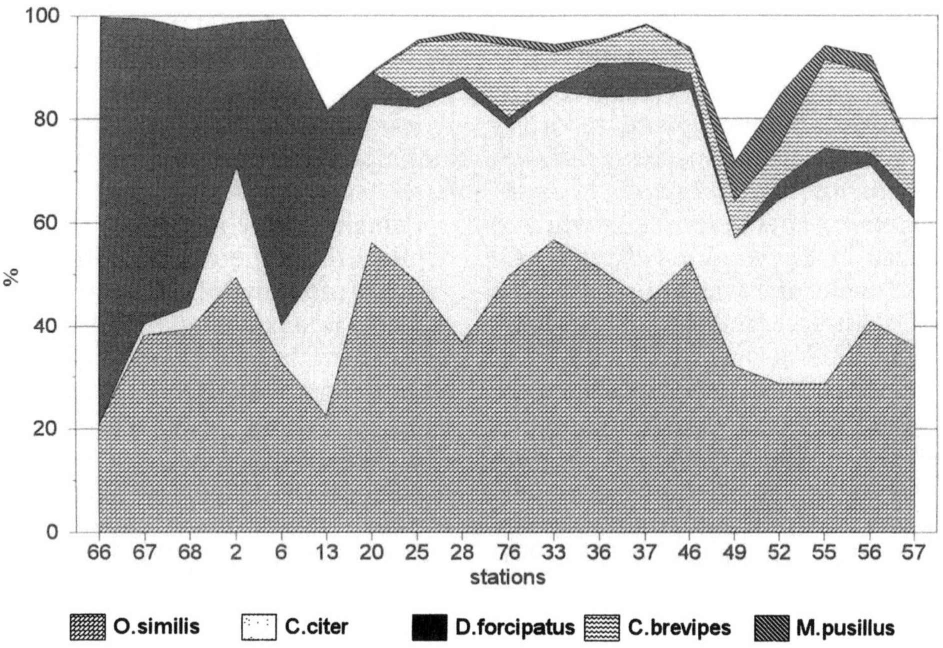

Fig. 1.18. Percentage composition of the five most abundant copepod species in the Straits of Magellan during the austral spring 1989. (Mazzocchi and Ianora 1991). *Microcalanus pusillus* has been identified as *M. pygmaeus* in the present study

This species was important at all stations but highest values were recorded in the central part of the Straits with values as high as 3000 ind. m^{-3} at one station. The second most abundant species, *C. citer*, with mean values of 26.2% of copepod numbers, was more abundant in the central and Pacific regions. *D. forcipatus*, with a mean relative abundance of 18.4%, was more common at Atlantic sites. Most other species had a much lower frequency of occurrence and were generally found at more western stations.

On the basis of these findings, Mazzocchi and Ianora (1991) subdivided the Straits into two geographic areas, each of which was defined by a different copepod assemblage. The species structure at Atlantic stations was very simple, with only two dominant species (*D. forcipatus* and *O. similis*). Towards the Pacific, the copepod assemblages became richer in terms of species composition due to the presence of a number of typically Antarctic species (e.g. *Paraeuchaeta antarctica*, *Scolecithricella minor*, *Oncaea curvata*) and others such as *Neocalanus tonsus*, *Eucalanus longiceps*, *Rhincalanus nasutus*, *Spinocalanus terranovae* (identified as *S. brevicaudatus* in the present study), and *Chiridius gracilis* that generally inhabit sub-surface or deep layers of slope and open oceanic waters. The same authors conclude that the community structure of the Straits is comparable to that reported for most other Chilean and Argentine shelf areas.

Of the other zooplankton groups, chaetognaths, euphausiids, amphipods and ostracods are the most abundant. Ten species of chaetognaths were identified during the austral spring 1989 (Ghirardelli et al. 1991, 1993). *Sagitta tasmanica* was the most abundant chaetognath species throughout the Straits, with highest numbers at depths of 150–200 m at the Pacific entrance. Other species characteristic of this region included *Eukrohnia hamata*, *Sagitta decipiens*, *S. planctonis*, *S. gazellae*, *S. minima*, *S. maxima* and *S. marri*. There was a drastic reduction in species diversity towards the easternmost sector with only one species, *S. tasmanica*, present at the Atlantic entrance.

Euphausiids also showed higher species diversity at Pacific stations up until Cape Froward (Guglielmo et al. 1991; Antezana et al. 1992). Highest numbers were recorded from 100–200 m. The species sampled included *Nematoscelis megalops*, *Stylocheiron maximum*, *Thysanoessa gregaria* and juvenile stages of *Euphausia vallentini*. Of these, only *E. vallentini* was sampled within the Straits. It was present in high numbers especially near the bottom.

Eleven species of amphipods were identified, seven of which were found only at Pacific stations (*Cyphocaris faurei*, *Primno macropa*, *Cyllopus magellanus*, *Hyperiella dilatata*, *Scina borealis*, *S. uncipes* and *Orchomene plebs*) (Guglielmo et al. 1991; Antezana et al. 1992). *Temisto gaudichaudi* was the most important species reaching maximum abundances in the area of Paso Ancho and the Atlantic. This was the only amphipod, together with *Hyperoche mediterranea*, sampled in this area.

Pelagic ostracods were collected at all depths throughout the Straits with the greatest abundances in the area of Paso Ancho (Benassi et al. 1994). Eight species were identified. Of these, *Discoconchoecia elegans* and *Obtusoecia antarctica* were the most common. *Loricoecia loricata* and *Metaconchoecia skogsbergi* were confined to Pacific areas whereas *Pseudoconchoecia serrulata* was present only at Pacific and Atlantic entrances to the Straits.

Note: A manual for the identification of Antarctic and sub-Antarctic copepods has recently been published by Razouls (1994). Unfortunately, it has not been possible to mention this work in our identification cards since it was published during the preparation of proofs of the present Atlas.

Chapter 2 Materials and Methods

2.1 Sampling Techniques

Zooplankton samples were collected in the late austral summer of 1991 (from February 20 to March 4) by the R/V *Cariboo* (Guglielmo et al. 1993). Nineteen stations were sampled from 30 to 1120 m depth along the main latitudinal axis of the Straits (52°–54°S latitude; 69°–73°W longitude) (Fig. 2.1). Station and sample data are given in Table 2.1.

Samples were taken with an EZNET BIONESS (Easy Zooplankton Net – Bedford Institute of Oceanography Net and Environmental Sensing System) (Sameoto et al. 1980). This is an opening and closing net sampler equipped with ten nets with a mouth opening of 0.25 m^2. The EZNET BIONESS was fitted with three types of plankton nets; one 1000-μm net followed by five 500-μm and four 230-μm nets alternately arranged on the main frame. For the present study, only the data from the 230-μm nets are reported.

The EZNET BIONESS continuously recorded temperature, salinity and light attenuance with a KMS II (ME Meerestichnik Elektronic GmbH) multiparametric probe. Also, oxygen concentration was measured with a calibrated oxygen electrode mounted on the KMS II. Flow velocity through the nets was monitored with external and internal acoustic doppler flowmeters (SM 21H-ME Meerestichnik Elektronic GmbH). Chlorophyll *a* concentration was measured with a Mie-Backscat Fluorometer (Haardt Optik Mikroelektronik) mounted on the EZNET BIONESS.

The EZNET BIONESS was towed at a speed of 1.5–2 m/s while it was slowly lowered along an oblique path to the desired depths, whereupon the nets were opened and closed by command from the ship. The volume of water filtered for each sample varied from 20 to 300 m^3 with greater volumes filtered at increasing depths due to the assumption that zooplankton abundances would be less at depth.

A series of discrete samples were collected from the surface to 900 m during both day and night. The depth layers were chosen so as to sample the water column above the thermocline in greater detail. Sampling stopped approximately 20–40 m from the bottom and, in many cases, only a few meters above the bottom. The distance from the bottom was calculated using a PSA 900 Data Sonic echosounder mounted on the EZNET BIONESS.

Samples were preserved in a 4% buffered formaldehyde-seawater solution. The entire sample was sorted for macrozooplankters (>1 cm in length) which were removed and identified. The remaining sample was then processed for the present copepod study. Copepods were counted and identified to species level, and distinguished as females, males or copepodite stage. Population abundances were estimated as the number of individuals per 100 m^3 of water filtered.

2.2 Terminology, Measurements and Line Drawings

The taxonomic descriptions are based on adult morphology. General body structures are described according to the terminology of Huys and Boxshall (1991). The abbreviations used in the descriptions are listed below:

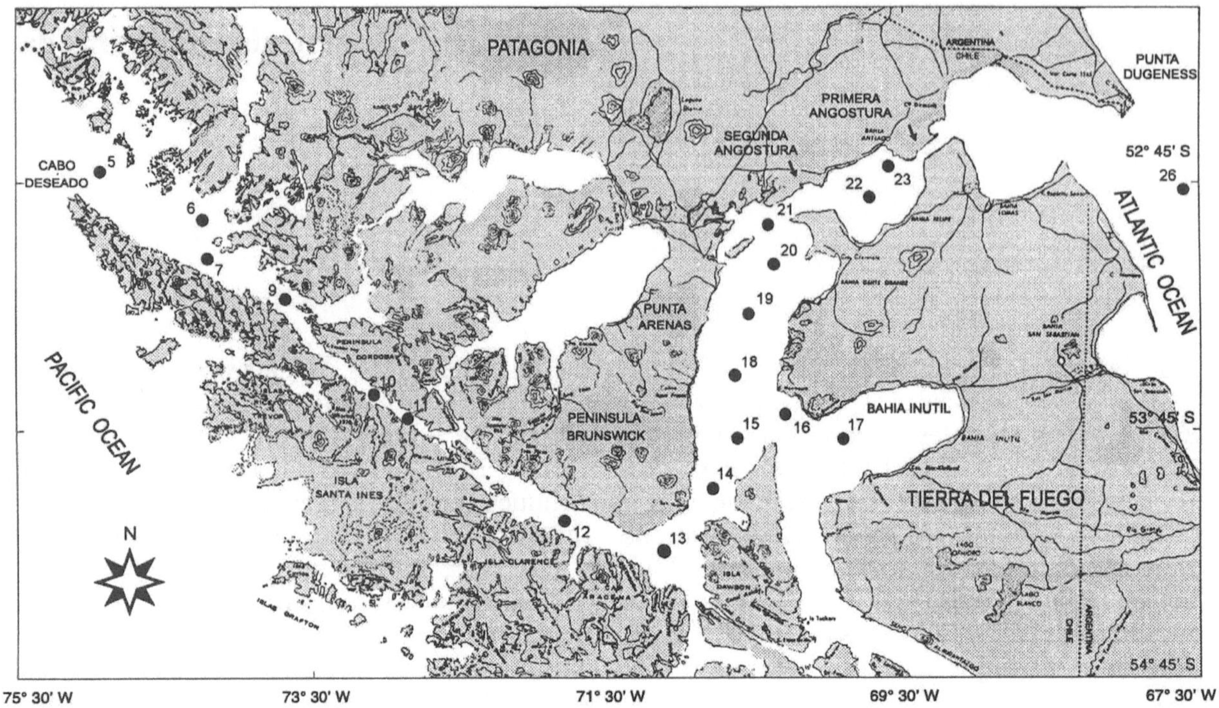

Fig. 2.1. Zooplankton sampling sites during the R/V *Cariboo* Cruise in the Straits of Magellan in February–March 1991

A1–A2	antennae 1–2
B1–B2	basal segments 1–2
P1–P4	swimming legs 1–4
Ri	endopod
Re	exopod
Ri1-Ri3	endopod segments (joints) 1–3
Rel-Re3	exopod segments (joints) 1–3
Th5	fifth thorax somite

All specimen measurements were made under a dissecting microscope with an ocular micrometer. The cephalothorax length is considered to be the distance between the anterior tip of the forehead and the posterior margin of the intersegmental membrane lying between the last metasomal somite and the genital somite. The abdomen length is the distance between the mid-dorsal anterior margin of the genital somite and the distal end of the caudal ramus. These lengths were used to determine both body length and proportional lengths of the cephalothorax and abdomen.

Parts were dissected from the animals and were permanently mounted in glycerine. Drawings of the body and appendages were made with the use of a camera lucida mounted on an Axiophot Zeiss microscope. Original ink figures were then processed by a computer graphic system (Apple MacIntosh Quadra with 20 MB RAM and 128 MB magneto-optic disk). The images were scanned through a UMAX 600 DPI scanner and then proportionally reduced and paged through an Aldus Free Hand drawing program.

2.3 Scanning Electron Micrographs

Preserved specimens were washed twice in distilled water and then left in distilled water for another 24 h. The next day, they were

Table 2.1. Station and sample data for the R/V *Cariboo* Cruise in the Straits of Magellan in February–March 1991

Station no.	Local date	Position				Local Time		Sonic depth (m)	Max depth Sampled (m)
		Start		End		Start	End		
		Lat. S	Long. W	Lat. S	Long. W				
5	23-02-91	52.42.90	74.28.90	52.42.30	74.30.80	16.41	17.13	220	200
6	24-02-91	52.51.10	73.54.30	52.52.60	73.57.60	9.25	10.34	550	500
7	24-02-91	53.02.00	73.39.70	53.01.00	73.42.40	16.16	17.23	640	600
9	25-02-91	53.13.30	73.17.60	53.11.50	73.20.40	11.20	12.38	1120	900
10	25-02-91	53.25.80	72.51.70	53.24.30	72.55.00	17.44	18.58	640	600
11	25-02-91	53.30.40	72.40.60	53.32.40	72.35.50	20.59	22.05	500	500
12	26-02-91	53.48.20	71.47.30	53.48.40	71.47.50	13.06	13.47	166	130
13	27-02-91	53.56.40	72.06.20	53.55.00	71.03.40	10.28	11.27	520	500
14	27-02-91	53.42.60	70.48.90	53.40.60	70.46.40	14.59	15.54	544	500
15	27-02-91	53.33.20	70.39.50	53.32.20	70.42.30	22.18	23.08	460	430
16	27-02-91	53.30.10	70.21.30	53.29.30	70.22.80	19.40	20.15	110	80
17	28-02-91	53.35.30	70.03.80	53.34.80	70.05.30	9.05	9.31	159	140
18	28-02-91	53.21.10	70.42.70	53.20.00	70.43.30	13.52	14.21	200	180
19	28-02-91	53.08.40	70.38.30	53.07.50	70.37.20	23.28	0.07	118	105
20	01-03-91	53.00.10	70.32.20	53.00.00	70.33.60	10.20	10.48	75	70
21	01-03-91	52.52.07	70.30.00	52.52.60	70.29.60	13.03	13.31	46	42
22	01-03-91	52.39.60	69.54.80	52.39.80	69.54.40	17.52	18.17	35	33
23	02-03-91	52.36.30	69.44.20	52.36.60	69.45.00	10.34	11.01	38	27
26	03-03-91	53.30.50	69.59.80	52.30.60	68.02.00	8.48	9.14	30	30

Total no. of stations	No. of samples			Total no. of samples	Max depth sampled (m)	Thickness of layer sampled (m)	
	230 µm	500 µm	1000 µm			Max	Min
19	76	95	19	190	900	200	3

post-fixed in a 2% OsO_4 solution for 1 h prior to dehydration in a graded alcohol series: 50% for 30 min, 70% for 30 min, 90% for 30 min, followed by three washings in 100% alcohol for 30 min each.

Specimens were critical point-dried with liquid CO_2 for 3–5 h. CO_2 was changed every hour. Specimens were then slowly brought to a temperature of 50 °C and a pressure of 50 atm. The pressure was gradually brought back to normal within 30 min and specimens were slowly cooled to ambient temperature. The specimens were mounted on stubs with adhesive tape and coated with gold palladium for 3 min at a setting of 2.2 kV and 10 mA of the sputtering apparatus.

Since SEM preparations required a large number of clean and intact specimens for detailed taxonomic description, SEM tables are lacking in some cases due to the paucity of material.

2.4 Identification Cards

The presentation of the species follows the taxonomic order proposed by Huys and Boxshall (1991) and Razouls (1993), which is

Table 2.2. Volumes of water filtered, total copepod numbers, temperature, salinity and fluorescence data for the different depth layers sampled at 19 stations during the R/V *Cariboo* Cruise in the Straits of Magellan in February–March 1991. Note that coventional Unit (F) for in vivo fluorescence measurements in the range 0–10 volts conforms to the range 0–50 mg/m³ for chlorophyll *a* units, with[a] resolution of 0.1 mg/m³ and an accuracy variability <10%

Station no.	Depth sampled (m)	Volume filtered (m^3)	Total copepod abundance (Ind/m^3)	Temperature (°C)	Salinity (psu)	Fluorescence (UF)
5	160–140	53.10	55.09	10.350	33.008	0.074
	115–100	32.60	71.87	10.559	32.931	0.070
	80–60	45.10	35.04	9.184	33.008	0.065
	40–20	32.20	42.22	9.629	30.193	0.166
6	455–400	120.80	14.46	8.163	33.260	0.046
	300–250	45.10	176.11	9.069	33.251	0.050
	200–150	45.90	1155.39	9.808	33.213	0.054
	100–50	99.10	107.55	10.090	31.203	0.124
7	500–400	79.80	154.90	8.204	33.256	0.043
	300–250	56.80	174.84	8.446	33.192	0.045
	200–150	63.70	542.71	9.486	33.113	0.052
	100–50	58.20	66.13	9.235	31.339	0.087
9	700–600	79.00	27.02	8.204	33.252	0.043
	400–300	62.30	54.89	8.393	33.236	0.046
	200–140	23.70	451.00	9.064	32.975	0.052
	100–50	44.60	253.38	8.720	31.493	0.092
10	500–400	191.00	35.62	8.244	33.242	0.043
	300–250	79.10	102.72	8.421	33.132	0.045
	200–150	73.00	169.76	8.596	32.803	0.052
	100–50	36.90	11351.17	8.901	30.479	0.187
11	400–300	302.70	37.95	8.272	33.222	0.045
	250–200	130.40	71.83	8.375	33.102	0.050
	150–100	94.70	169.45	8.526	32.557	0.067
	50–25	32.80	6202.61	9.342	29.094	0.263
12	100–50	49.60	66.90	7.998	30.605	0.150
	24–0	38.40	2278.94	8.464	30.183	0.210
	130–100	20.90	121.34	7.738	30.763	0.084
	50–25	47.00	327.07	8.223	30.448	0.287
13	400–300	249.80	300.39	6.926	31.112	0.053
	250–200	48.80	284.95	7.225	31.062	0.058
	150–100	68.70	216.13	7.765	30.812	0.066
	50–25	38.60	2197.65	8.385	30.382	0.233
14	400–300	156.30	417.69	6.883	31.082	0.052
	250–200	65.90	66.35	7.406	30.989	0.054
	150–100	85.00	1243.10	7.447	30.780	0.077
	90–50	56.90	250.55	8.332	30.525	0.114
15	300–200	155.40	174.86	6.922	31.004	0.054
	150–100	36.00	223.83	7.614	30.695	0.066
	80–60	26.00	261.99	8.258	30.537	0.093
	40–20	26.10	1710.05	8.594	30.358	0.194
16	60–30	55.80	114.29	8.590	30.482	0.215
	20–0	36.10	20.25	9.548	30.199	0.526
	80–60	30.70	2685.15	7.619	30.617	0.071
	40–20	22.40	725.60	8.946	30.425	0.248

Table 2.2. *Continued*

Station no.	Depth sampled (m)	Volume filtered (m^3)	Total copepod abundance (Ind/m^3)	Temperature (°C)	Salinity (psu)	Fluorescence (UF)
17	120–100	41.40	1332.82	7.021	30.647	0.063
	80–60	43.90	5434.84	7.836	30.597	0.083
	40–30	20.00	172.80	9.003	30.406	0.291
	20–10	18.70	130.83	9.298	30.341	0.530
18	160–120	92.00	625.18	7.037	30.914	0.053
	100–80	43.40	150.49	7.225	30.804	0.055
	60–40	39.20	23.24	8.221	30.612	0.128
	20–10	14.50	22.84	9.237	30.453	0.548
19	75–50	46.30	229.88	8.534	30.552	0.115
	25–0	29.70	6629.14	8.996	30.403	0.309
	100–75	57.70	524.85	8.074	30.635	0.093
	50–25	42.30	922.64	8.834	30.500	0.211
20	60–40	35.60	1350.20	8.836	30.495	0.219
	20–0	23.20	6.18	9.079	30.380	0.417
	80–75	52.10	2053.20	8.699	30.525	0.186
	40–20	20.30	818.00	8.988	30.487	0.256
21	20–0	39.20	583.14	9.121	30.517	0.250
	40–20	66.00	1307.93	9.149	30.522	0.229
	0–40	77.40	916.17	9.178	30.522	0.243
	40–20	55.90	1008.39	9.190	30.522	0.227
22	32–29	34.10	359.54	9.828	30.552	0.233
	30–20	17.20	226.02	9.805	30.542	0.233
	20–10	51.70	184.75	9.788	30.543	0.233
	10–0	24.10	51.57	9.761	30.539	0.236
23	20–10	34.10	79.28	10.208	30.754	0.241
	0–30	46.70	50.83	10.217	30.754	0.330
	20–10	28.30	89.83	10.219	30.763	0.250
	0–25	46.10	89.15	10.225	30.760	0.302
26	30–20	15.40	9959.94	9.302	32.654	0.243
	26–20	28.90	19016.25	9.295	32.654	0.229
	10–4	33.30	368.77	9.307	32.648	0.246
	10–4	23.50	40.26	9.299	32.651	0.222

based not only on the morphology of the species but also the group phylogeny. However, to facilitate consultation of identification cards, copepod species have also been arranged in alphabetical order in Appendix 1.

When possible, both female and male specimens are described, reporting only those morphological characters which we feel are the most useful for a rapid and precise species identification. A somewhat longer description is devoted to species which present greater difficulty in their identification. Three of the species collected (*Ischnocalanus tenuis*, *Oncaea mediterranea* and *Euterpina acutifrons*) are not reported due to extreme paucity and heavy damage of the available material.

The synonymy at the beginning of each identification card is very concise and is not intended to be exhaustive. In these, we refer only to papers with species illustrations, preferably taking into account more recent publications and those which we feel are more accessible in libraries of various marine biology institutions.

Line drawings have been prepared only of those anatomical parts which are essential for the correct identification of the species.

SEM micrographs are intended as companion illustrations to line drawings, often in substitution of these since they reveal many anatomical details not always evident at the light microscope.

The data on species spatial distributions are expressed as numbers 100^{-3}. The geographic positions of all stations are plotted in distribution maps. Each station includes all samples collected at that specific geographic position with no consideration for the depths at which different samples were collected.

The vertical distribution of each species, given at the end of each identification card, was obtained by grouping data for three geographic areas which take into consideration the bottom topography and hydrography of the Straits. These are referred to in the text as:

PACIFIC AREA, from the Pacific entrance to Carlos III Island (stations: 5, 6, 7, 9, 10, 11);

CENTRAL AREA, from Carlo III Island to Pelicano Point (stations: 12, 13, 14, 15, 16, 17, 18, 19);

ATLANTIC AREA, from Pelicano Point to the Atlantic entrance (stations: 20, 21, 22, 23, 26).

Illustrations for vertical distribution data were compiled by considering only stations where a given species had been sampled at least once. Distribution maps show the mean for a given species sampled at the same depth throughout the Pacific, Central and Atlantic areas. Temperature (T) and fluorescence (F) profiles are shown for a key station within each area: St. 9 for the Pacific area, St. 15 for the Central area and St. 20 for the Atlantic area. The conventional unit (F) for in vivo fluorescence in the range 0–10 volts corresponds to 0–50 mg/m^3 for Chl *a* with a resolution of 0.1 mg/m^3 and an accuracy variability of less than 10%.

Temperature and fluorescence values for each station and each depth are reported in Table 2.2 together with values for salinity and total copepod abundance.

Species abundance data for the stations and depths sampled during the *Cariboo Cruise* are reported in Appendix 2.

Chapter 3 Systematic Account

Order CALANOIDA

Family Augaptilidae

3.1 *Augaptilus glacialis* Sars 1900

Augaptilus glacialis Sars, 1900: p. 88, pls. 26, 27; 1924–1925: p. 254, pl. 76 figs. 1–16 – Brodskii, 1950: p. 368, fig. 258 – Vervoort, 1951: p. 144, figs. 80, 81 – 1957: p. 138, fig. 131 – Tanaka 1964b: p. 77, fig. 212 – Tanaka and Omori, 1971: p. 259, fig. 5a.
Augaptilus zetesios Wolfenden, 1902: p. 369, pl. 3.

Body length –
females: 4.51–5.19 mm (15 specimens)
males: 4.04–4.39 mm (10 specimens)

Female. Cephalothorax oblong and ovate, 2.51 times longer than wide. Last thoracic somite narrowly rounded at distal corners. A1 exceeds end of furca by distal 4–5 segments. Proportional lengths of cephalothorax and abdomen 69:31. Abdomen 3-segmented. Proportional lengths of abdominal somites and furca 39:11:17:33 = 100. Genital somite slightly asymmetrical. Caudal rami 5 times longer than wide and strongly divergent. Anal somite and furca naked on dorsal side, coarsely covered with fine hairs on ventral surface. Mandibular blade with two sets of large teeth of equal length, one single smaller tooth and another single large tooth with serrated fringe. Inner marginal spine of P5 Re2 reaching base of 3rd inner marginal seta of Re3.

Male. Cephalothorax 2.6 times longer than wide. Proportional lengths of cephalothorax and abdomen 71:29. Left A1 geniculated. Abdomen 5-segmented. Proportional lengths of abdominal somites and furca 16:13:13:11:14:33 = 100. Ventral side of caudal rami bearing long hairs. P5 slightly asymmetrical; distal joint of right exopod bearing two spines of unequal length, the larger spine slightly longer than Re3. Inner distal margin of B2 of both legs bearing small bump with cluster of short blunt denticles. This latter feature is reported only for right B2 by Brodskii (1950).

Remarks

It seems to be remarkably difficult to separate *A. glacialis* from *A. longicaudatus* (Claus 1863) as discussed by Vervoort (1965b). The lengths reported for *A. g.* females from different areas are always larger than 4 mm, while for *A. l.* they are generally smaller (Tanaka and Omori 1971). Otherwise, a comparison of the data available for *A. l.* suggests that this species is greatly variable in length; Farran (1908) reported an extremely wide length range for females from the Irish Atlantic slope (3.6–6.1 mm). In *A. l.* females, the caudal rami and anal somite are covered by spiniform hairs; in *A. g.* the furca may be completely naked, but usually there are fine hairs along the internal borders. Our specimens are in the size range usually reported for *A. g.*; the caudal rami and anal somite of females are coarsely covered with fine hairs on the ventral side (no spiniform hairs as in *A. l.*) and the genital somite is only slightly asymmetrical without the remarkable swelling on the left side (in dorsal view) characterizing *A. l.* females (as shown also by Tanaka and Omori (1971, p. 260). In the specimens from the Straits of Magellan, as also in the specimens analyzed by Vervoort (1965b), the female does not have the blunt, coarsely serrulated internal marginal spine of the P5 Re2 as figured by Brodskii (1950, fig. 258) but this spine has short hairs along its length. This spine reaches the base of the 3rd inner marginal seta of the Re3, while in *A. l.* it reaches the 2nd marginal seta as reported by Tanaka (1964b). Apparently, only the males of *A. glacialis* and *A. longicaudatus* can easily be separated by characters of the P5: only in *A. l.* males does the left exopodite end with a very long apical spine.

Distribution

This species is primarily reported for the North Atlantic-Arctic region (Sars 1925; Brodskii 1950; Johnson 1963), penetrating far to the south in the Atlantic Ocean, apparently absent in the North Pacific and scarcely represented in the Antarctic (Vervoort 1951). In the Arctic and Atlantic Oceans, the species occurs in sub-surface waters and at intermediate depths, but in the temperate and tropical Atlantic the occurrence of *A. glacialis* is exclusively confined to cold, deep waters (Vervoort 1965b). Tanaka and Omori (1971) recorded this species in the Izu Region, central Japan, at depths of 0–1200 m. In the present study, *A. glacialis* was present only at Pacific stations in the Straits of Magellan with maximum abundances from 400–500 m.

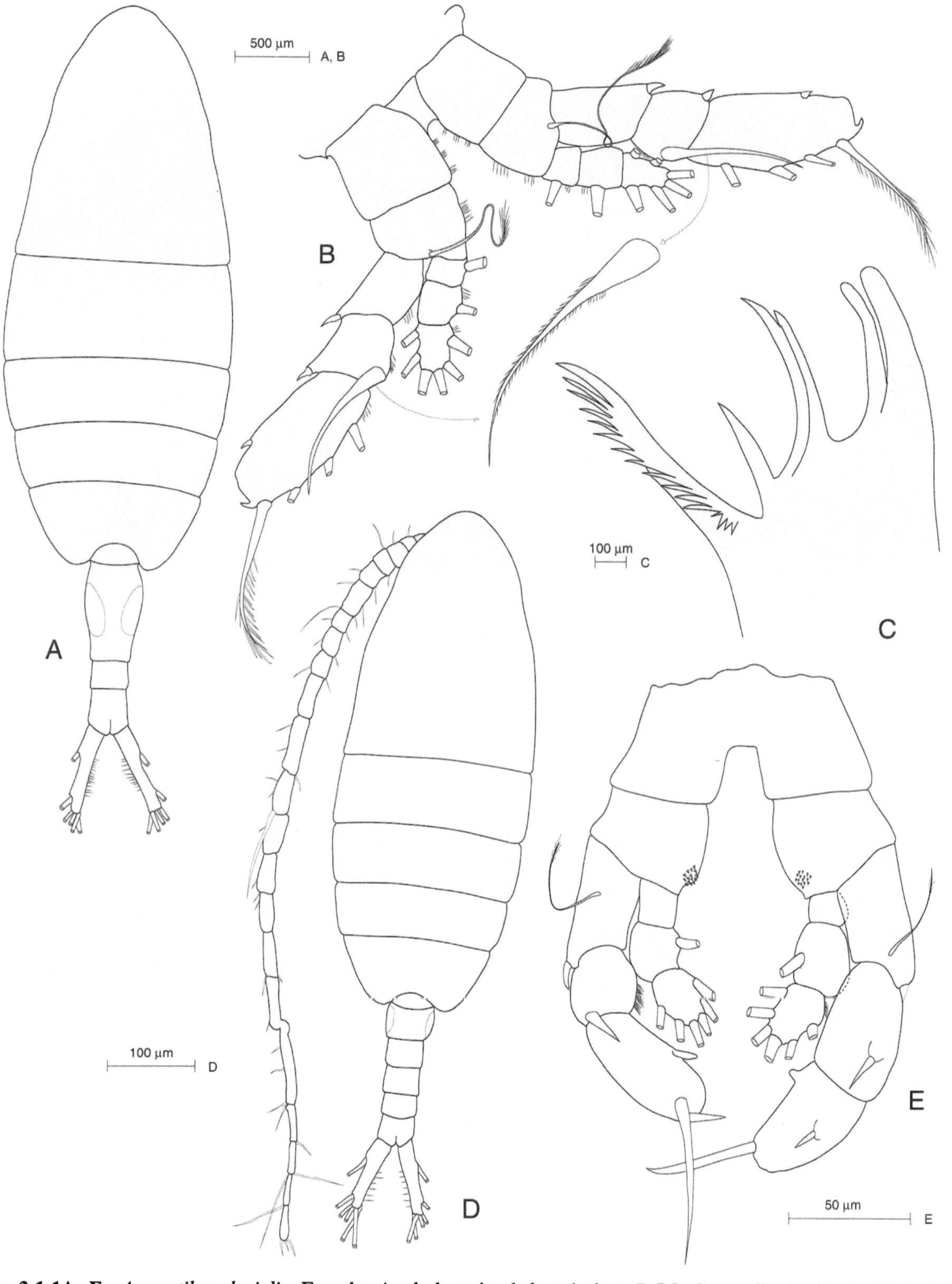

Fig. 3.1.1A–E. *Augaptilus glacialis*. Female: **A** whole animal dorsal view; **B** P5; **C** mandibular blade. Male: **D** whole animal dorsal view; **E** P5

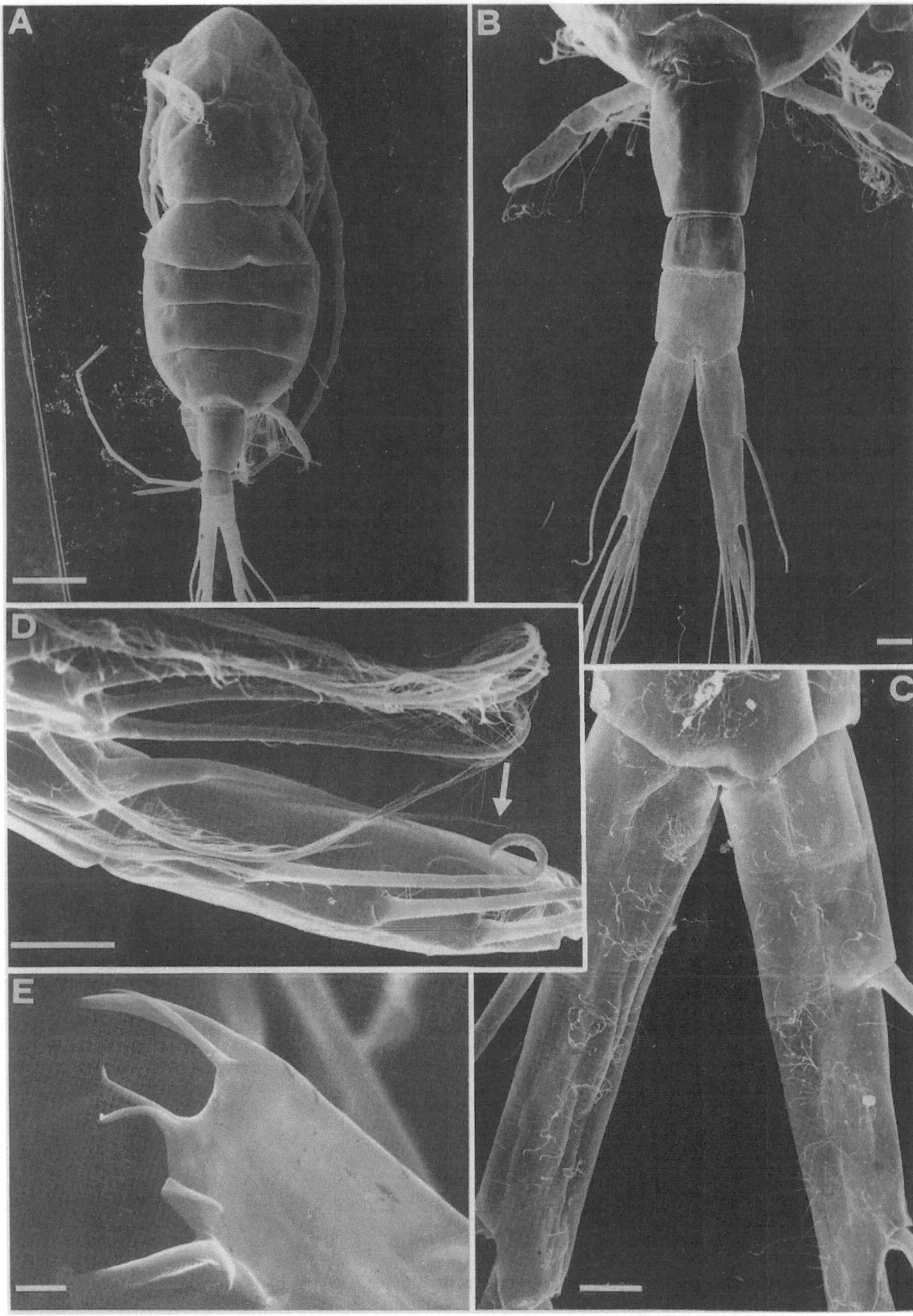

Fig. 3.1.2A–E. *Augaptilus glacialis*. Female: **A** whole animal dorsal view; **B** abdomen dorsal view; **C** furca ventral view; **D** Re3 of P_5 with marginal spine (*arrow*) of Re2; **E** terminal part of mandibular blade. *Bars*: **A** 500 μm; **B** 6 μm; **C**, **D** 50 μm; **E** 10 μm

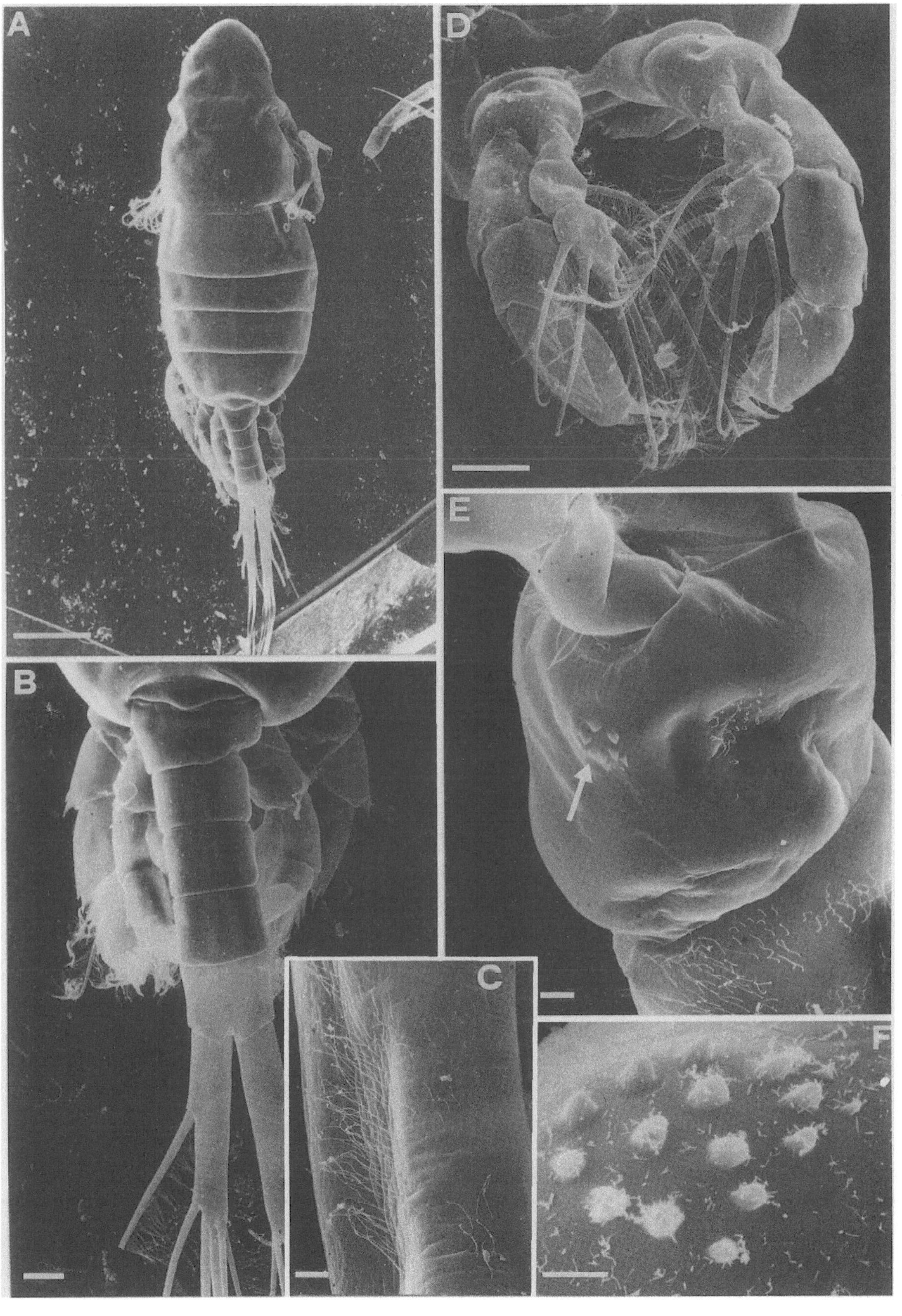

Fig. 3.1.3A–F. *Augaptilus glacialis*. Male: **A** whole animal dorsal view; **B** abdomen dorsal view; **C** ventral surface of furca; **D** P5; **E** B_2 of right P5; **F** detail of blunt denticles on B2 of right P5 (arrowed in E). *Bars*: **A** 500 µm; **B**, **D** 100 µm; **C**, **E** 10 µm; **F** 5 µm

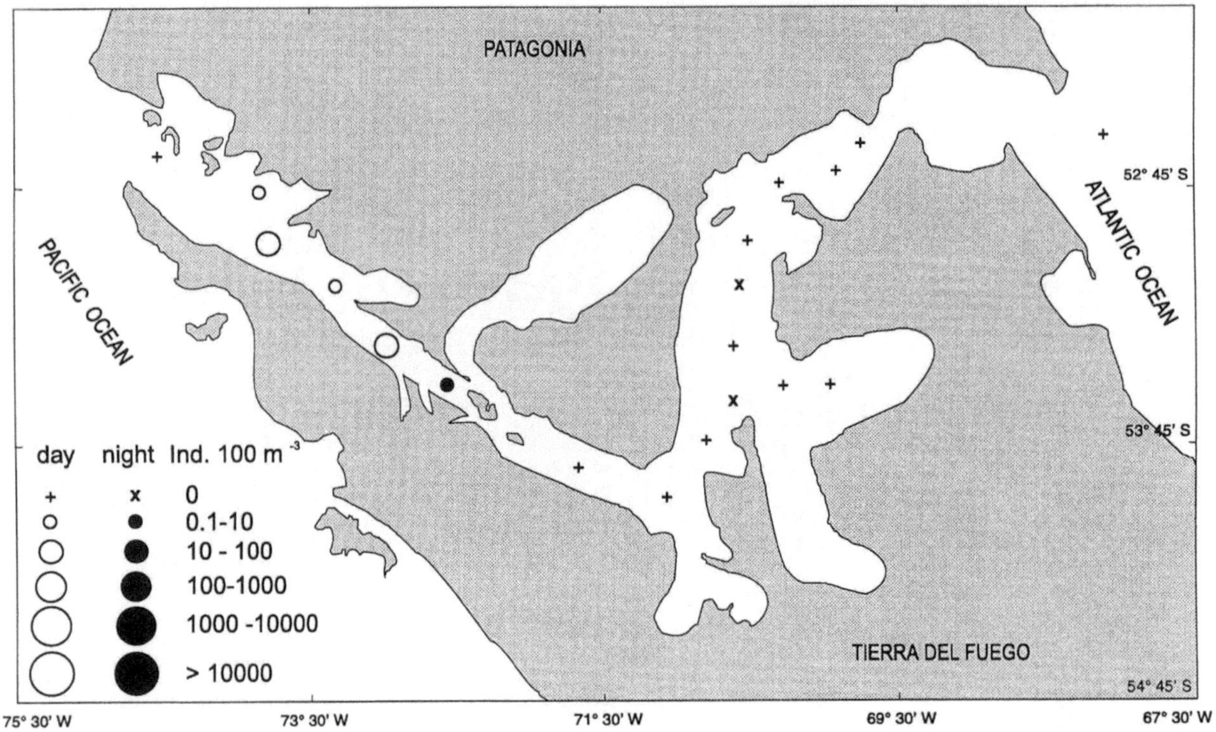

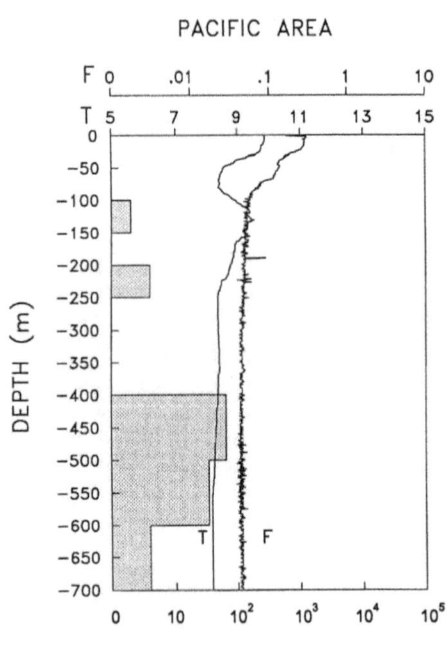

Depth range of occurrence= **100 - 700** m
Population core = **400 - 500** m
Temperature (T) = **8.20 - 9.80** °C
Oxygen = **348.4 - 437.2** µmol dm^{-3}
Chl fluorescence (F) = **0.04 - 0.39**

Fig. 3.1.4. *Augaptilus glacialis*. Distribution in the Straits of Magellan in February–March 1991

3.2 *Haloptilus acutifrons* Giesbrecht 1892

Hemicalanus acutifrons Giesbrecht, 1892: p. 384, pl. 3 fig. 11, pl. 27 figs. 4, 12, 18, 26, pl. 42 figs. 12, 20.
Hemicalanus spinifrons Sars, 1900: p. 95, pl. 28 figs. 1–12.
Haloptilus acutifrons Sars, 1924–1925: p. 250, pl. 74 figs. 1–11 – Brodskii, 1950: p. 364, fig. 255 – Grice, 1962: p. 223, pl. 25 figs. 1–8 – Owre and Foyo, 1967: p. 81, figs. 523–525.

Body length –
females: 4.35–4.66 mm (6 specimens)

Female. Body transparent. Cephalothorax 2.8 times longer than wide. Proportional lengths of cephalothorax and abdomen 84:16. Forehead produced into long, strong spine, slightly bent downward, with subtle filamentous rostrum. A1 exceeds end of body by 5–7 distal joints. Mandibular blade with strong distal tooth and group of 3 much smaller teeth on opposite margin. Proportional lengths of abdominal somites and furca 39:11:9:14:27 = 100. P5 biramous, symmetrical; last exopodal joint with 2 terminal spines of different length. Longer of 2 spines slightly shorter than same joint. Endopod not reaching distal margin of Re2.

Remarks

Haloptilus acutifrons resembles *H. oxycephalus*, the latter of which is often recorded from Antarctic and sub-Antarctic seas (Park 1988). As discussed by Park (1988), the two species are practically indistinguishable; the only useful anatomical detail to determine *H. acutifrons* is the relatively shorter cephalic spine. *H. acutifrons* is also smaller than *H. oxycephalus*; both species, however, show a wide size range depending on the sampling region and latitude. Our specimens show a size range overlapping that of *H. oxycephalus*, but the shape of the forehead is exactly as always reported for *H. acutifrons*.

Distribution

Oceanic, cosmopolitan species with a wide geographical distribution range both in the Northern and Southern hemispheres. Farran (1929) recorded *Haloptilus acutifrons* off New Zealand during the "Terra Nova Expedition". It was found at a depth of 400–1000 m in the central part of the Arctic Ocean, and below 1000 m in Greenland waters; it can also occur at shallower depths such as 100 m or 50 m, and even at the surface (Brodskii 1950). This species was also recorded from 0–2000 m in the Mediterranean Sea (Scotto di Carlo et al. 1984). Vervoort (1957) summarized records of its geographical distribution and Roe (1972) discusses its vertical distribution. In the present study, *H. acutifrons* was only found from 100–400 m at two stations in the Pacific area of the Straits of Magellan.

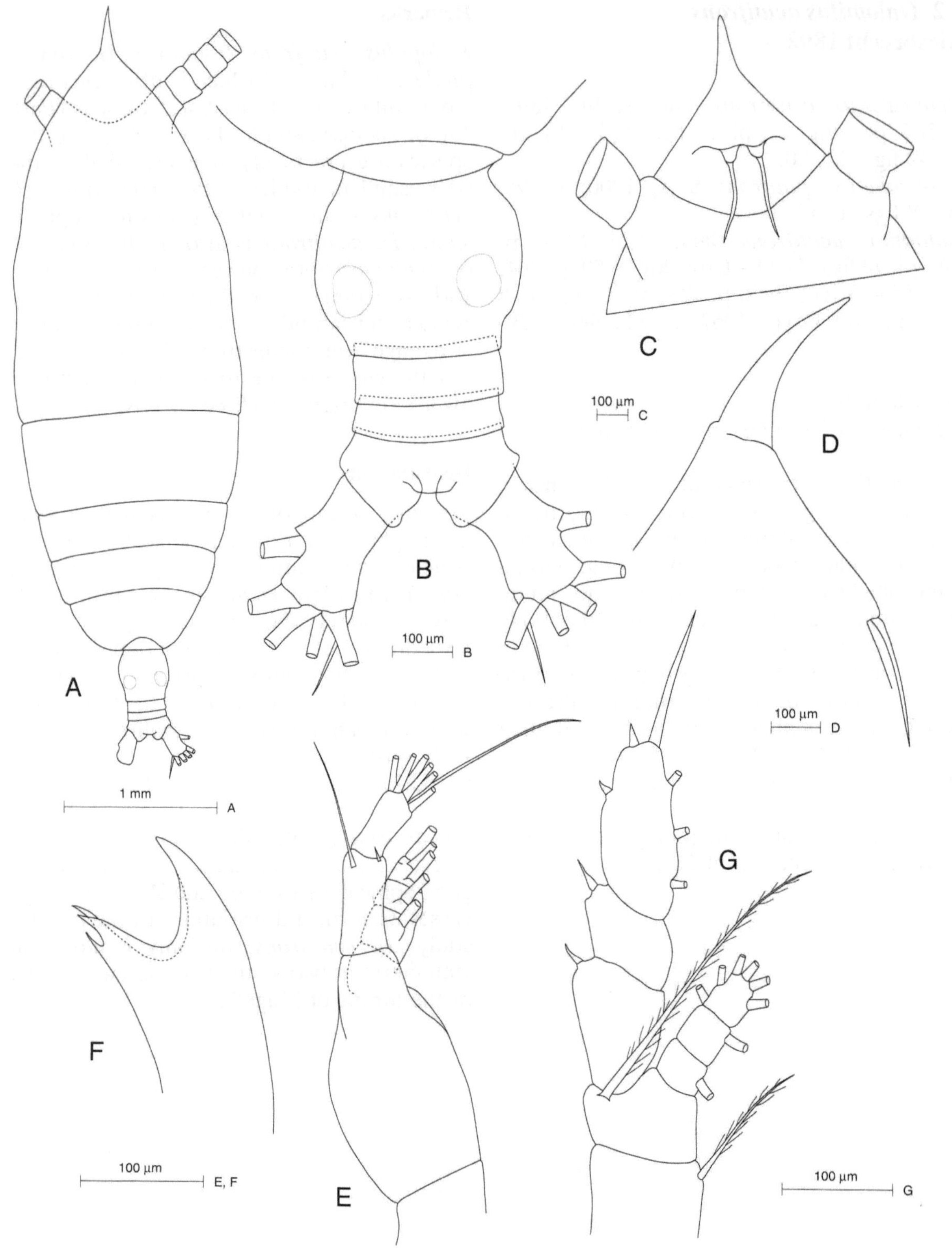

Fig. 3.2.1A–G. *Haloptilus acutifrons*. Female: **A** whole animal dorsal view; **B** abdomen dorsal view; **C** pointed forehead with filiform rostrum ventral view; **D** pointed forehead with filiform rostrum lateral view; **E** mandibular palp; **F** mandibular blade; **G** P5

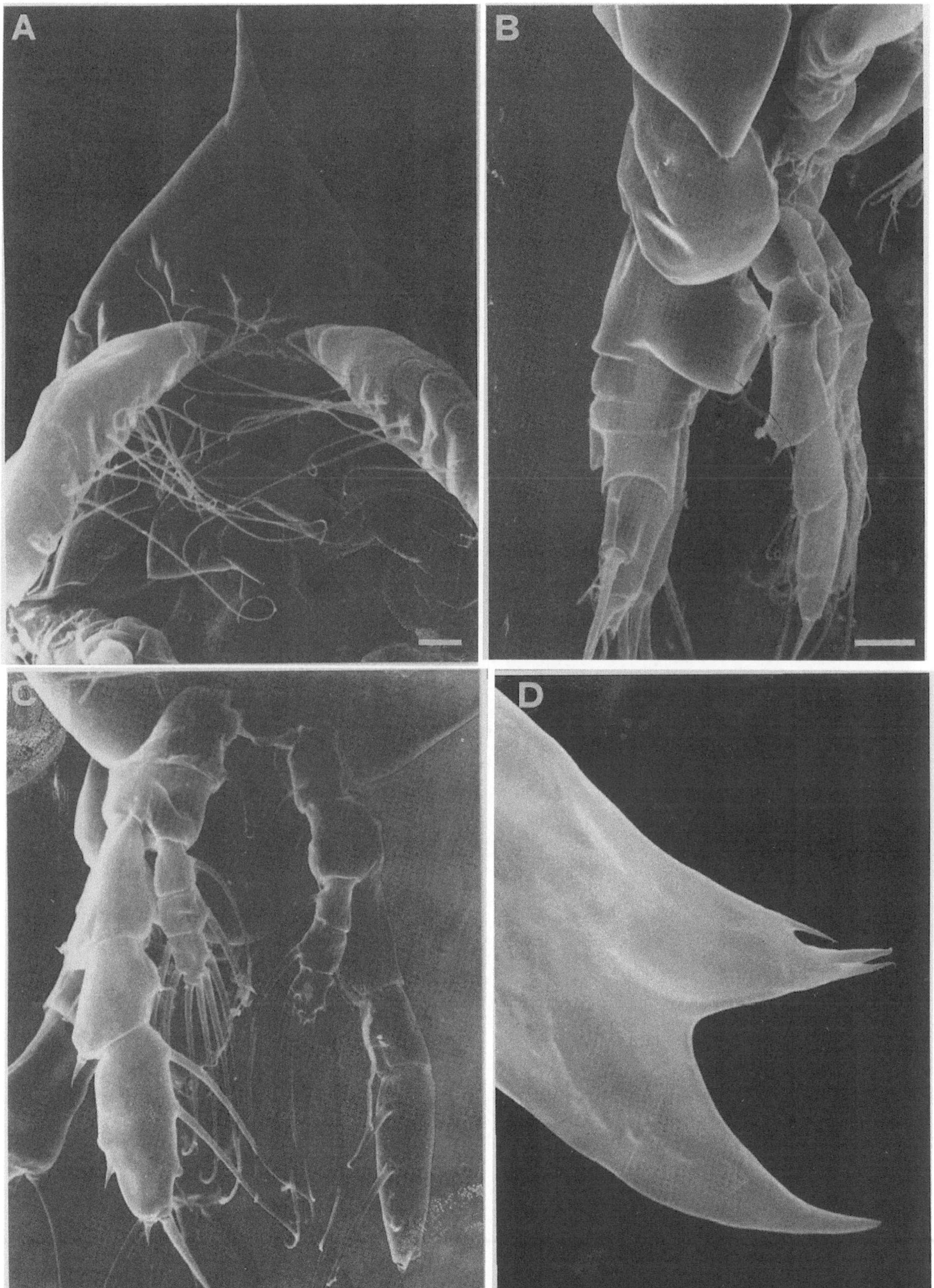

Fig. 3.2.2A–D. *Haloptilus acutifrons*. Female: **A** forehead ventral view; **B** abdomen lateral view; **C** P5 anterior view; **D** terminal part of mandibular blade. *Bars*: **A**, **B**, **C** 100 µm; **D** 10 µm

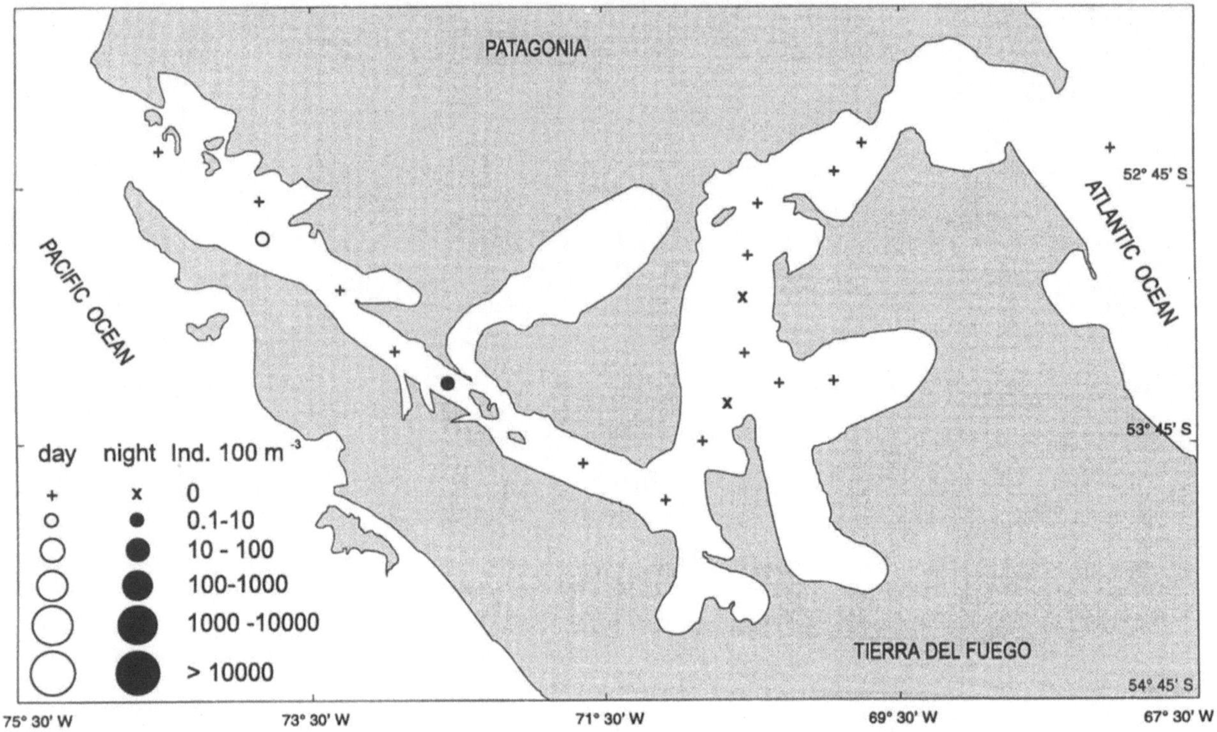

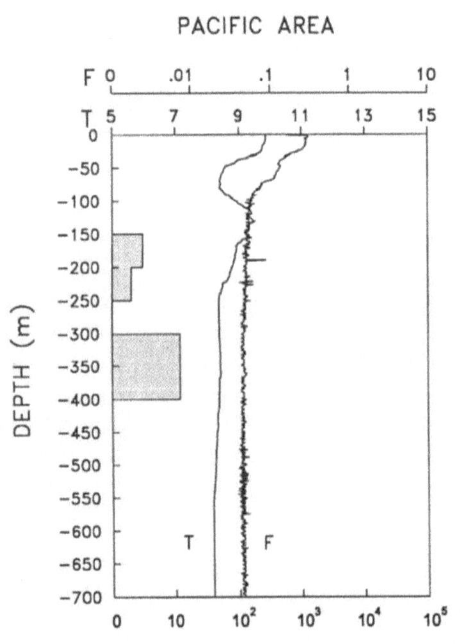

Depth range of occurrence = **150 - 400** m
Population core = **300 - 400** m
Temperature (T) = **8.20 - 9.80** °C
Oxygen = **348.4 - 437.2** µmol dm^{-3}
Chl fluorescence (F) = **0.04 - 0.39**

Fig. 3.2.3. *Haloptilus acutifrons*. Distribution in the Straits of Magellan in February–March 1991

3.3 *Heterorhabdus spinifrons* Claus 1863

Heterochaeta spinifrons Claus, 1863: p. 183, pl. 32 – Giesbrecht, 1892: p. 372, pl. 20 figs. 1, 3, 11, 16, 19, 31, pl. 39 figs. 42, 43, 51, 52, 54.
Heterorhabdus spinifrons Esterly, 1905: p. 183, fig. 37a,b; 1906: p. 77, pl. 14 fig. 93 – Sars, 1925: p. 227, pl. 62 figs. 9–12 – Brodskii, 1950: p. 351, fig. 246 – Grice, 1962: p. 222, pl. 24 figs. 10, 11 – Tanaka, 1964a: p. 13, fig. 179a–e – Mazza, 1965: p. 401, figs. 2c,f,g, 3–5, 6a – Arcos, 1975: p. 22, pl. 8 figs. 68–70.

Body length –
females: 3.49–3.82 mm (12 specimens)
males: 3.43–3.59 mm (9 specimens)

Female. Large species, easily recognized for its prolonged forehead ending with a sharp spine. Rostrum bearing two long, fine filaments. Postero-lateral corners of last thoracic somite rounded. A1 exceeds caudal rami by about 2 segments. Proportional lengths of cephalothorax and abdomen 68:32; proportional lengths of abdominal somites and furca 42:15:12:12:19 = 100. In dorsal view, posterior margins of abdominal somites bordered by small spinules. At light microscope, ventral surface of genital somite fringed with row of small denticles on either side of genital aperture. At SEM, these "denticles" appear as tubules. Caudal rami asymmetrical; left ramus longer, bearing very long strong inner marginal seta. Inner spine of P5 Re2 as long as Re3. Terminal spine of Re3 about 1/3 length of the joint.

Male. General shape of body as female, slightly smaller. Cephalothorax 2.3 times longer than wide. Proportional lengths of cephalothorax and abdomen 75:32; proportional lengths of abdominal somites and furca 23:19:16:11:8:23 = 100. Posterior margins of abdominal somites ornated with spinules on dorsal side, as in female. Caudal rami asymmetrical, with short hairs on dorsal side. P5 asymmetrical. Both left and right B2 with prolonged inner borders forming ear-shaped expansion ornated with numerous fine hairs. Re2 of right leg dilated; inner border bearing bilobed expansion ornated with tufts of long hairs. Re3 about twice as long as second exopodal joint with pointed end bearing long, fine, apical spine. Re2 of left leg with large triangular expansion; Re3 ending in one long spine and bearing two short, strong spines.

Remarks

There are two other *Heterorhabdus* species closely resembling *H. spinifrons*: *H. sub-spinifrons* described by Tanaka (1964a) for specimens from deep waters of Sagami Bay, Japan, and *H. spinifer* described by Park (1970) for specimens from the Caribbean Sea and Gulf of Mexico. Our specimens differ from the descriptions given for the above species in several anatomical details. Also, they are bigger in size compared to *H. spinifer* (1.6–1.7 mm) and do not bear small processes on either side above the genital aperture characterizing *H. sub-spinifrons* females.

Distribution

H. spinifrons is reported from temperate and tropical regions of the Atlantic, Pacific and Indian Oceans and the Mediterranean Sea. It has a very wide depth range of occurrence from just below the surface in some areas to more than 1000 m in other oceans (Sewell 1947; Grice and Hülsemann 1965; Roe 1972; Scotto di Carlo et al. 1984). The species has previously been found both off the Chilean coast (Fagetti 1962) and in the fjord region of southern Chile (Arcos 1974). In the present study, *H. spinifrons* was found from 100–500 m in the central Pacific area and at one station from 100–150 m in the Central area of the Straits of Magellan.

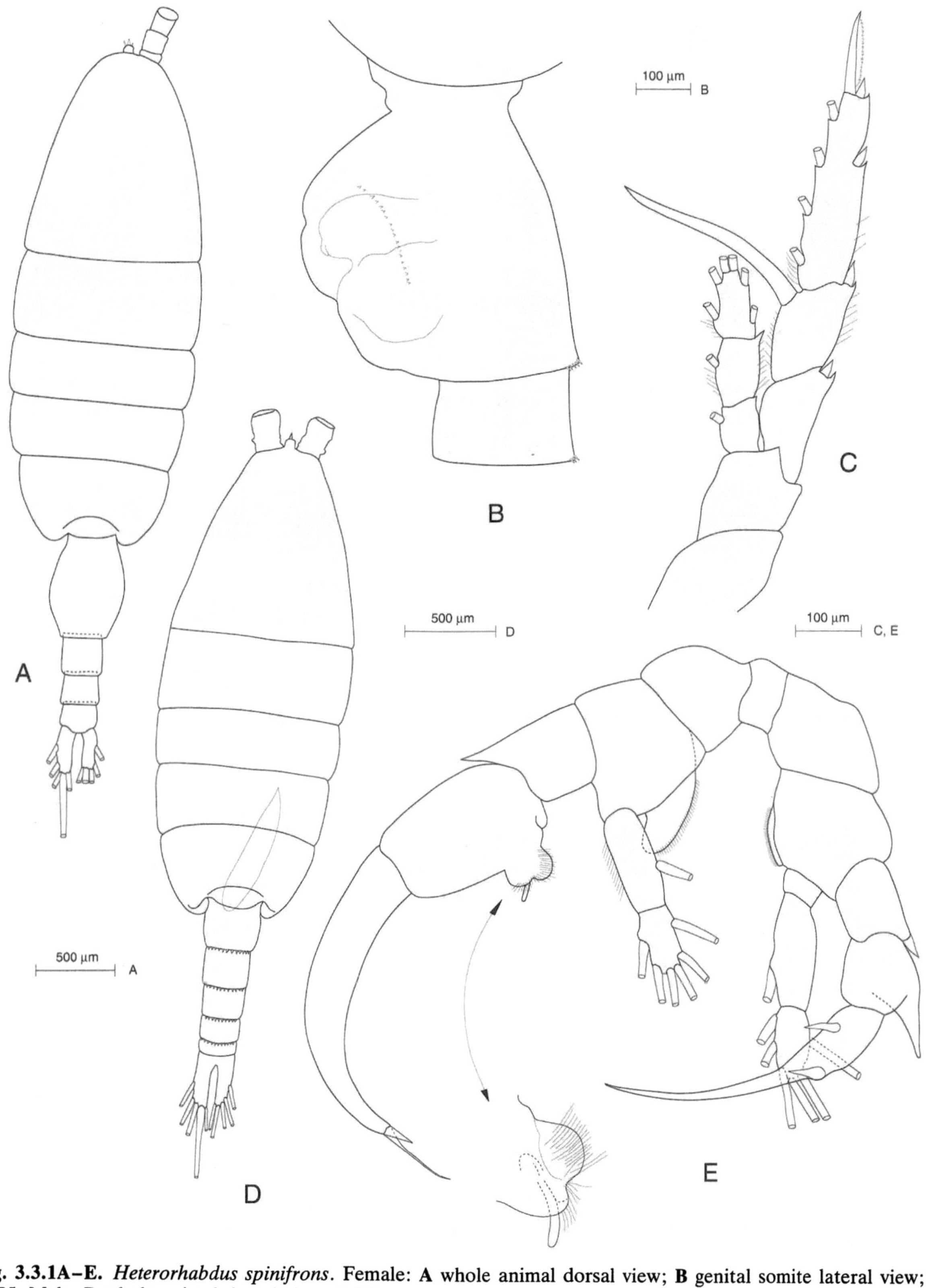

Fig. 3.3.1A–E. *Heterorhabdus spinifrons*. Female: **A** whole animal dorsal view; **B** genital somite lateral view; **C** P5. Male: **D** whole animal dorsal view; **E** P5

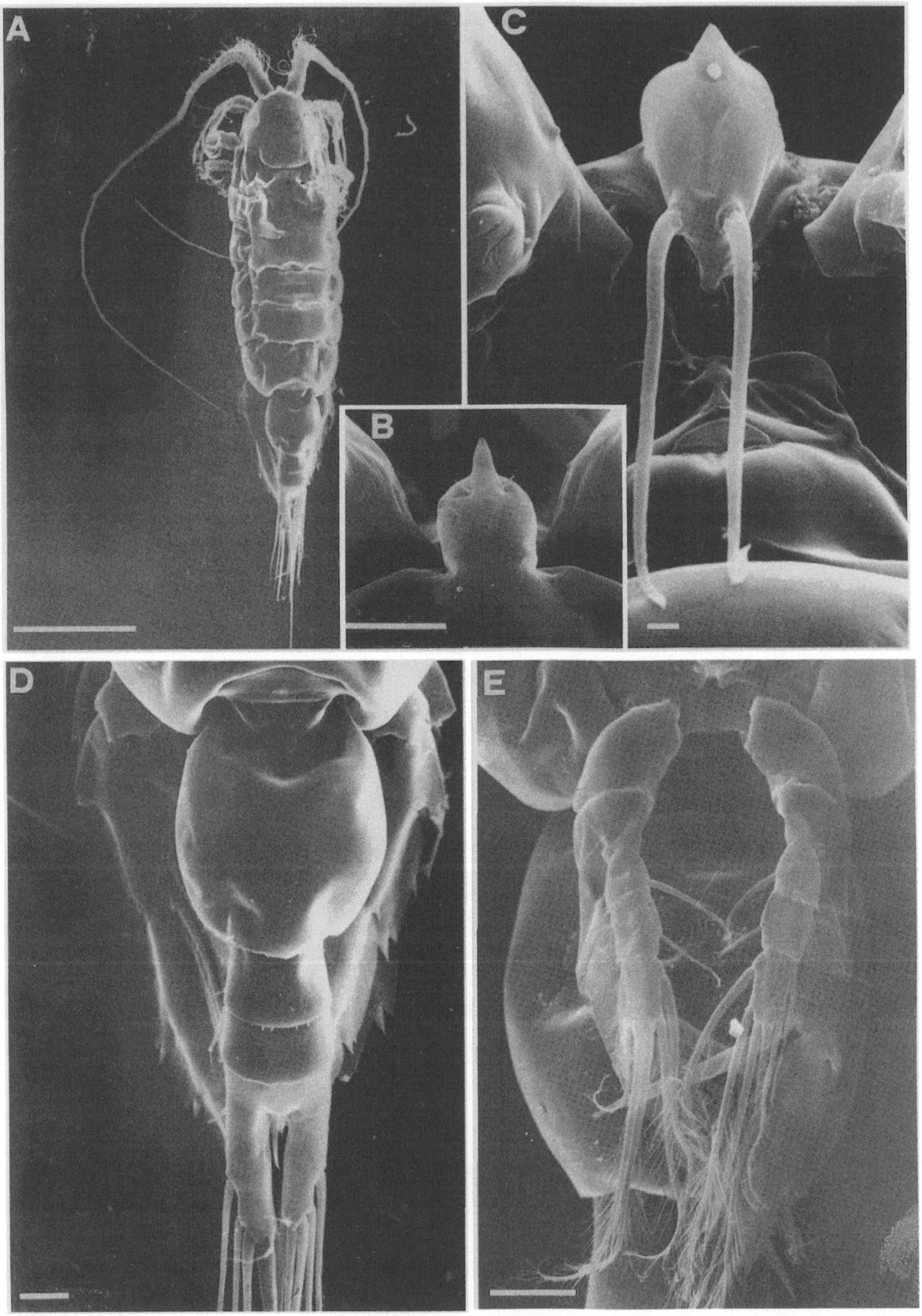

Fig. 3.3.2A–E. *Heterorhabdus spinifrons*. Female: **A** whole animal dorsal view; **B** dorsal view of the forehead with characteristic protuberance ending with a sharp spine; **C** rostrum with two long subtle filaments; **D** abdomen dorsal view; **E** P5. *Bars*: **A** 1 mm; **B** 50 μm; **C** 5 μm; **D**, **E** 100 μm

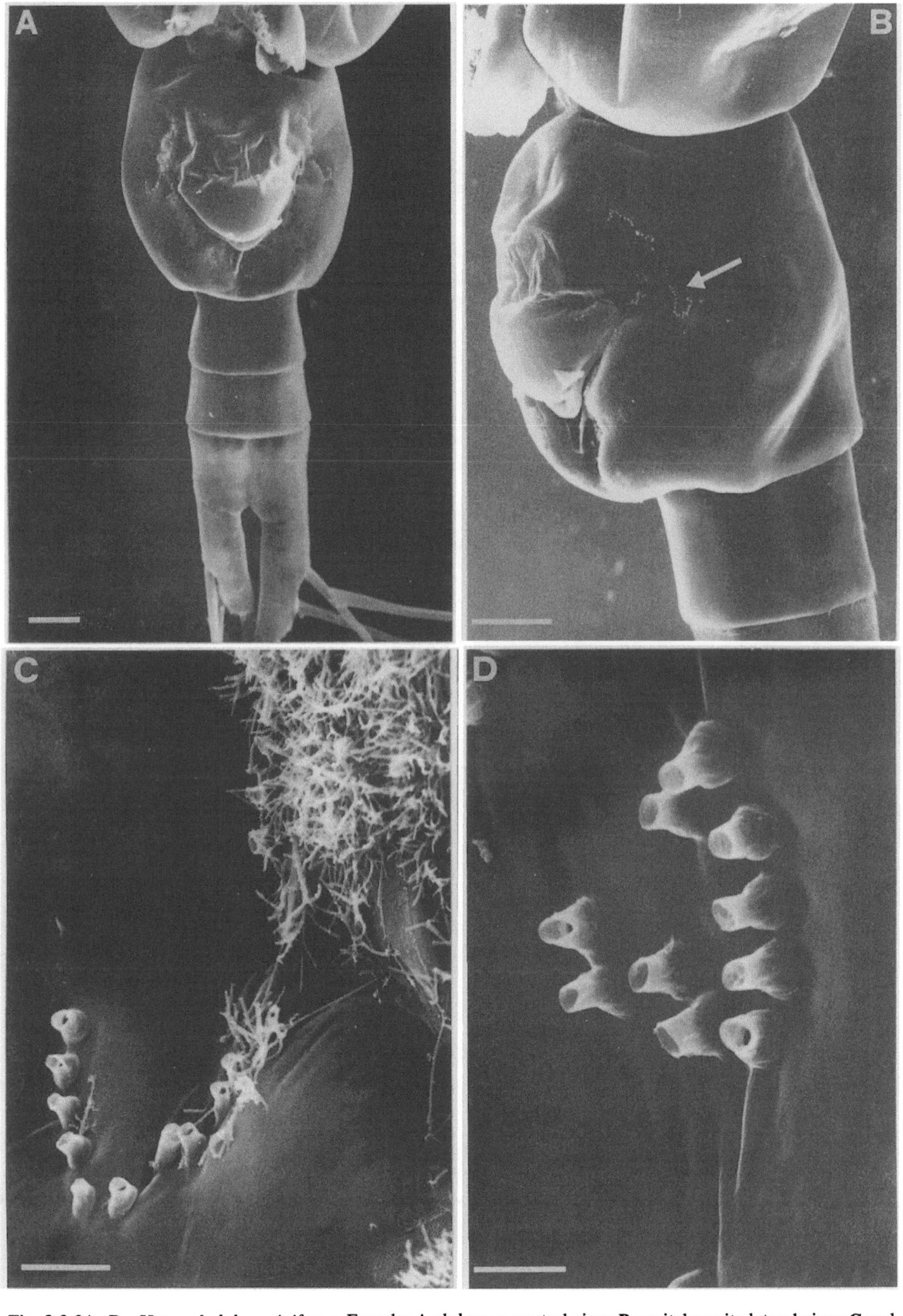

Fig. 3.3.3A–D. *Heterorhabdus spinifrons*. Female: **A** abdomen ventral view; **B** genital somite lateral view; **C** and **D** detail of integumental cone-like pores on either side of genital aperture (*arrow* in **B**). *Bars*: **A**, **B** 100 µm; **C** 10 µm; **D** 5 µm

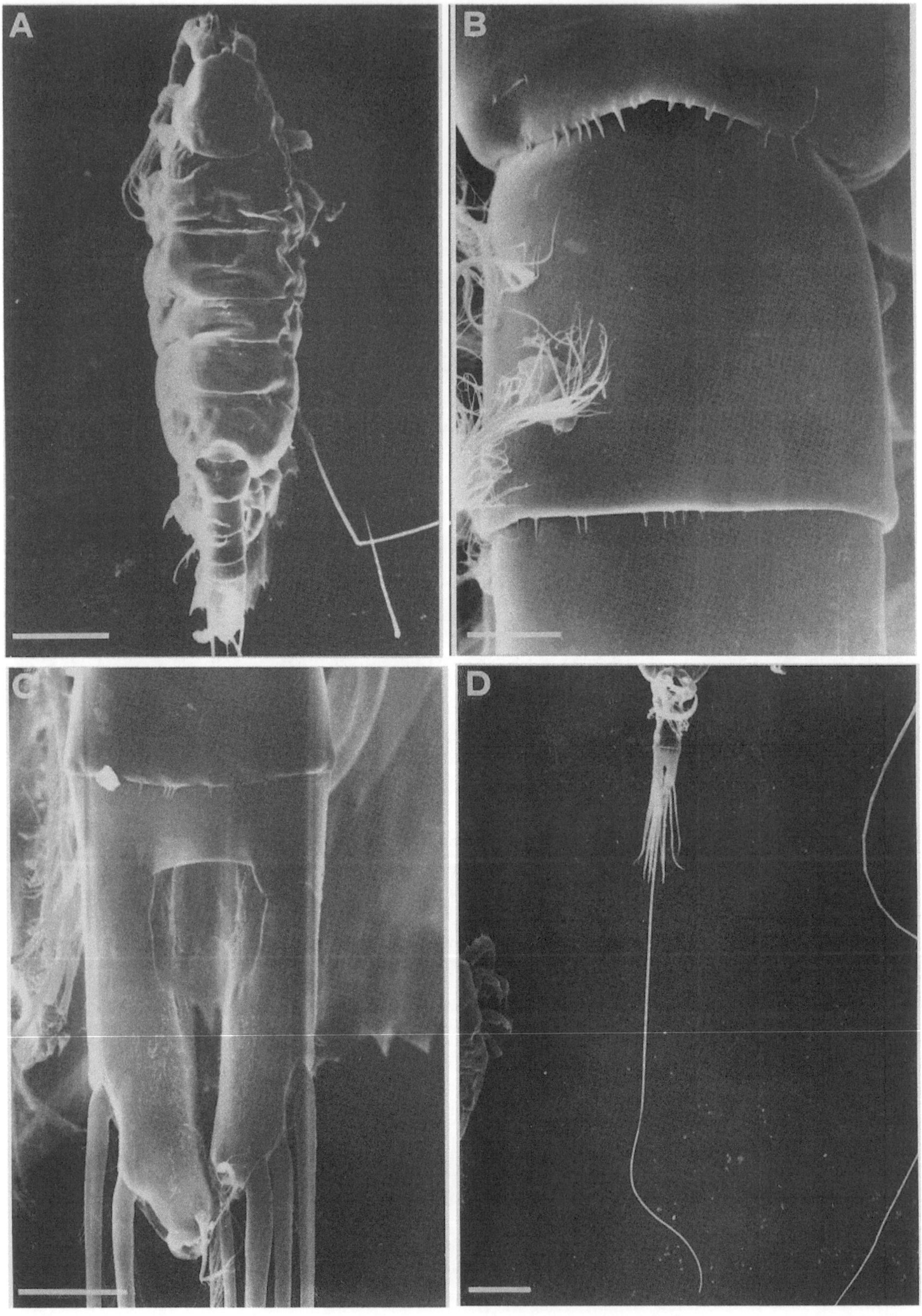

Fig. 3.3.4A–D. *Heterorhabdus spinifrons*. Male: **A** whole animal dorsal view; **B** posterior margins of abdominal somites; **C** asymmetrical caudal rami dorsal view; **D** characteristic long spine on left caudal ramus, ventral view. *Bars*: **A**, **D** 500 μm; **B** 50 μm; **C** 100 μm

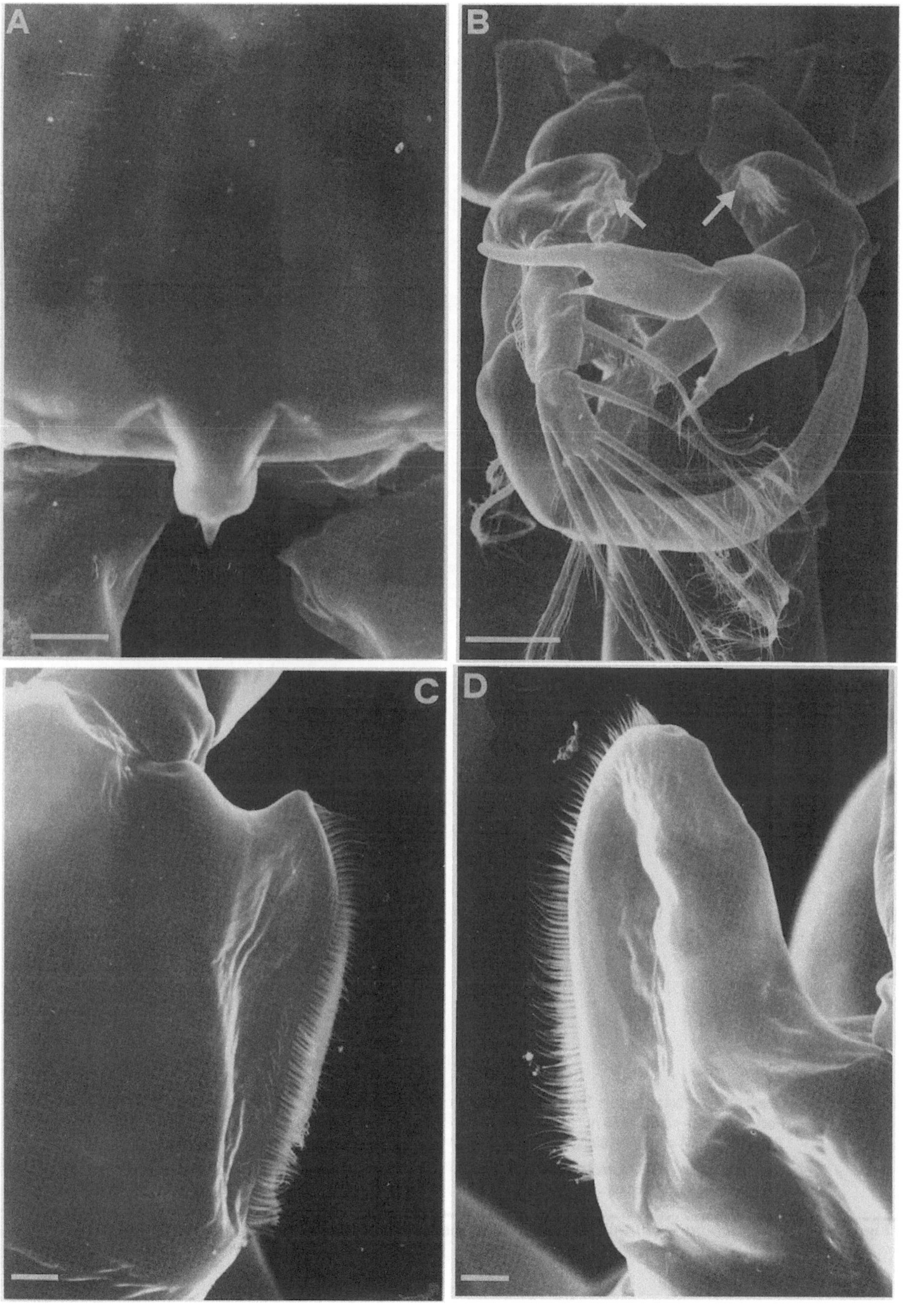

Fig. 3.3.5A–D. *Heterorhabdus spinifrons*. Male: **A** protuberance on forehead ending with a short spine, dorsal view seen from above; **B** P5; **C** and **D** ear-like lobes on inner margins of P5 B2 (*arrows* in **B**). *Bars*: **A** 50 μm; **B** 100 μm; **C**, **D** 10 μm

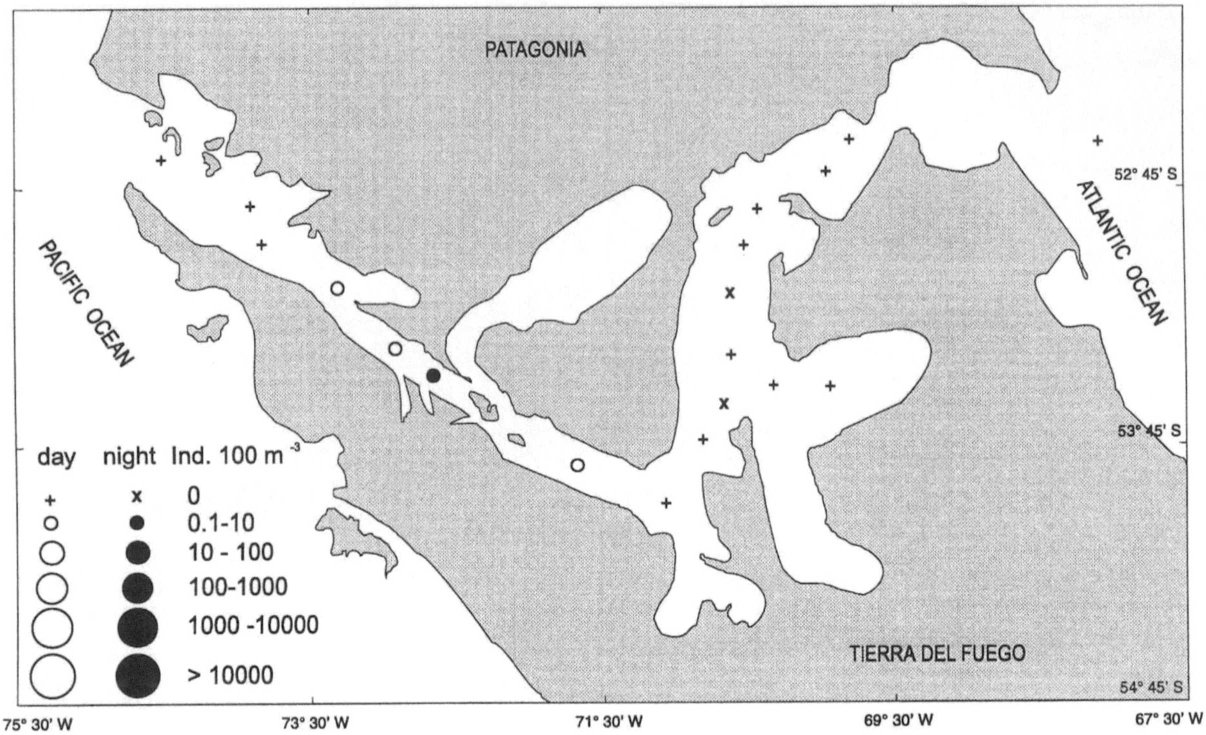

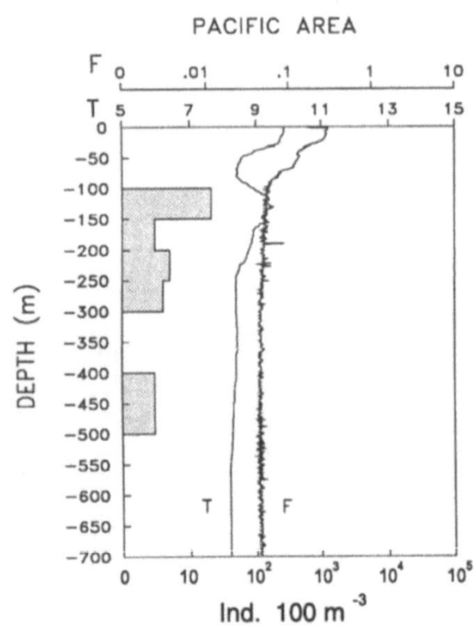

Depth range of occurrence = **100 - 500** m
Population core = **100 - 150** m
Temperature (T) = **8.20 - 9.80** °C
Oxygen = **348.4 - 437.2** µmol dm^{-3}
Chl fluorescence (F) = **0.04 - 0.39**

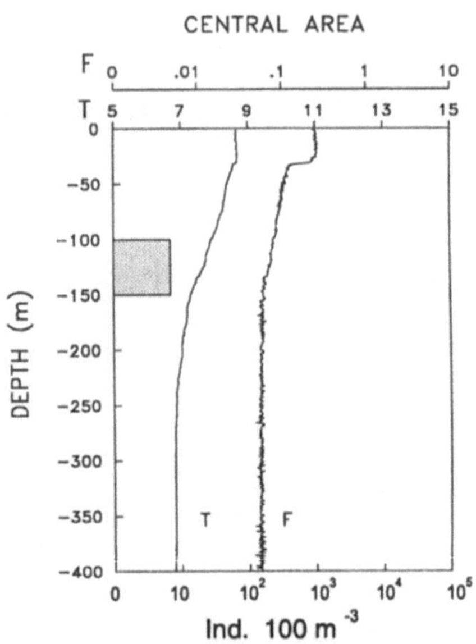

Depth range of occurrence = **100 - 150** m
Population core = **100 - 150** m
Temperature (T) = **6.70 - 9.01** °C
Oxygen = **346.1 - 407.3** µmol dm^{-3}
Chl fluorescence (F) = **0.05 - 0.27**

Fig. 3.3.6. *Heterorhabdus spinifrons*. Distribution in the Straits of Magellan in February–March 1991

3.4 *Lucicutia clausi* Giesbrecht 1889

Leuckartia clausi Giesbrecht, 1889: p. 812; 1892: p. 359, pl. 19 figs. 5, 6, 12–14, 24, 27, pl. 37 fig. 37.
Lucicutia clausi Giesbrecht and Schmeil, 1898: p. 111 – Tanaka, 1963: p. 38, fig. 166 – Hülsemann, 1966: p. 709, figs. 12, 13, 87–91, 123–126 – Park, 1970: p. 517, figs. 211–219 – Ali-Khan and Ali-Khan, 1982: p. 265, figs. 1–4.

Body length –
females: 2.02–2.25 mm (20 specimens)
males: 1.90–2.14 mm (16 specimens)

Female. Cephalothorax 2.3 times longer than wide. Head with poorly developed lateral processes that are rounded and minute in form. Al 25-segmented, about 2 segments longer than end of furca. Proportional lengths of cephalothorax and abdomen 65:35; proportional lengths of abdominal somites and furca 31:10:10:12:37 = 100. Genital somite longer than wide in dorsal view, with conspicuous ventral protrusion. Anal somite longer than preceeding. In P1, endopod 2-jointed and exopod 3-jointed. In P2–P5, both endopods and exopods 3-jointed. In P2–P4, number and position of setae is the same. P5 Re3 with 1 apical and 2 lateral spines and 3 internal setae. Apical spines of exopods progressively smaller from P1 to P5 where they are about 1/4 length of last joint.

Male. General body shape as in female. Proportional lengths of cephalothorax and abdomen 60:40; proportional lengths of abdominal somites and furca 15:13:12:12:13:35 = 100. P1 Ri 2-jointed. In left P5, the inner distal corner of B2 bearing protrusion with a cluster of denticles varying in number. In right P5, inner margin of B2 with small tooth-like projection; Ri2 leaf-like with 6 setae. Last joint of right exopod elongated and recurved, with long hairs on inner side.

Remarks

According to Giesbrecht (1892), the spine-like protrusions on the sides of the head of this species can vary enormously in shape, from straight to turned backwards. They can also be greatly reduced or lacking altogether. In our specimens, these protrusions are scarsely developed and are similar to those reported by Hülsemann (1966) for the tropical Atlantic. Our specimens differ from the latter forms, however, since they are more slender and have longer caudal rami. They are also similar to those reported by Ali-Khan and Ali Khan (1982) for the Indian Ocean but in this case the caudal rami are even shorter than those described for the tropical Atlantic. In lateral view, the head protrudes over the first thoracic somite in almost all our specimens.

Distribution

L. clausi has been recorded in offshore cold, temperate and warm waters of all oceanic regions (Hülsemann 1966) and also in the Mediterranean Sea (Scotto di Carlo et al. 1984) and in Chilean waters (Fagetti 1962). It has been reported as occurring from the surface to 3000 m (de Decker and Mombeck 1965), but other authors report its occurrence at shallower depths up to 1000 m (Deevey and Brooks 1977). Maximum abundances for this species in the Mediterranean are from 150–300 m (Scotto di Carlo et al. 1984). In the present study, *L. clausi* was present from 50–500 m throughout the Pacific area of the Straits of Magellan, with maximum abundances from 400–500 m.

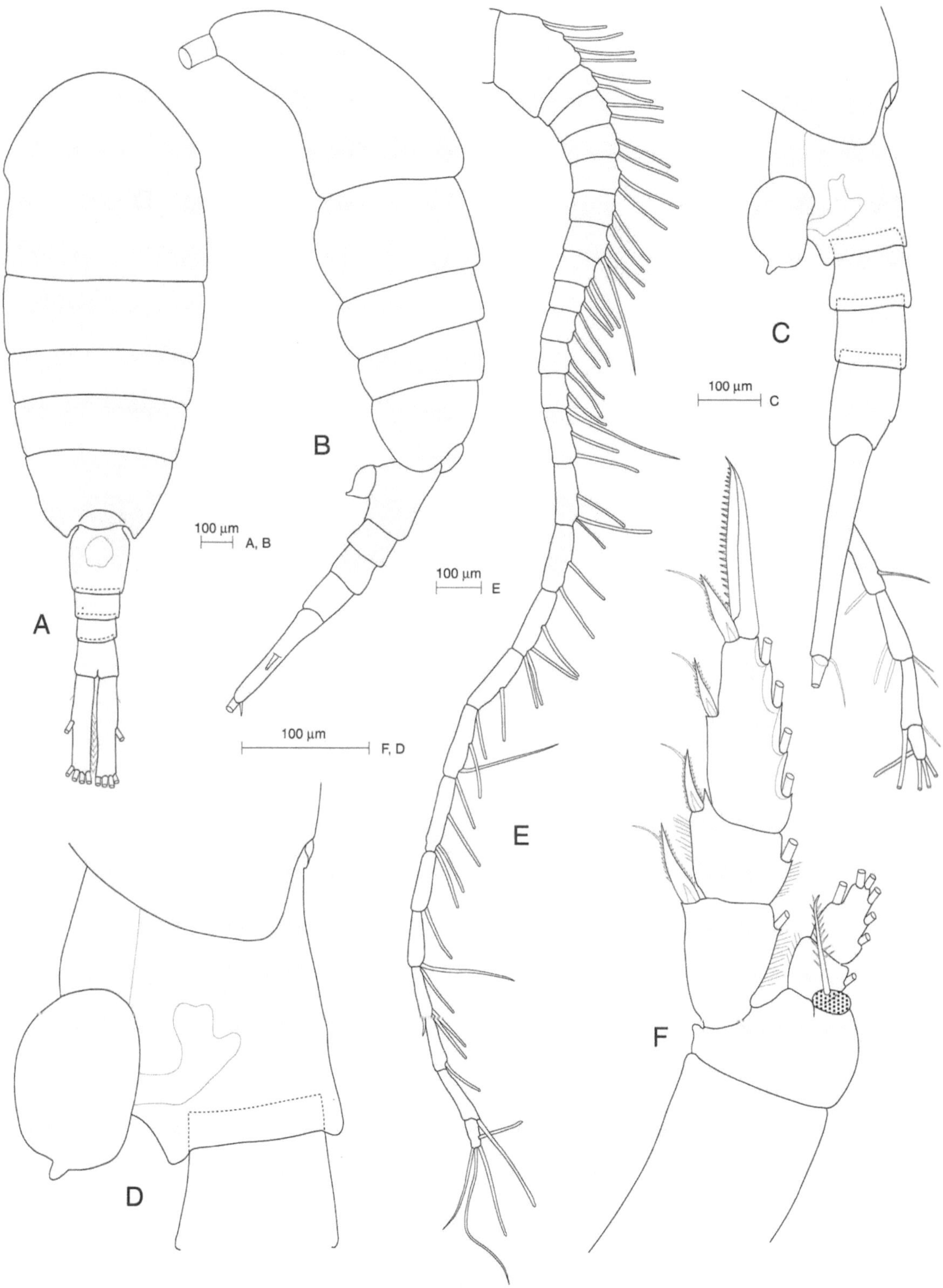

Fig. 3.4.1A–F. *Lucicutia clausi*. Female: **A** whole animal dorsal view; **B** whole animal lateral view; **C** abdomen lateral view; **D** genital somite with conspicuous ventral protrusion in lateral view; **E** A1; **F** P1

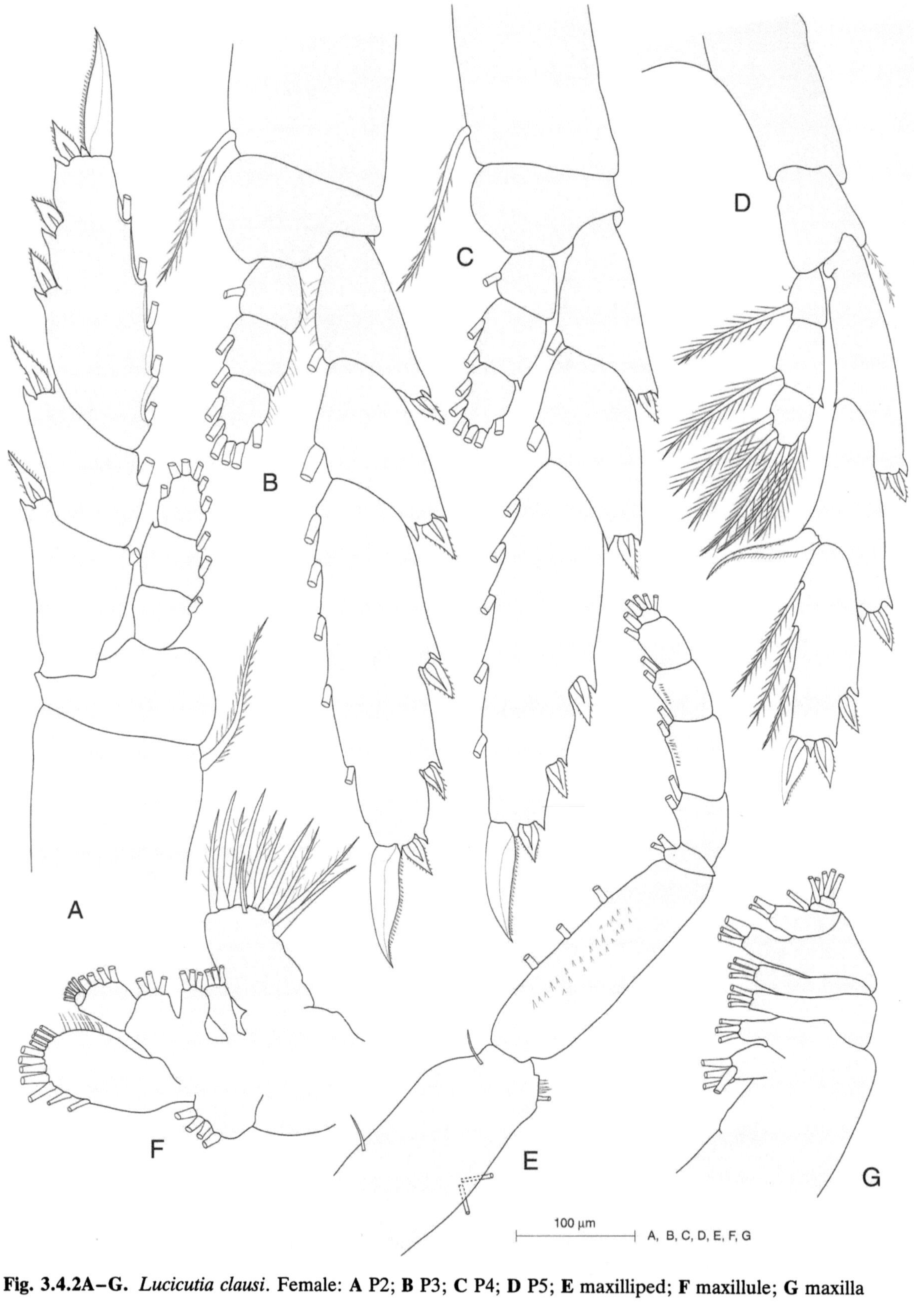

Fig. 3.4.2A–G. *Lucicutia clausi*. Female: **A** P2; **B** P3; **C** P4; **D** P5; **E** maxilliped; **F** maxillule; **G** maxilla

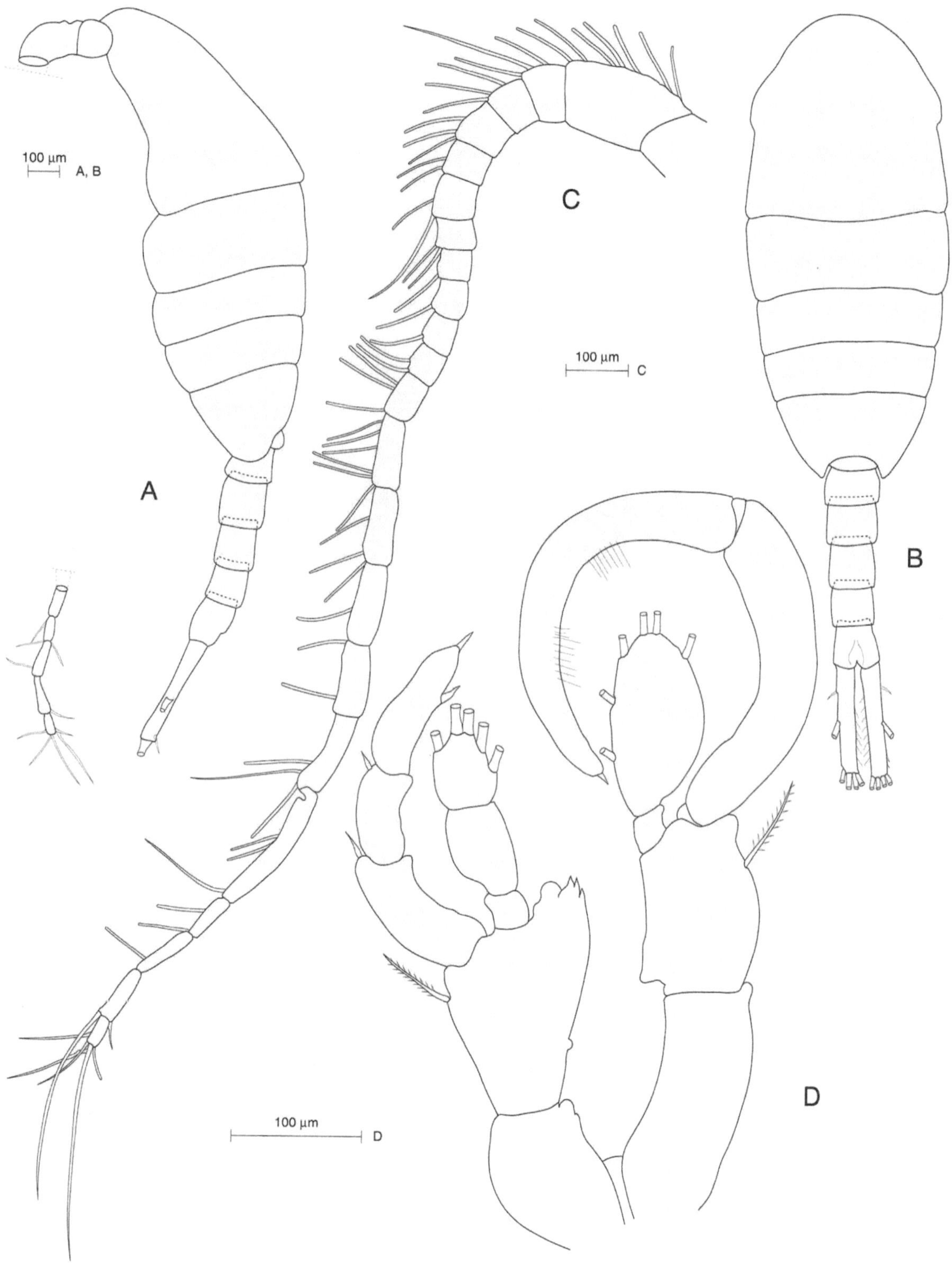

Fig. 3.4.3A–D. *Lucicutia clausi*. Male: **A** whole animal lateral view; **B** whole animal dorsal view; **C** A1; **D** P5

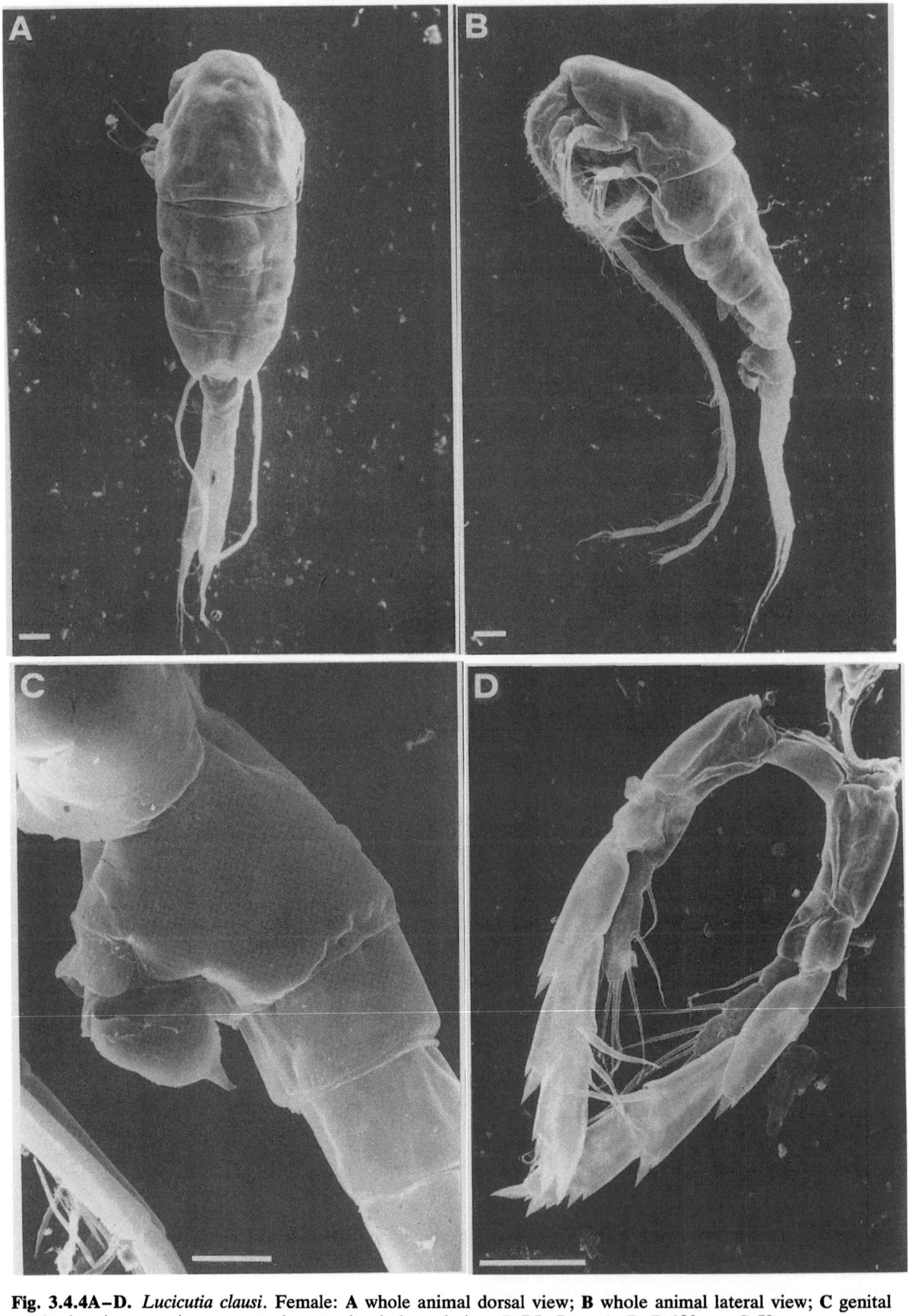

Fig. 3.4.4A–D. *Lucicutia clausi*. Female: **A** whole animal dorsal view; **B** whole animal lateral view; **C** genital somite bearing conspicuous ventral protrusion in lateral view; **D** P5. *Bars*: **A**, **B**, **D** 100 µm; **C** 50 µm

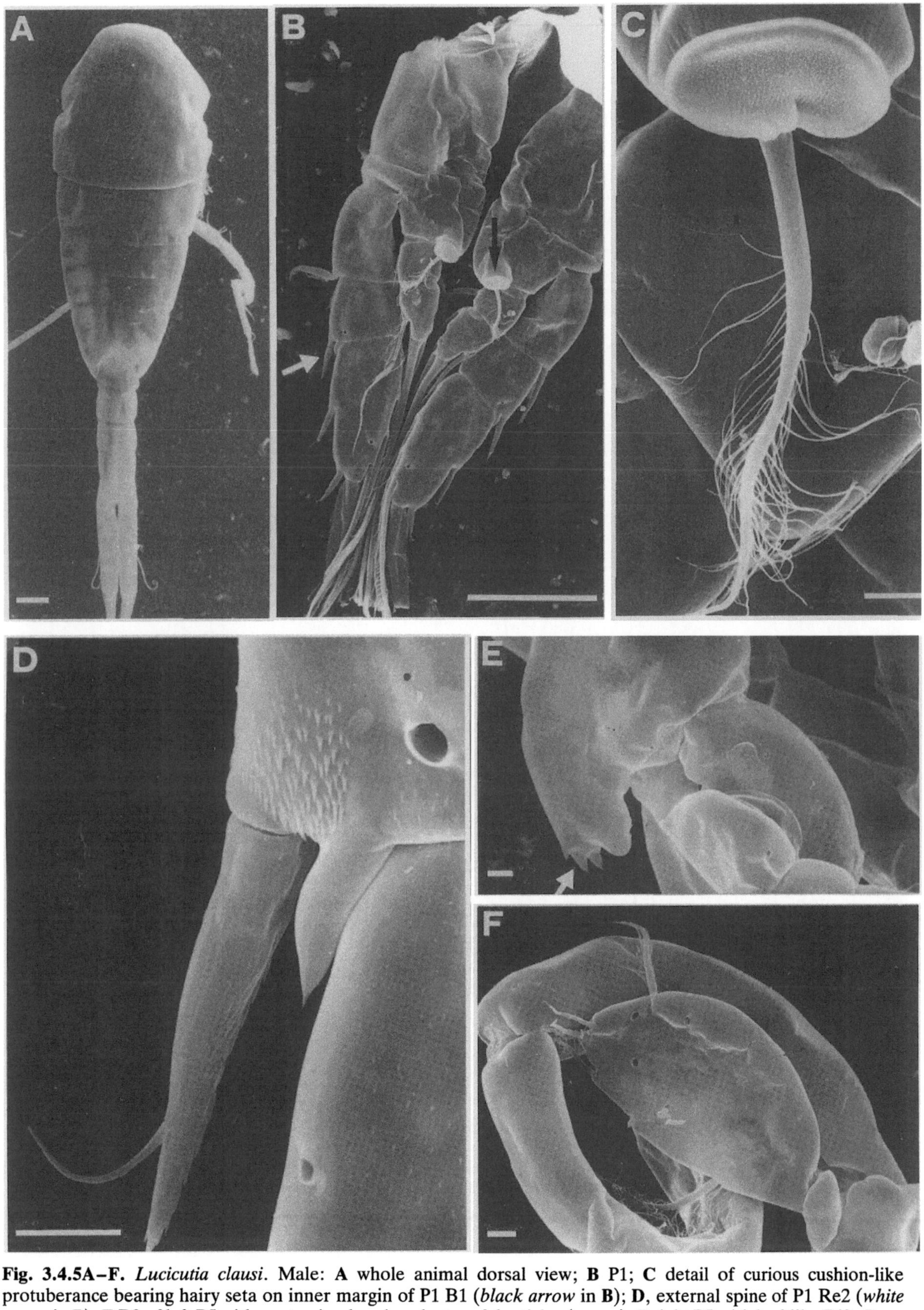

Fig. 3.4.5A–F. *Lucicutia clausi*. Male: **A** whole animal dorsal view; **B** P1; **C** detail of curious cushion-like protuberance bearing hairy seta on inner margin of P1 B1 (*black arrow* in **B**); **D**, external spine of P1 Re2 (*white arrow* in **B**); **E** B2 of left P5 with protrusion bearing cluster of denticles (*arrow*); **F** right P5 with leaf-like Ri2. *Bars*: **A** 100 μm; **C** 50 μm; **B**, **D**, **E**, **F** 10 μm

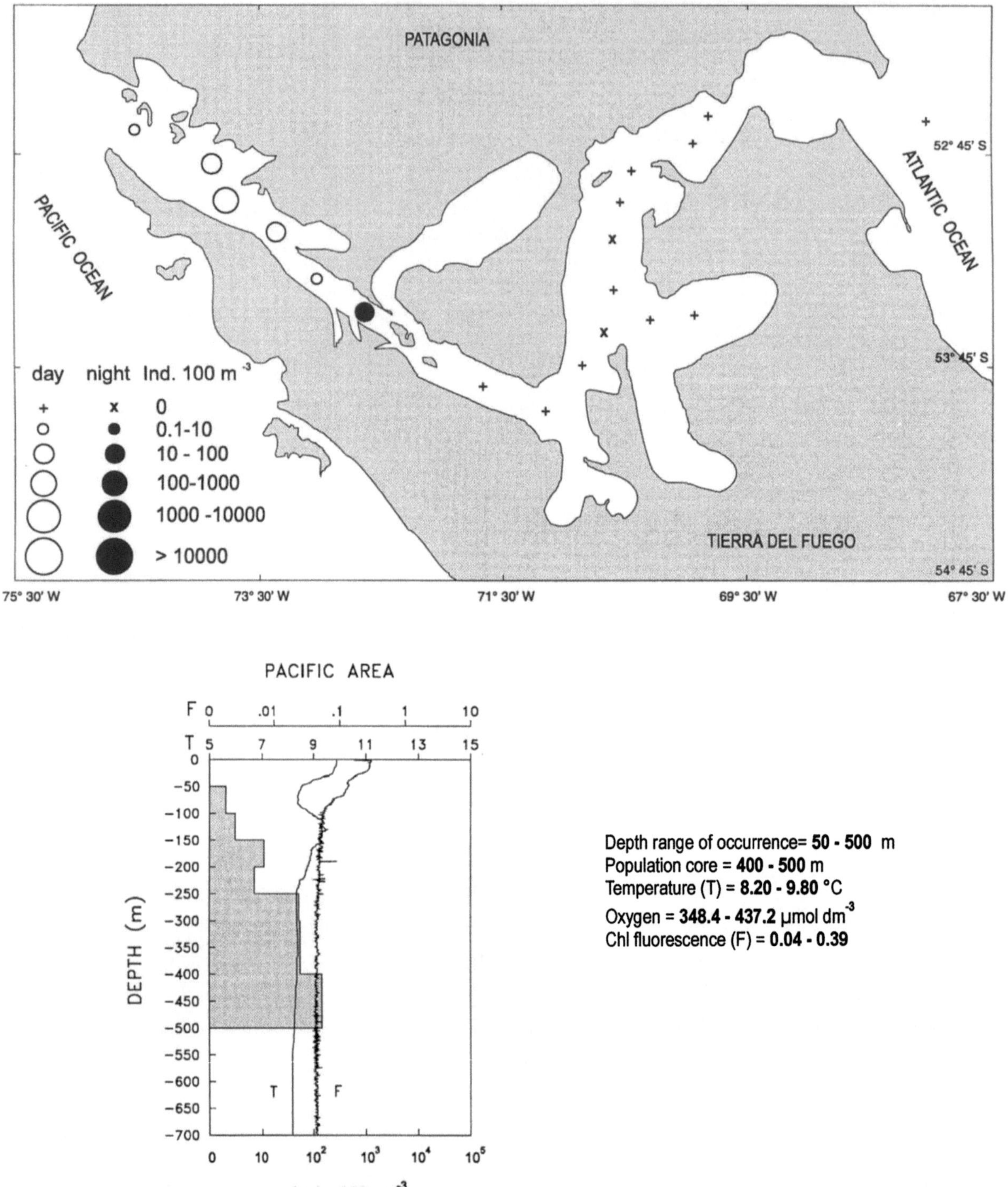

Fig. 3.4.6. *Lucicutia clausi.* Distribution in the Straits of Magellan in February–March 1991

3.5 *Metridia lucens* Boeck 1864

Metridia lucens Boeck, 1864: p. 14 – Sars, 1903: p. 113, pl. 77 – Brodskii, 1950: p. 294, fig. 200 – Ramirez, 1969: pl. 68, pl. 12 figs. 84, 86, 87, 89.
Paracalanus hibernicus Brady and Robertson, 1873: p. 126, pl. 8 figs. 1–3.
Metridia hibernica Giesbrecht, 1892: pp. 61, 341, pl. 33 figs. 2, 12, 16, 22, 28, 36, 39.

Body length –
females: 2.62–3.21 mm (25 specimens)
males: 1.80–2.00 mm (20 specimens)

Female. Cephalothorax elongated and ovate. Head convex on dorsal side and strongly curved in lateral view. Posterior corners of last thoracic somite with small acute apex. A1 somewhat longer than last thoracic somite. Proportional lengths of cephalothorax and abdomen 65:35; proportional lengths of abdominal somites and furca 41:24:16:19 = 100. P5 uniramous, 3-jointed, sometimes with defined 4th segment. Distal joint bearing apical setae and one rather long spine on outer margin (in rare cases there are 2 spines). Second joint generally with similar outer spine. P5 highly variable for the number of joints and the presence or absence of an outer spine on 2nd joint. P5 characterized by relative length of inner apical seta with respect to distal joint, i.e. this seta clearly longer than two others.

Male. General shape as female, but dorsal side of head smoother in lateral view. Proportional lengths of cephalothorax and abdomen 63:37; proportional lengths of abdominal somites and furca 14:19:18:18:13:18 = 100. P5 asymmetrical. Third joint of right P5 bearing spiniform process on inner side as long or slightly longer than terminal joint. Process denticulated distally along inner margin. Distal joint of right leg larger than corresponding joint of left leg. Left P5 with 2nd joint bearing thick bundle of hairs on distal inner margin; 3rd joint with rather long spine and round, lamellar process. Fourth joint bearing 2 strong spines on inner margin.

Remarks

The specimens examined closely resemble *M. lucens* but show a marked variability in many morphological features, especially in size, the lateral profile of the female head, female P5, and length of the spiniform process on right P5 of the male. Interestingly, some of our male specimens had P5 with reversed rami. Brodskii (1950) described *M. pacifica*, closely allied to *M. lucens*, as a new species from northern cold waters of the Pacific. According to Brodskii's taxonomic key, our females concord with *M. lucens*. On the other hand, males have caudal rami that are longer than the anal somite and the left ramus of the P5 bears 2 spines on the inner margin as in *M. pacifica*. This is why this species was identified as *M. pacifica* in a previous study on copepods of the Straits of Magellan (Mazzocchi and Ianora 1991). In fact, the systematic literature is contradictory and insufficient to clearly distinguish between the two species. Such uncertainties emerge also from the work of Tanaka (1963), who found specimens resembling *M. pacifica* but reported the female P5 of *M. lucens* in his fig. 5c (p. 22).

Distribution

M. lucens is a very abundant oceanic boreal and temperate species, widespread in the North Atlantic (Giesbrecht 1892; Sars 1903; Wilson 1932). It has been found south of New Zealand to within the Antarctic Circle (Farran 1929). It has also been reported off the coasts of Argentina (Ramirez 1970, 1971) and Chile (Marin and Antezana 1985; Hirakawa 1989). Mazzocchi and Ianora (1991) found this species at Pacific stations in the Straits of Magellan, where it was reported as *M. pacifica*. The occurrence of this species in the Pacific has been questioned by a number of authors (Brodskii 1950; Tanaka 1963) since it may have been confused with *M. pacifica*. The species undergoes pronounced diel vertical migrations (Wilson 1932). In the present study, *M. lucens* was present throughout the study area up to the entrance of the Primera Angostura. It was distributed at all depths but with maximum abundances from 150–200 m in the Pacific area and from 100–150 m in the Central area of the Straits of Magellan.

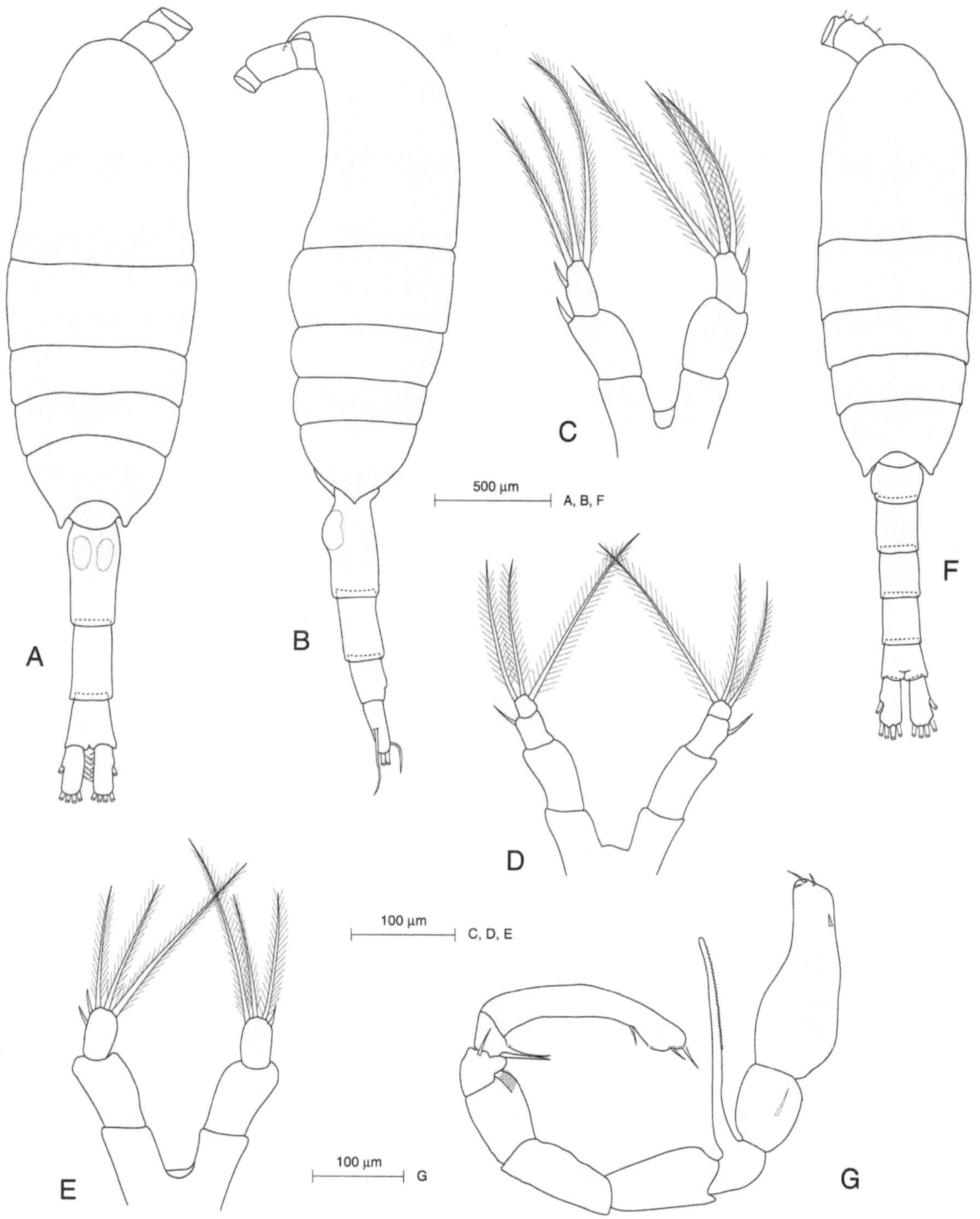

Fig. 3.5.1A–G. *Metridia lucens*. Female: **A** whole animal dorsal view; **B** whole animal lateral view; **C**, **D**, **E** P5 showing variations in number of joints and spines on outer margins. Male: **F** whole animal dorsal view; **G** P5

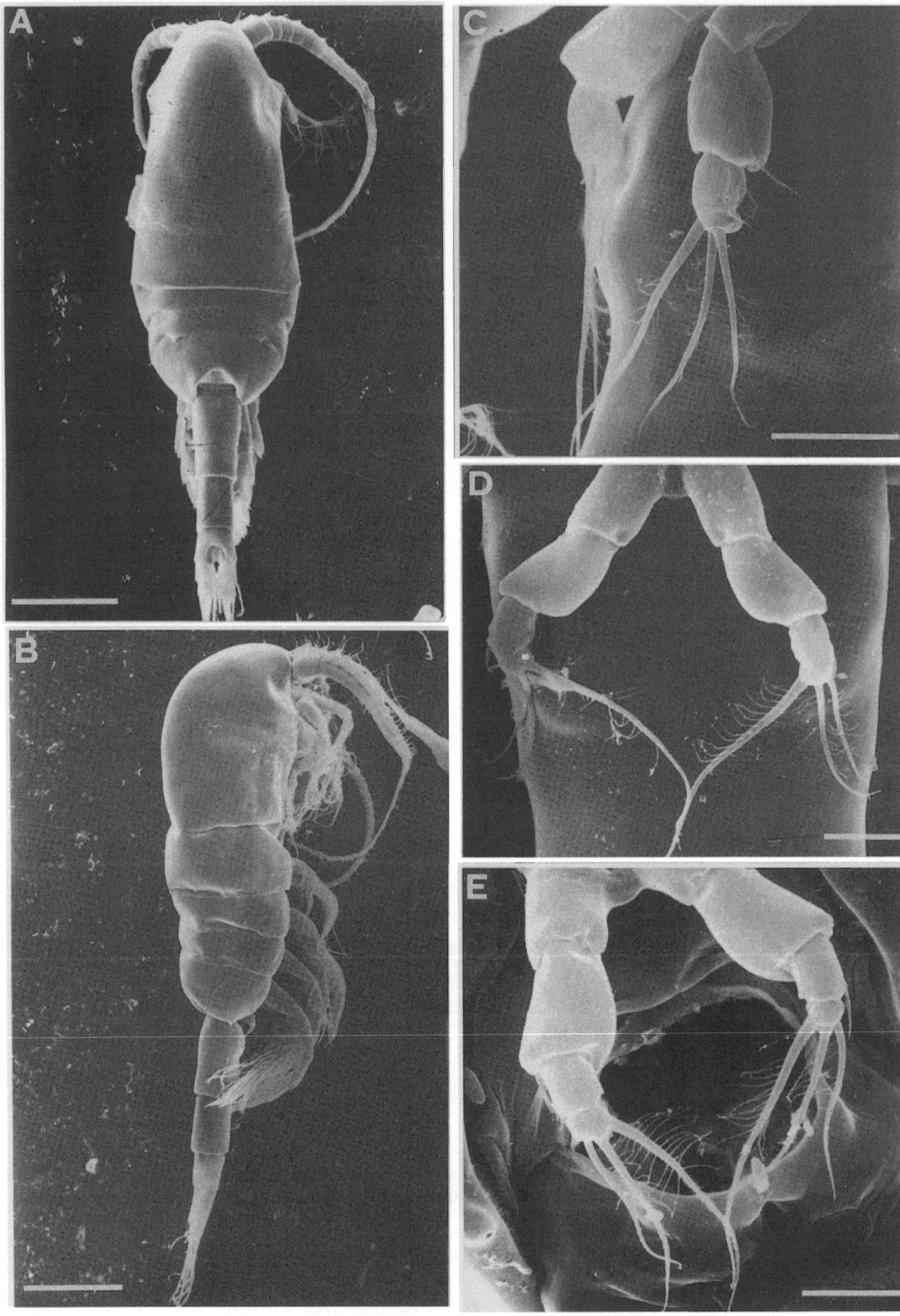

Fig. 3.5.2A–E. *Metridia lucens*. Female: **A** whole animal dorsal view; **B** whole animal lateral view; **C**, **D**, **E** variations in the number of joints in the P5. *Bars*: **A**, **B** 500 µm; **C**, **D**, **E** 50 µm

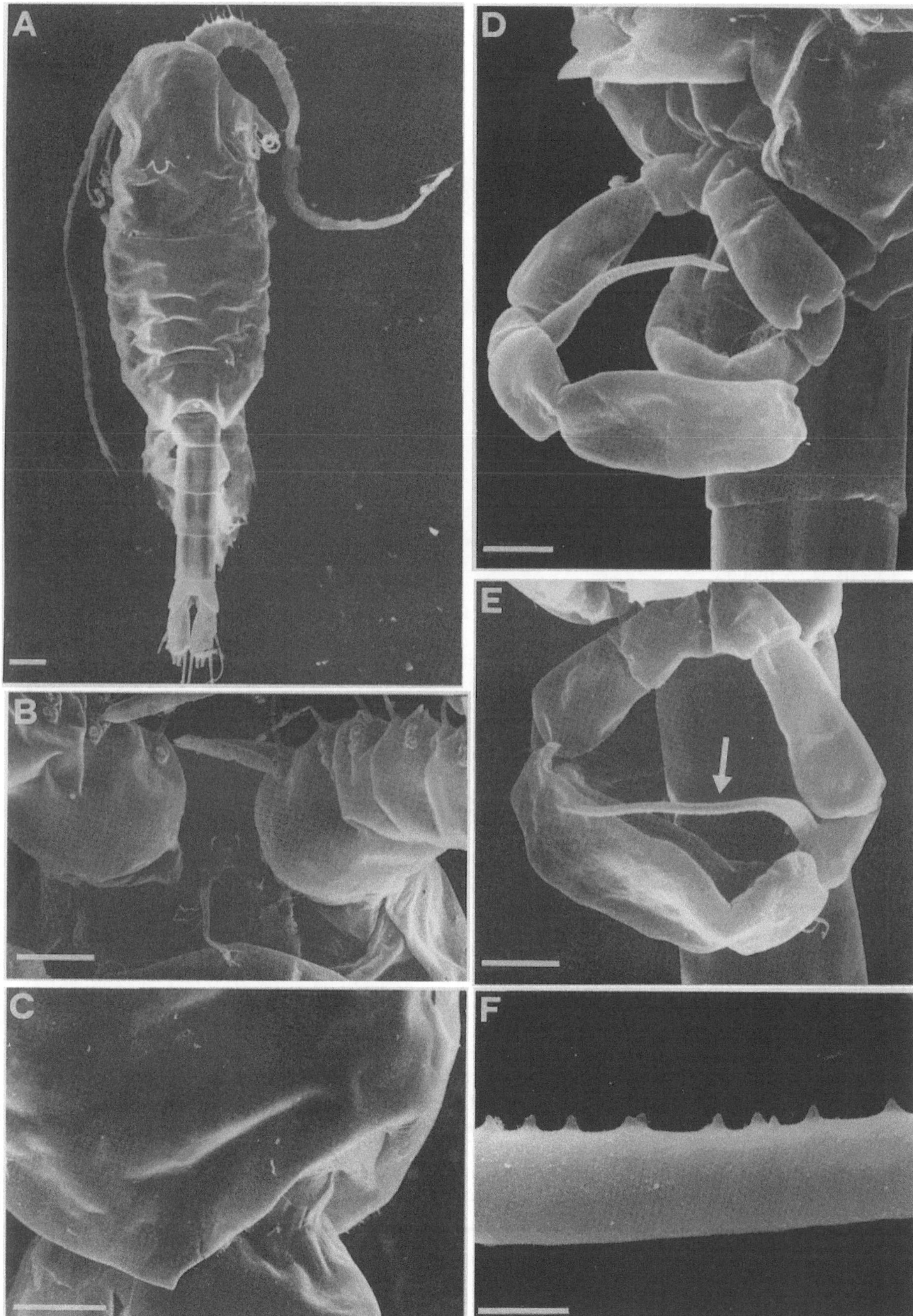

Fig. 3.5.3A–F. *Metridia lucens*. Male: **A** whole animal dorsal view; **B** forehead with rostral filaments; **C** left postero-lateral corner of last thoracic somite; **D**, **E** P5 with reversed rami; **F** detail of denticles on spiniform process of P5 (*arrow* in **E**). *Bars*: **A** 100 µm; **B**, **C**, **D**, **E** 50 µm; **F** 5 µm

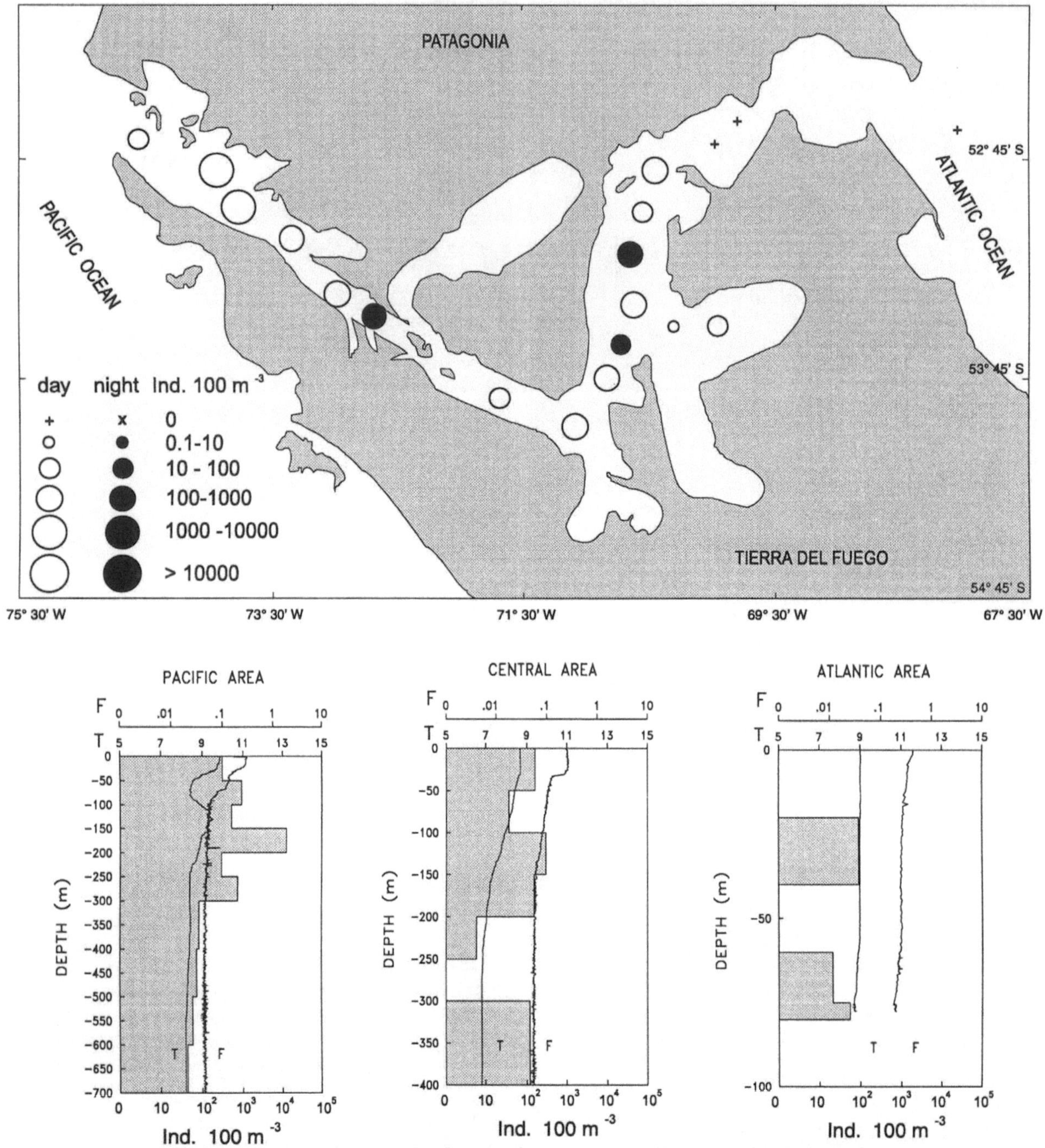

Depth range of occurrence = **0 - 700** m
Population core = **150 - 200** m
Temperature (T) = **8.20 - 9.80** °C
Oxygen = **348.4 - 437.2** μmol dm^{-3}
Chl fluorescence (F) = **0.04 - 0.39**

Depth range of occurrence = **0 - 400** m
Population core = **100 - 150** m
Temperature (T) = **6.70 - 9.01** °C
Oxygen = **346.1 - 407.3** μmol dm^{-3}
Chl fluorescence (F) = **0.05 - 0.27**

Depth range of occurrence = **0 - 80** m
Population core =20 **- 40** m
Temperature (T) = **8.68 - 9.02** °C
Oxygen = **390.4 - 410.9** μmol dm^{-3}
Chl fluorescence (F) = **0.16 - 0.44**

Fig. 3.5.4. *Metridia lucens*. Distribution in the Straits of Magellan in February–March 1991

3.6 *Pleuromamma robusta* Dahl 1893

Pleuromma robustum Dahl, 1893: p. 104
Pleuromamma robusta Sars, 1903: p. 115, pls. 78, 79 – Steuer, 1933: p. 11, figs. 44, 45 – Ramirez, 1969: p. 73, figs. 95, 96.
Pleuromamma robusta f. *antarctica* Steuer, 1931: p. 7, fig. 2; 1933: p. 11, figs. 56, 57 – Vervoort, 1951: p. 123, figs. 68, 69.

Body length –
females: 3.45–3.82 mm (5 specimens)
males: 3.14–3.27 mm (2 specimens)

Female. Body robust. Cephalothorax 2.5 times longer than wide. Proportional lengths of cephalothorax and abdomen 68:32. Frontal part of head with elongated protrusion ending in rostrum composed of two filiform processes. Pigmented spot on right side of cephalothorax. Postero-lateral corners of thorax rounded. Abdomen symmetrical. Proportional length of abdominal somites and furca 41:24:17:18 = 100. Genital somite symmetrical in dorsal view, with a pronounced ventral swelling in lateral view. Anal somite produced into distinct tooth-like corner on each side with three pores at tip; dorsal distal margin bordered with a row of denticles and concave area ornated with short setae. Caudal rami slightly asymmetrical, left one slightly longer; inner edge of both rami hairy; dorsal surface of left ramus with row of long setules along midline. P5 with 3 free symmetrical joints. First joint with hairy spine at extreme outer corner. Second joint with long hairs along inner border and few hairs along outer edge; outer edge bears one short naked spine. Terminal joint with long hairs along both borders. This joint ends with one long and two much shorter spines covered with thorns. Between shorter spines are two short tubules. These tubules appear as small spines at the light microscope. They are not generally reported in figures from other authors; only Vervoort's (1951) drawings of the P5 of *P. robusta* f. *antarctica* are exactly like ours.

Male. Same shape as female. Basal segment of Al armed with flattened spatulate aesthetascs. Left Al geniculated. Cephalothorax 2.6 times longer than wide. Proportional lengths of cephalothorax and abdomen 67:33. Pigmented spot on right side of body in the two specimens we collected. Abdomen symmetrical and slender; proportional lengths of abdominal somites and furca 15:18:16:19:16 = 100. Third abdominal somite with a clutch of long hairs on posterior margin of left side. Anal somite and caudal rami with same ornamentation as in female. P5 asymmetrical; left ramus longer than right with terminal joint narrow and oblong; terminal joint of right ramus considerably widened, laminate and securiform. According to Wilson (1932), the reversed rami of the P5 with respect to other species of the genus is a major character of the male of this species.

Remarks

This species was briefly mentioned by Dahl (1893) as occurring from 100 to 1500 m in the Atlantic Ocean, but the author gives no detailed description or figures. Steuer (1931) divided *Pleuromamma robusta* into two forms: f. *typica* [= *P. robusta robusta* (Dahl 1893)] and f. *antarctica* [= *P. robusta antarctica* (Steuer 1931)]. This latter form differs from the typical one mainly by the scantiness of the hairs covering the abdominal somites and furca, both in females and males. The A1 in females is ornated with strong hooks on the basal joints in the f. *typica*; in the f. *antarctica*, the proximal joints of A1 have no spines or teeth, but the various setae have a prominent base. In the f. *antarctica* male no spinules, tubes or hairs are found along the edge of the anal operculum. Our specimens show intermediate characters: the females lack hook-like teeth on the A1 and the P5 is like the drawings of Vervoort (1951) for *P. robusta* f. *antarctica*; both sexes have an ornamentation on the anal operculum that should be lacking in the f. *antarctica*. Furthermore, they are slightly smaller than normally reported for *P. robusta* f. *typica* Steuer, 1931.

Distribution

P. robusta is widely distributed, inhabiting intermediate depths but also occurring at the surface. Farran (1929) reported it as a common species in a small area south of New Zealand between 51° and 53°S. Sars (1903) found a few specimens north of the Faroe Islands (200–400 m) and from other areas of Norway. Ramirez (1969, 1971) reported the f. *typica* in offshore Argentinian waters from 0–600 m. Ramirez and Dinofrio (1976) sampled it in the Antarctic sector of Argentina. Björnberg (1963) found it from 400–600 m off Brazil in waters of Antarctic origin. Mackintosh (1934) reported it from circumpolar Antarctic areas and Hardy and Gunther (1935) near South Georgia from 0–1000 m. Vervoort (1951) found the f. *antarctica* in samples from 0–100 m in the Atlantic sector of the Antarctic. Vervoort (1965a) recorded *P. robusta robusta* in Atlantic waters of intermediate depth and also at the surface, but in lower numbers. Fagetti (1962) cites *P. robusta* in her catalogue of the planktonic copepods of Chile. Roe (1972) provides detailed information on the vertical distribution of *P. robusta*, without separation between the two forms. This author reported *P. robusta* as the least common of the *Pleuromamma* species from the Sound cruise but fairly common in intermediate waters, with maximum day depths at 400 m and extensive vertical migrations at night up to 25 m. In the present study, *P. robusta* was present only from 150–700 m in the Pacific area of the Straits of Magellan.

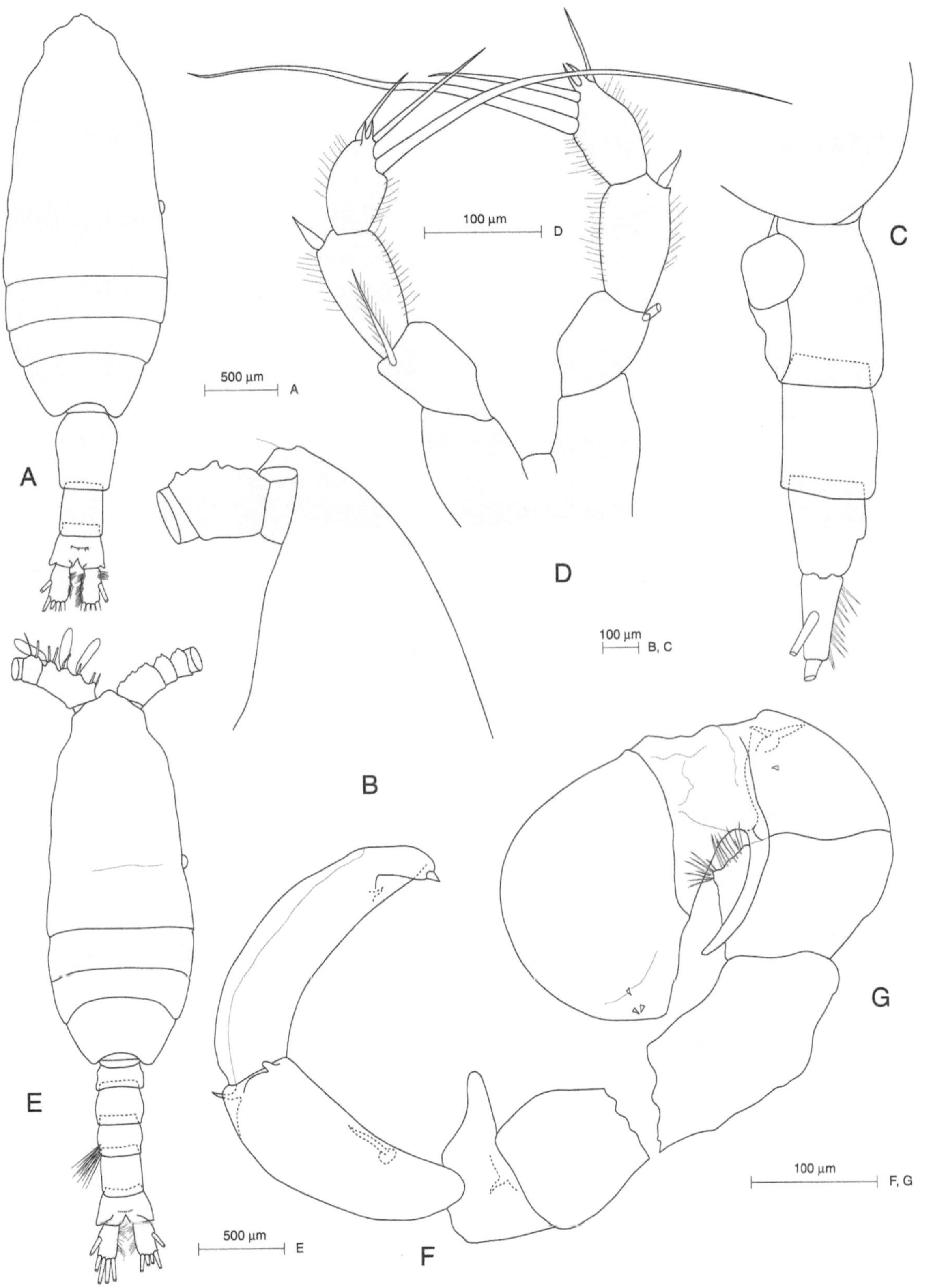

Fig. 3.6.1A–G. *Pleuromamma robusta.* Female: **A** whole animal dorsal view; **B** head lateral view; **C** abdomen lateral view; **D** P5. Male: **E** whole animal dorsal view; **F** left P5; **G** right P5

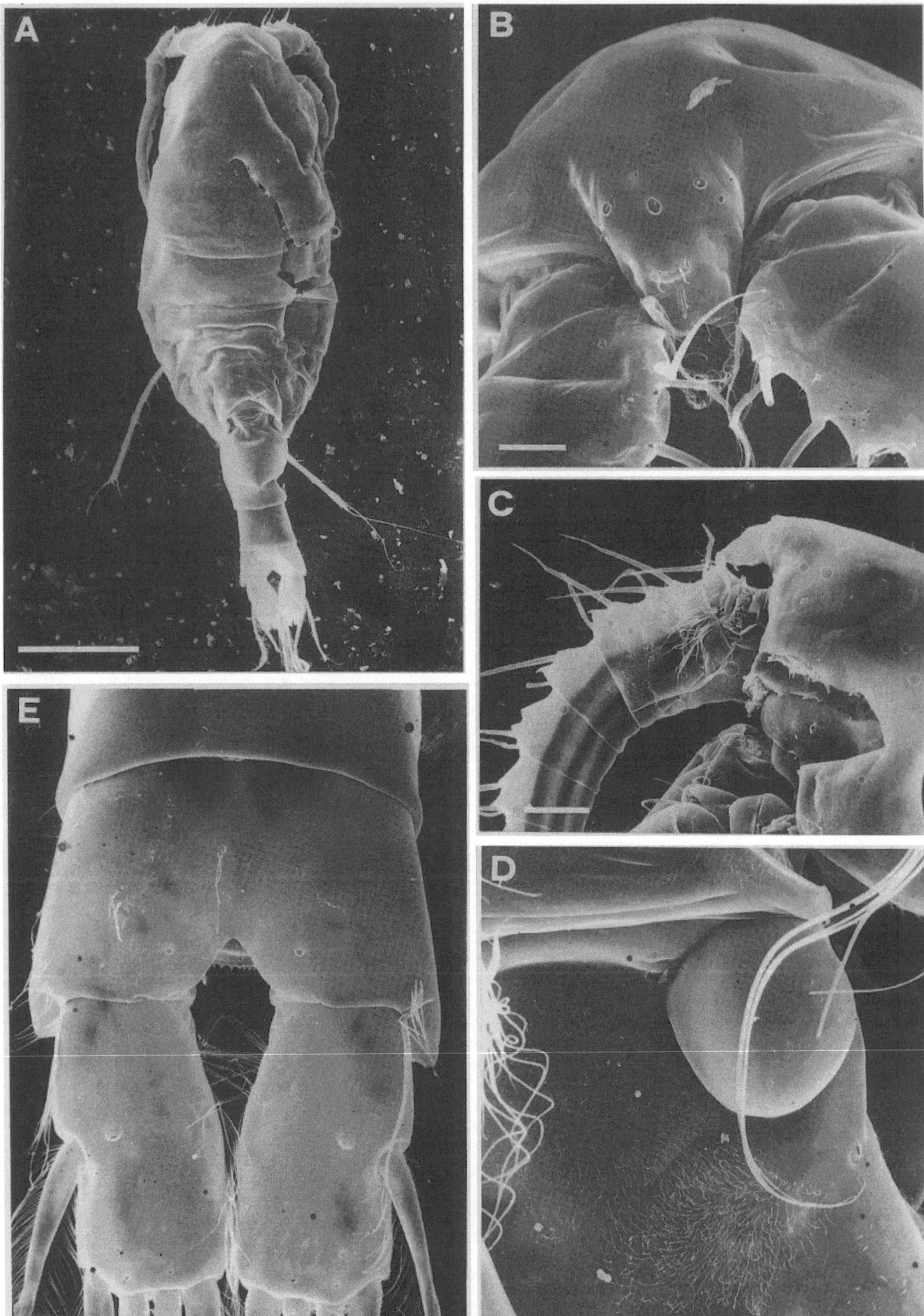

Fig. 3.6.2A–E. *Pleuromamma robusta*. Female: **A** whole animal dorsal view; **B** forehead bearing sensillory pores; **C** head lateral view showing elongated protrusion of forehead ending in rostrum; **D** pigment spot; **E** anal somite and furca ventral view. *Bars* **A** 500 μm; **B**, **D** 50 μm; **C**, **E** 100 μm

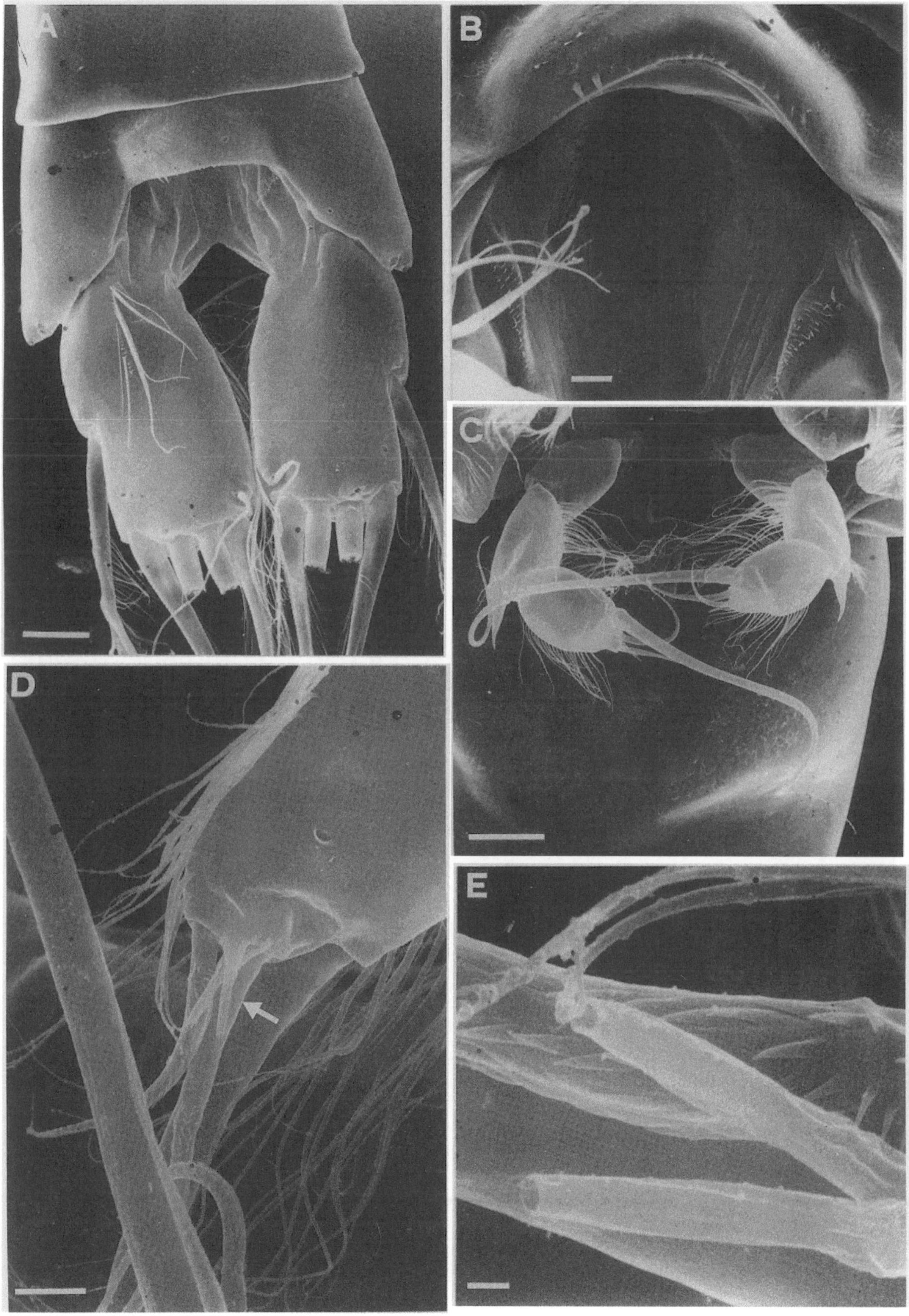

Fig. 3.6.3A–E. *Pleuromamma robusta*. Female: **A** anal somite and furca dorsal view; **B** detail of dorsal margin of anal somite; **C** P5; **D** detail of terminal edge of last joint of P5; **E** tubules on P5 (*arrow* in **D**); *Bars*: **A**, **C** 50 μm; **B**, **D** 10 μm; **E** 1 μm

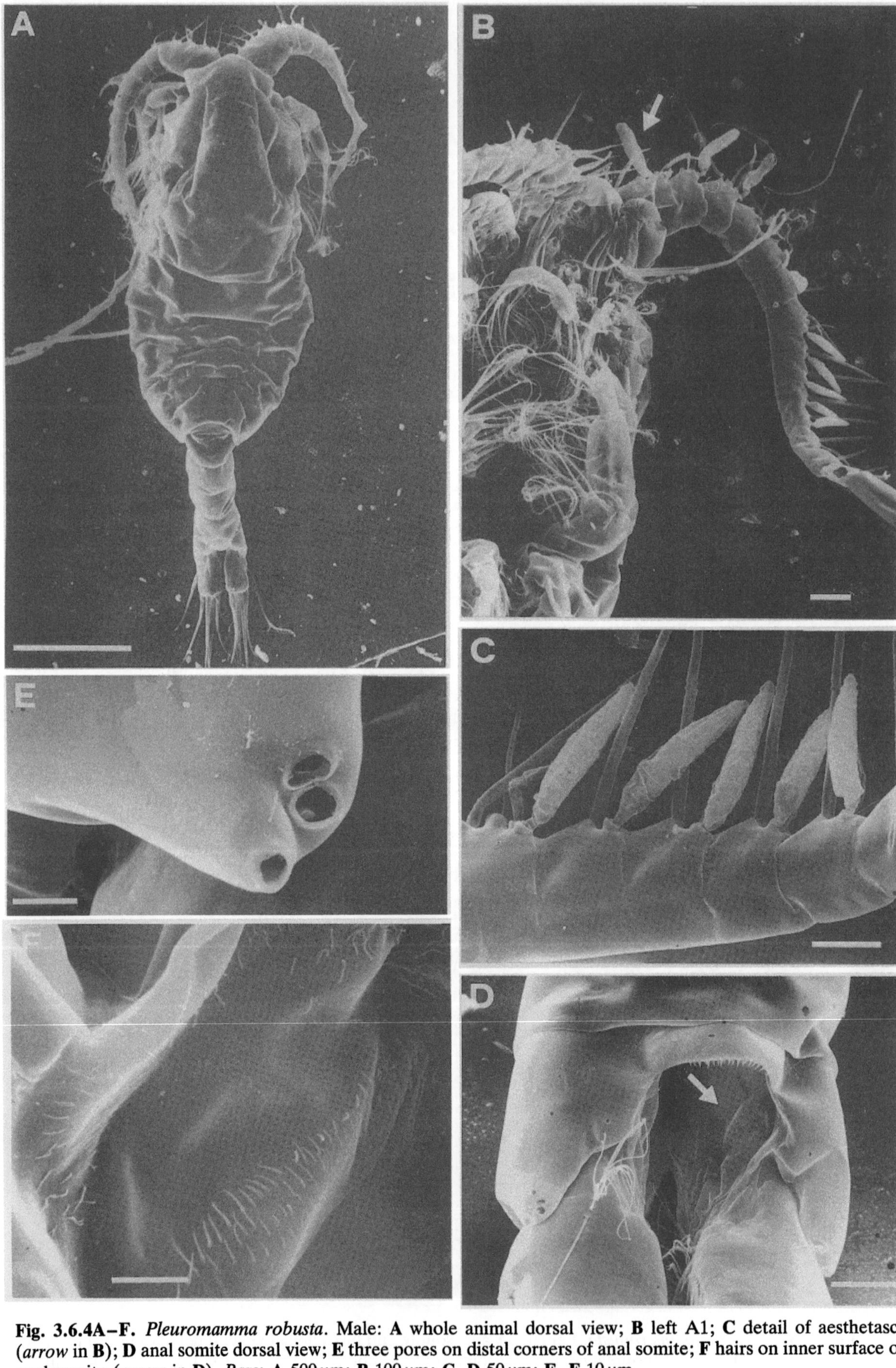

Fig. 3.6.4A–F. *Pleuromamma robusta*. Male: **A** whole animal dorsal view; **B** left A1; **C** detail of aesthetascs (*arrow* in **B**); **D** anal somite dorsal view; **E** three pores on distal corners of anal somite; **F** hairs on inner surface of anal somite (*arrow* in **D**). *Bars*: **A** 500 µm; **B** 100 µm; **C**, **D** 50 µm; **E**, **F** 10 µm

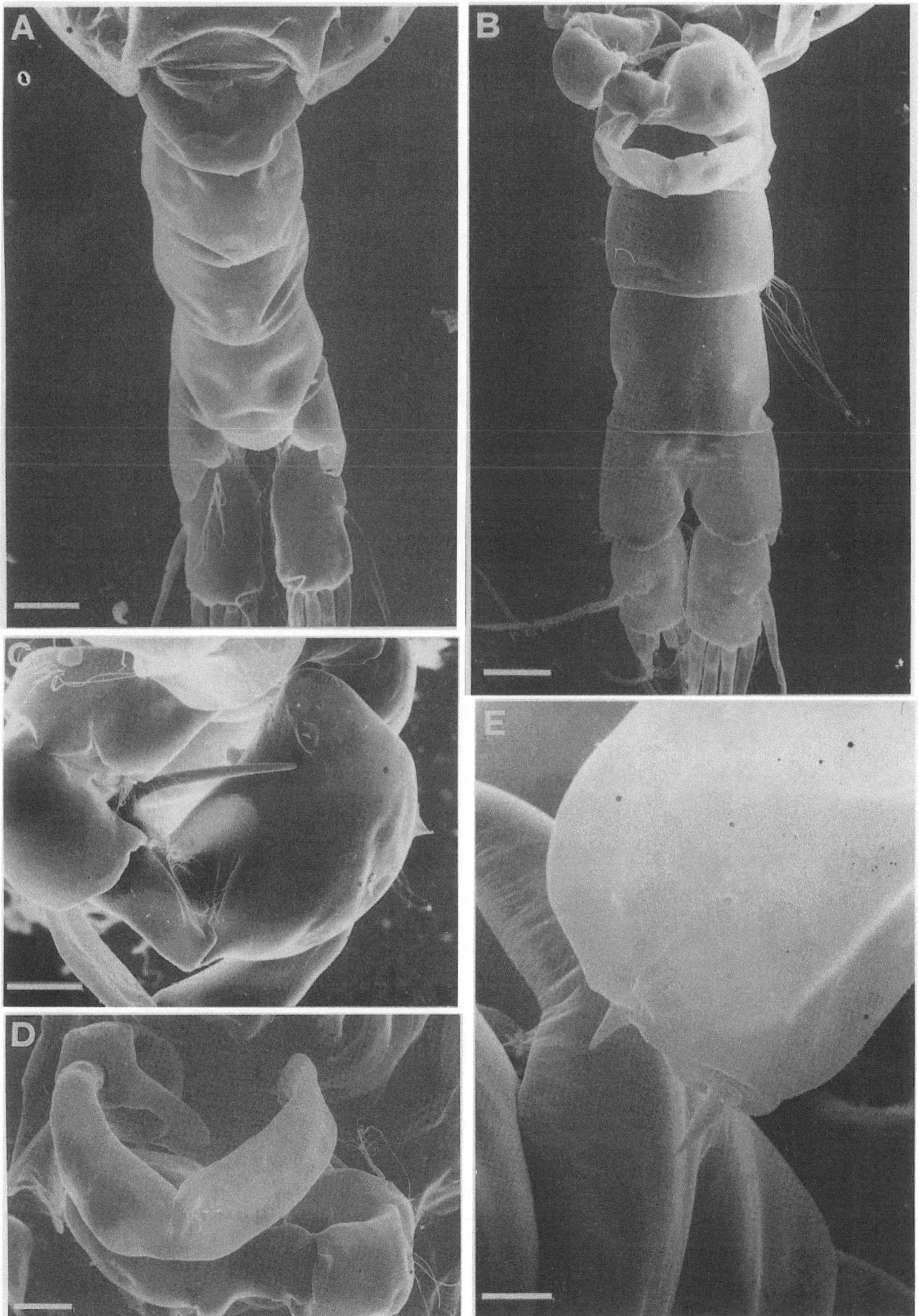

Fig. 3.6.5A–E. *Pleuromamma robusta*. Male: **A** abdomen dorsal view; **B** abdomen ventral view; **C** right P5; **D** P5 with left ramus seen from above; **E** detail of terminal joint of right P5 bearing two short spines. *Bars*: **A**, **B** 100 μm; **C**, **D** 50 μm; **E** 10 μm

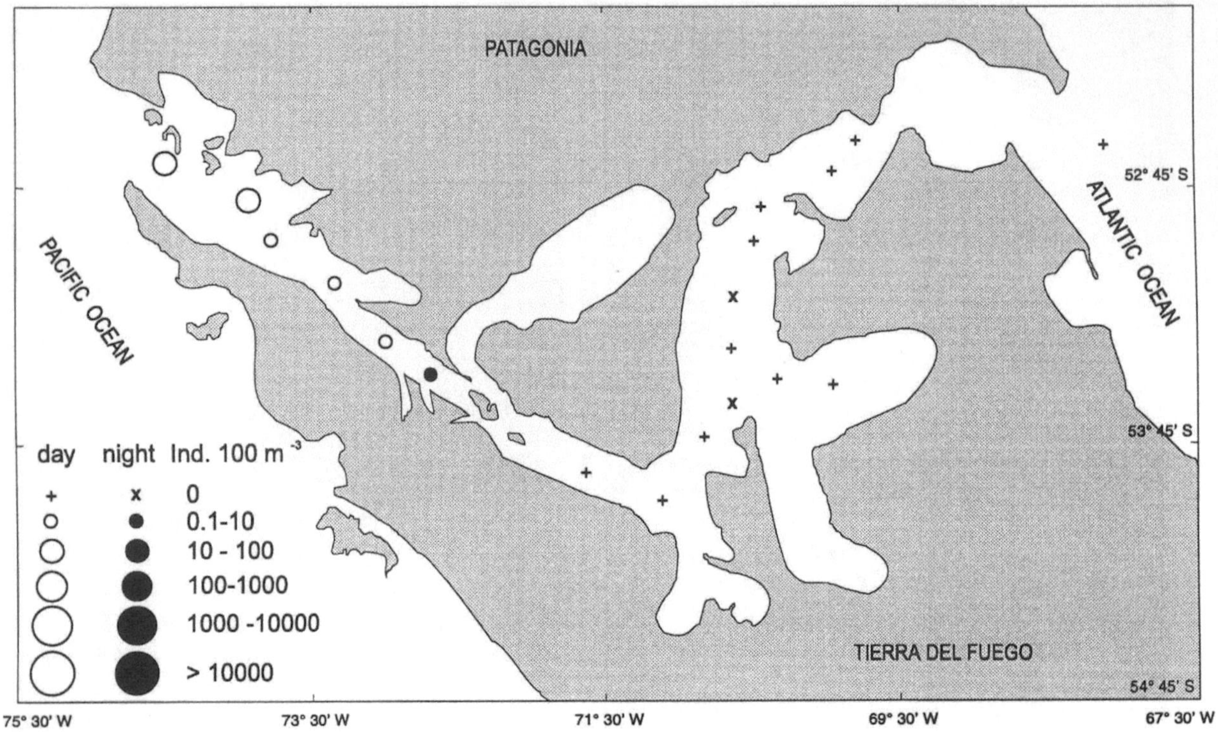

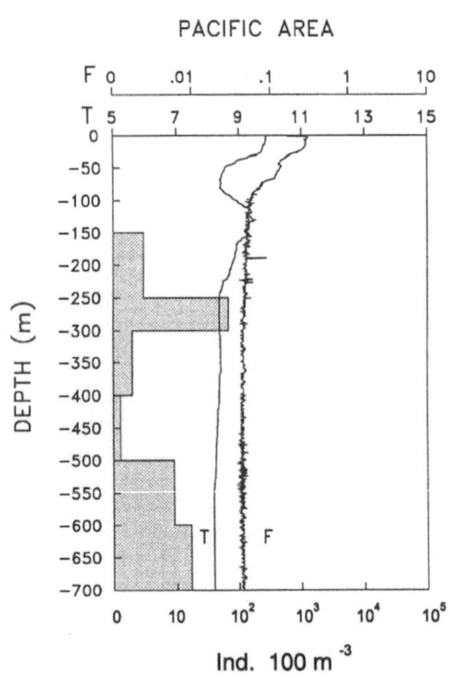

Depth range of occurrence = **150 - 700** m
Population core = **250 - 300** m
Temperature (T) = **8.20 - 9.80** °C
Oxygen = **348.4 - 437.2** µmol dm^{-3}
Chl fluorescence (F) = **0.04 - 0.39**

Fig. 3.6.6. *Pleuromamma robusta*. Distribution in the Straits of Magellan in February–March 1991

3.7 *Acartia tonsa* Dana 1849

Acartia tonsa Dana, 1849: p. 26 – Giesbrecht, 1892: p. 508, pl. 30 figs. 7, 24, 34, pl. 43 figs. 6, 10 – Wilson, 1932: p. 160, fig. 109 – Brodskii, 1950: p. 426, fig. 302 – Gonzalez and Bowman, 1965: p. 257, fig. 11c–f – Ramirez, 1966: p. 19, figs. 8–14 – Vidal, 1968: p. 46, pl. 17 figs. 12–14, pl. 18 fig. 8 – Ramirez, 1969: p. 84, pl. 16 fig. 135 – Arcos, 1975: p. 23, pl. 9 figs. 71–74 – Sabatini, 1990: p. 53, figs. 37–48.
Acartia (*Acanthacartia*) *tonsa* Steuer, 1915: p. 396.
Acartia tonsa var. *cryophylla* Björnberg, 1963: p. 64, fig. 34.

Body length –
females: 1.10–1.27 mm (30 specimens)
males: 0.98–1.12 mm (20 specimens)

Female. Body slender. Cephalothorax 3.4 times longer than wide; rounded posterior corners of last thoracic somite ornated with short spinules. Proportional lengths of cephalothorax and abdomen 80:20. Abdomen 3-segmented; proportional lengths of abdominal somites and furca 46:16:18:20 = 100. Genital somite with rows of small spinules on both dorsal and ventral sides. Dorsal posterior margins of abdominal somites naked. P5 basal joint with round process on inner distal margin. Distal joint consisting in strong, characteristic basally swollen spine, serrated at beginning of distal half and ornated with small scales on thin distal end. Basal joint with long, feathered seta on outer margin, longer than terminal spine.

Male. Same body shape as female. Right A1 geniculated. Proportional lengths of cephalothorax and abdomen 78:22. Posterior corners of last thoracic somite ornated with few spinules. Abdomen 5-segmented. Proportional lengths of abdominal somites and furca 16:29:18:8:14:15 = 100. Hairs and small spines on dorsal margin of first somite. Spinules sparsely distributed on dorsal surface and posterior margin of second abdominal somite. Caudal rami ornated with spinules of different size on dorsal surface. P5 uniramous, asymmetrical, with median B1 common to both right and left joints. Right ramus 4-jointed and left ramus 3-jointed. In right leg, penultimate joint with wide process on inner margin; last joint swollen in the middle, strongly turned inwards and with a hook-like apex. Last joint of left leg bearing spine and a characteristic lamellar projection resembling the head of a bird, with numerous, very small spinules on surface. At basis of this projection, cluster of long hairs and row of bristles.

Remarks

Both females and males of this species are highly variable in the ornamentation of the abdominal somites. This variability may reflect biogeographic differences. Our specimens closely resemble those described by Björnberg (1963) as *A. tonsa* var. *cryophylla* due to the remarkably reduced ornamentation of the abdominal somites compared to the typical *A. tonsa*.

Distribution

The species is widely distributed in bays and inshore areas of both Pacific and Atlantic coasts of North and South America. It has recently been reported also for brackish water lagoons in the Mediterranean Sea (Gaudy and Viñas 1985). *A. tonsa* has been recorded for both the coasts of Chile (Vidal 1968; Arcos 1975; Hirakawa and Zama 1985; Marin and Antezana 1985; Peterson et al. 1988) and Argentina (Ramirez 1969; Sabatini 1990). It was previously recorded in the Straits of Magellan but only at Pacific and central stations (Mazzocchi and Ianora 1991). In the present study, *A. tonsa* was entirely absent from the Pacific area but was present in subsurface waters of the Central and Atlantic areas of the Straits.

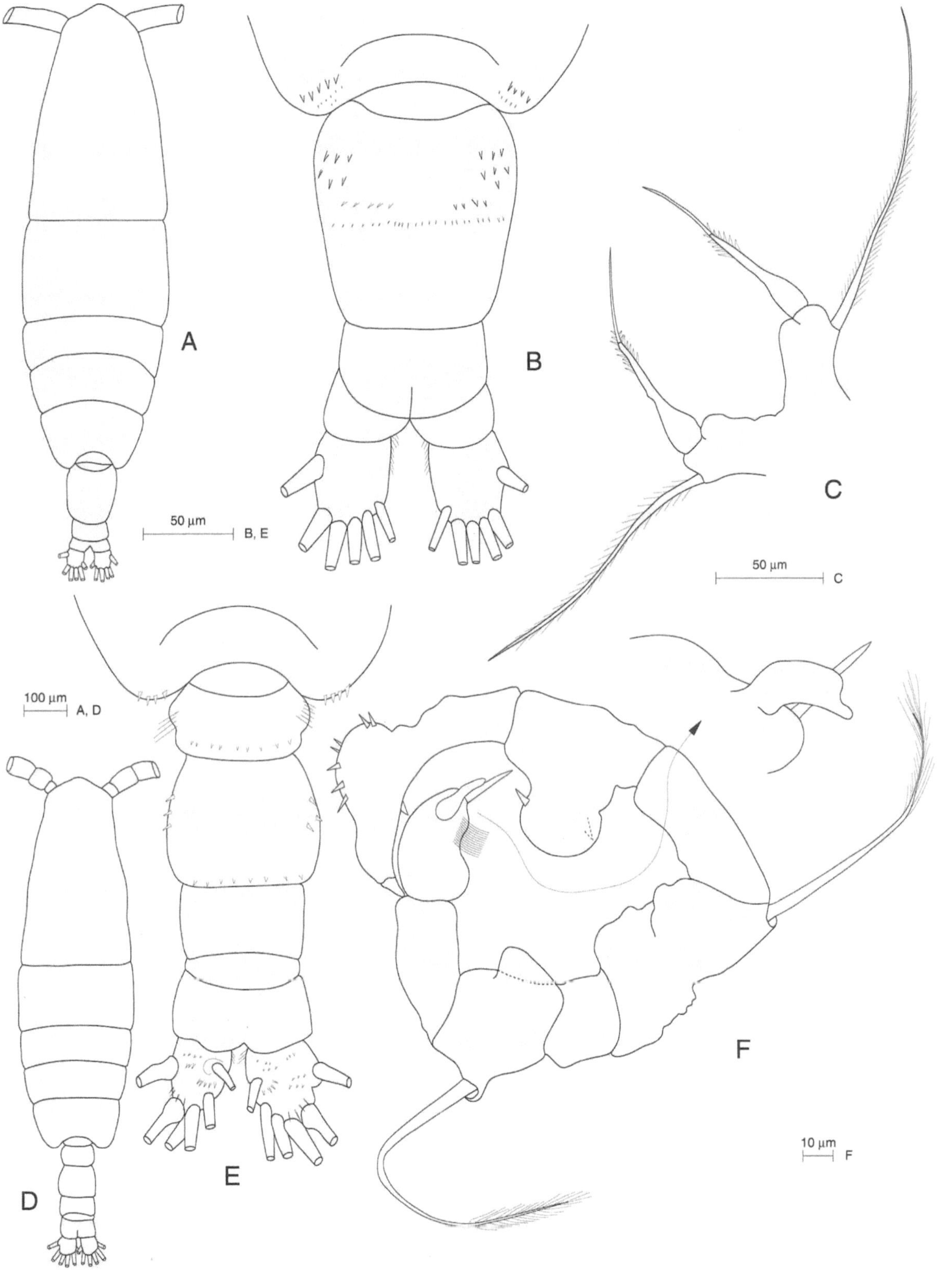

Fig. 3.7.1A–F. *Acartia tonsa*. Female: **A** whole animal dorsal view; **B** abdomen dorsal view; **C** P5. Male: **D** whole animal dorsal view: **E** abdomen dorsal view; **F** P5

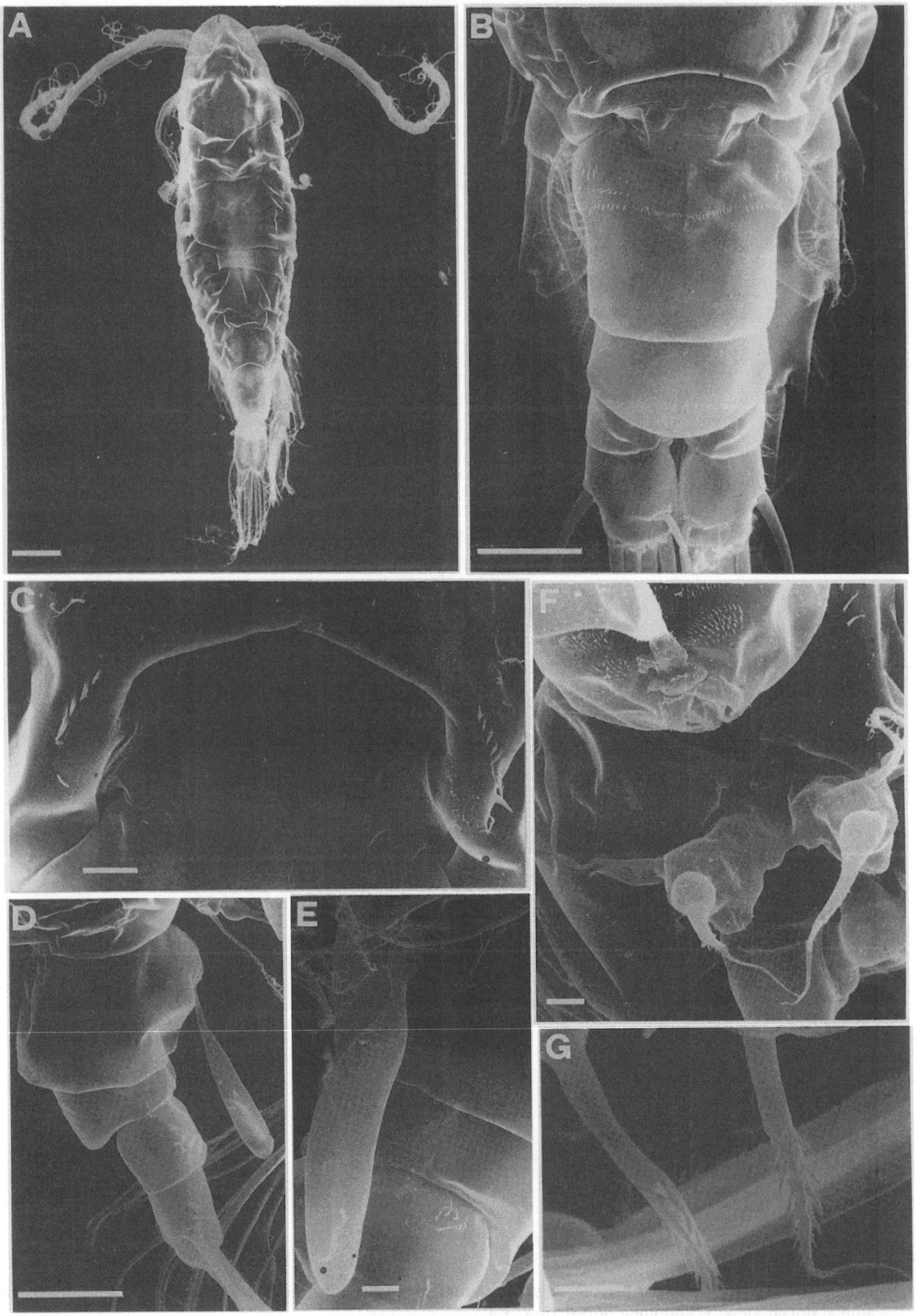

Fig. 3.7.2A–G. *Acartia tonsa*. Female: **A** whole animal dorsal view; **B** abdomen dorsal view with rows of small spinules on genital somite; **C** detail of inner posterior margins of last thoracic somite; **D** abdomen lateral view with spermatophore attached to genital somite; **E** detail of spermatophore; **F** P5; **G** detail of distal end of P5. *Bars*: **A** 100 µm; **B**, **D** 50 µm; **C**, **E**, **F**, **G** 10 µm

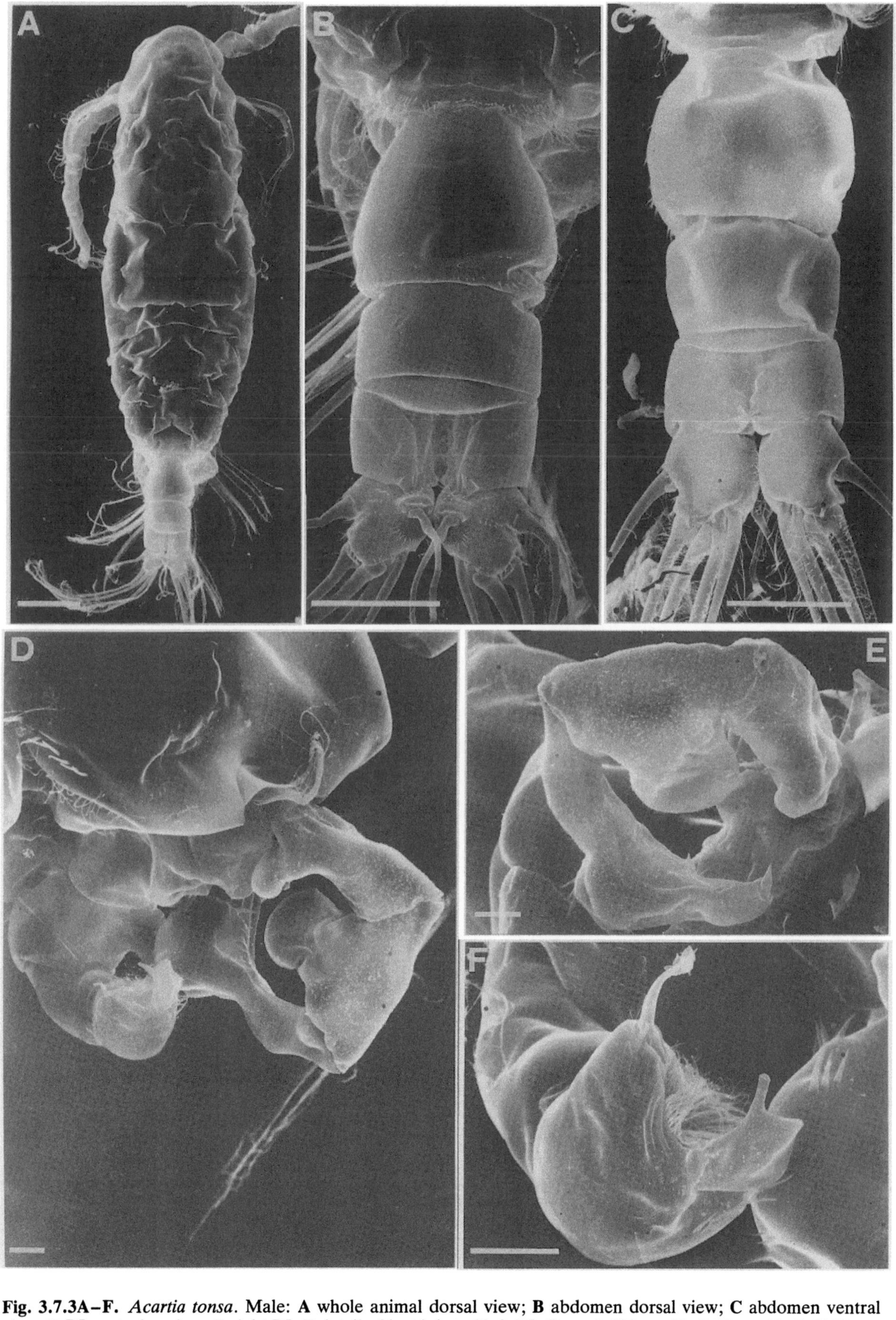

Fig. 3.7.3A–F. *Acartia tonsa*. Male: **A** whole animal dorsal view; **B** abdomen dorsal view; **C** abdomen ventral view; **D** P5 posterior view; **E** right P5; **F** detail of last joint of left P5. *Bars*: **A** 100 µm; **B**, **C** 50 µm; **D**, **E**, **F** 10 µm

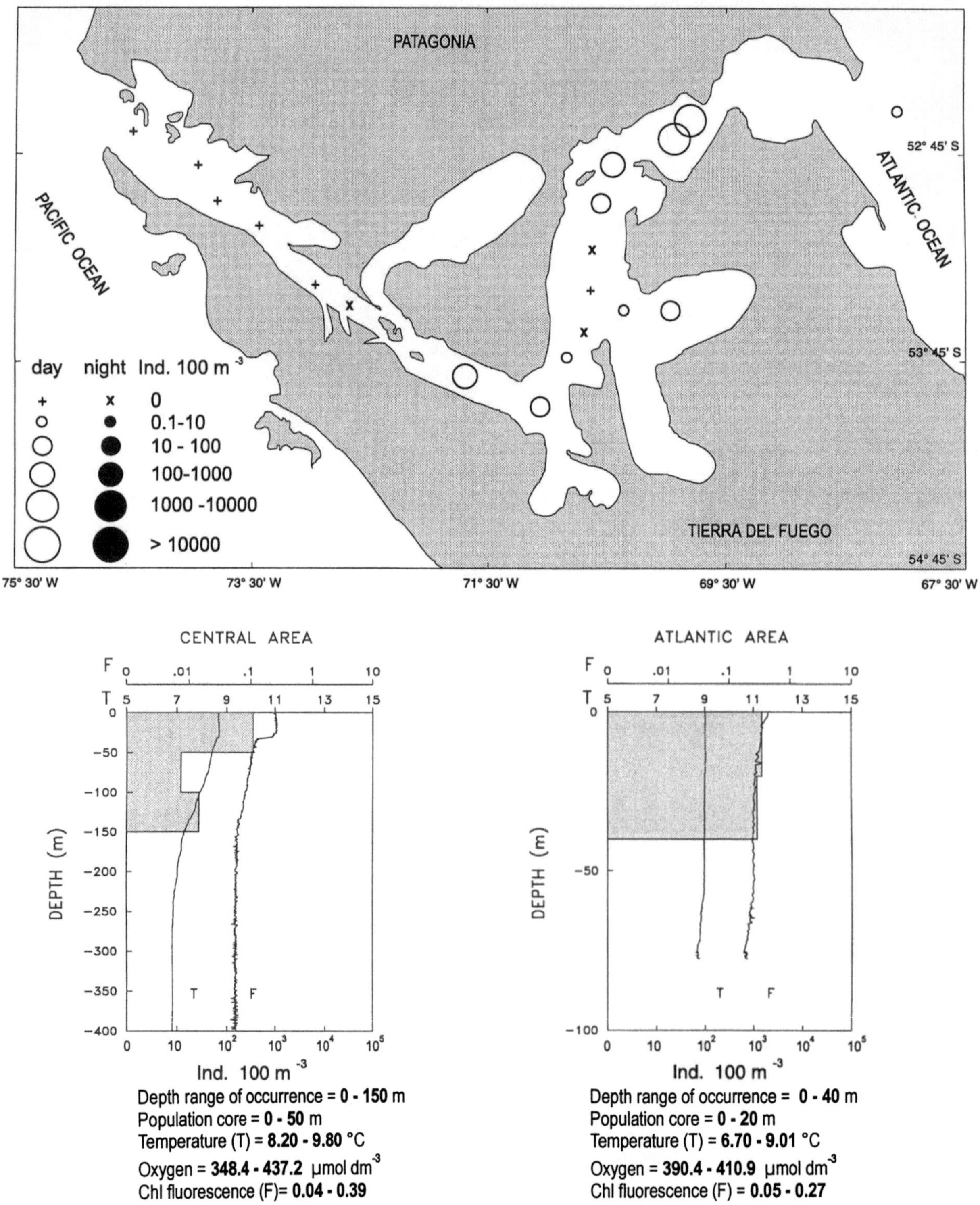

Fig. 3.7.4. *Acartia tonsa*. Distribution in the Straits of Magellan in February–March 1991

3.8 *Candacia cheirura* Cleve 1904

Candacia cheirura Cleve, 1904: p. 180, fig. 29a–e – Farran, 1929: p. 273, fig. 29 – Grice, 1963: p. 175, figs. 15, 16 – Ramirez, 1971: p. 87, pl. 2 figs. 4, 7, 11.

Body length –
females: 2.27–3.12 mm (7 specimens)
males: 2.18–2.84 mm (16 specimens)

Female. All the characteristic features of the genus. Cephalothorax 2.5 times longer than wide, with blunt head. Posterior corners of last thoracic somite with symmetrical points. First cephalothorax somite bears flap-like projection along mid-line of dorsal side with two small sensilla at base. Proportional lengths of cephalothorax and abdomen 76:24; proportional lengths of abdominal somites and furca 44:26:15:15 = 100. Lateral margins of genital somite rounded and symmetrical in dorsal view. Second abdominal somite bears a remarkable ventral posteriad protuberance, not visible in dorsal view. Terminal segment of P5 with 2 spine-like points on external margin and 3 at distal end; the middle point longest of three.

Male. General shape as female. Cephalothorax 2.6 times longer than wide. Proportional lengths of cephalothorax and abdomen 76:24. Right A1 geniculated. Flap-like projection on dorsal side of first thoracic somite, as in female. Posterior corners of thorax pointed and asymmetrical; left corner ends in a spiniform process; right corner with characteristic hazelnut-swelling ornated with small knobs, not visible at light microscope. Proportional lengths of abdominal somites and furca 18:24:21:13:10:14 = 100. In dorsal view, genital somite bears process on right side ending in slender point. Right P5 ending in a chela; distal joint of left P5 longer than penultimate joint and ornated with 5 spines and rows and bundles of hairs.

Distribution

Cleve (1904) described this species from collections west of South Africa in the Benguela Current at depths of 250–350 m. Farran (1929) recorded it between the Auckland and Campbell Islands, south of New Zealand in a sector influenced by sub-Antarctic waters. Farran supposed that it is a normal inhabitant of the upper mesoplankton in the circumpolar current known as the West Wind Drift from which the Benguela Current originates. Other areas of occurrence include the South Atlantic, South Pacific and South Indian Oceans. Vervoort (1957) recorded it in central Antarctic waters and suggested that this species drifts from the circumpolar current towards sub-Antarctic sectors. Ramirez (1971) recorded only one specimen of *C. cheirura* between 42° and 45°S on the South Argentine continental shelf influenced by sub-Antarctic waters. In the present study, *C. cheirura* was only present from 50–500 m in the Pacific area and from 200–250 m at a few stations in the Central area of the Straits of Magellan.

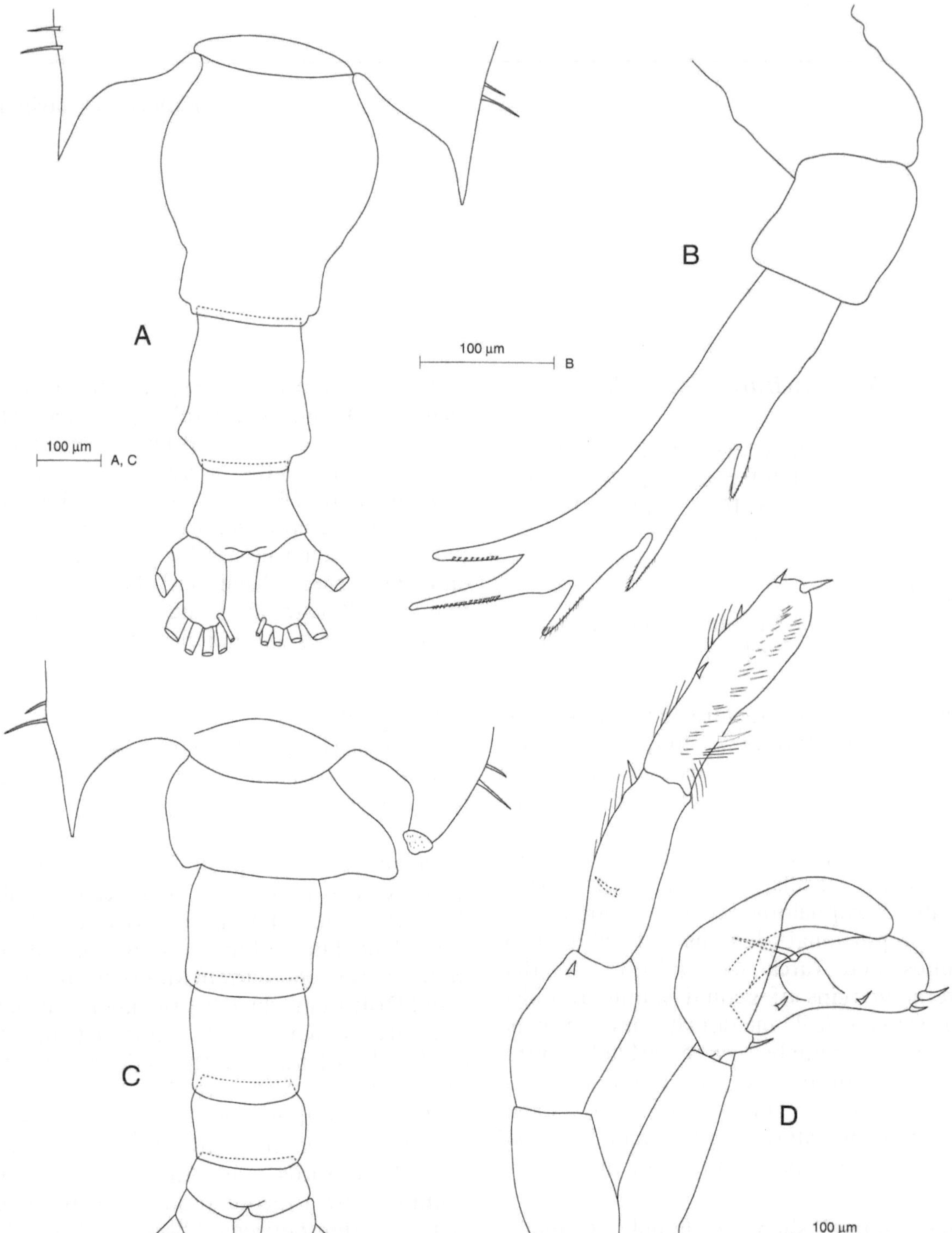

Fig. 3.8.1A–D. *Candacia cheirura*. Female: **A** abdomen dorsal view; **B** P5. Male: **C** abdomen dorsal view; **D** P5

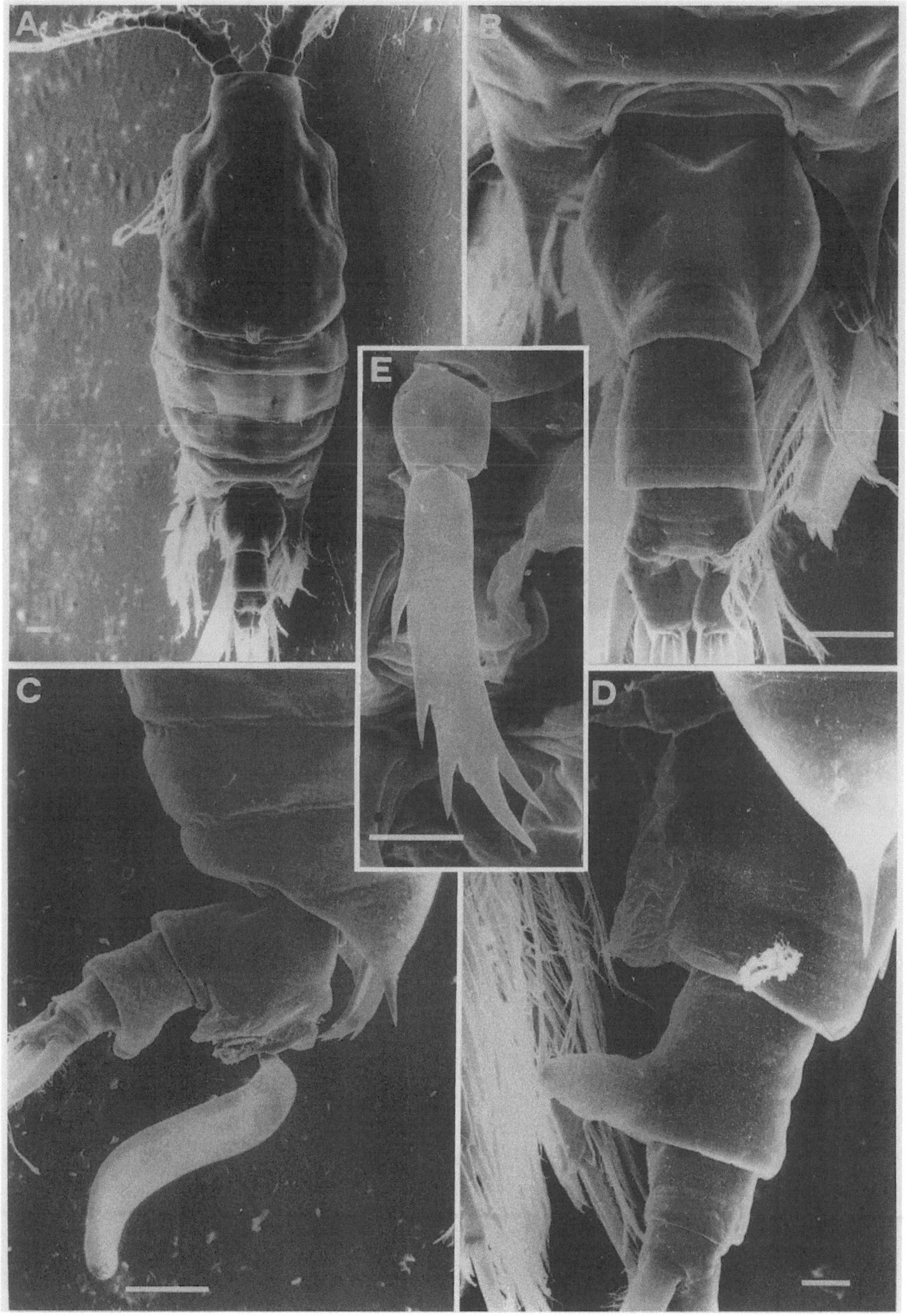

Fig. 3.8.2A–E. *Candacia cheirura*. Female: **A** whole animal dorsal view; **B** abdomen dorsal view; **C** abdomen bearing spermatophore in lateral view; **D** large protuberance on left side of second abdominal somite; **E** P5 *Bars*: **A**, **B**, **C**, **D** 100 µm; **E** 50 µm

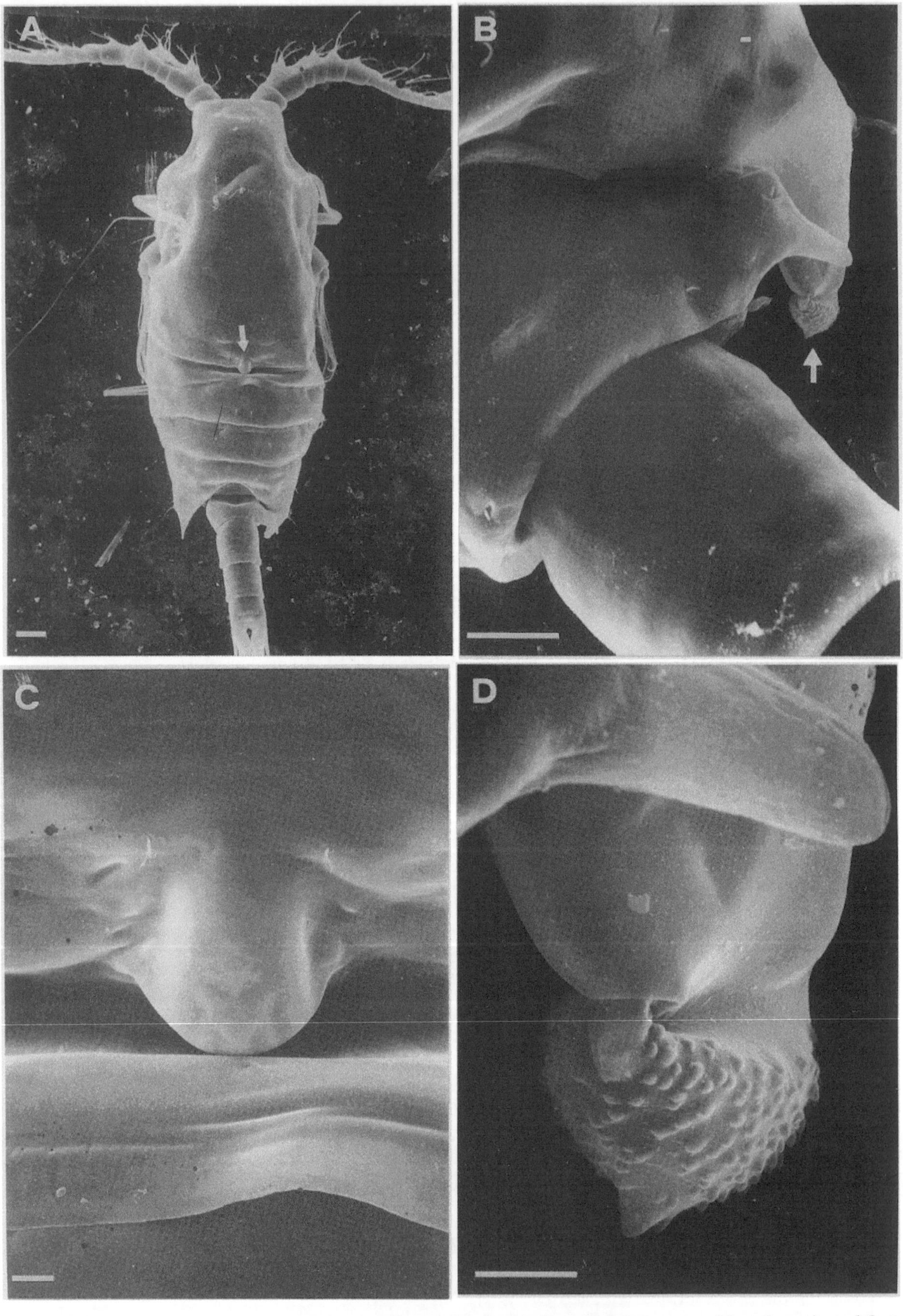

Fig. 3.8.3A–D. *Candacia cheirura*. Male: **A** whole animal dorsal view; **B** right corner of last thoracic and first abdominal somites; **C** detail of flap-like projection on dorsal side of first thoracic somite (*arrow* in **A**); **D** detail of acorn-like swelling on right corner of thorax (*arrow* in **B**). *Bars*: **A** 100 µm; **B** 50 µm; **C, D** 10 µm

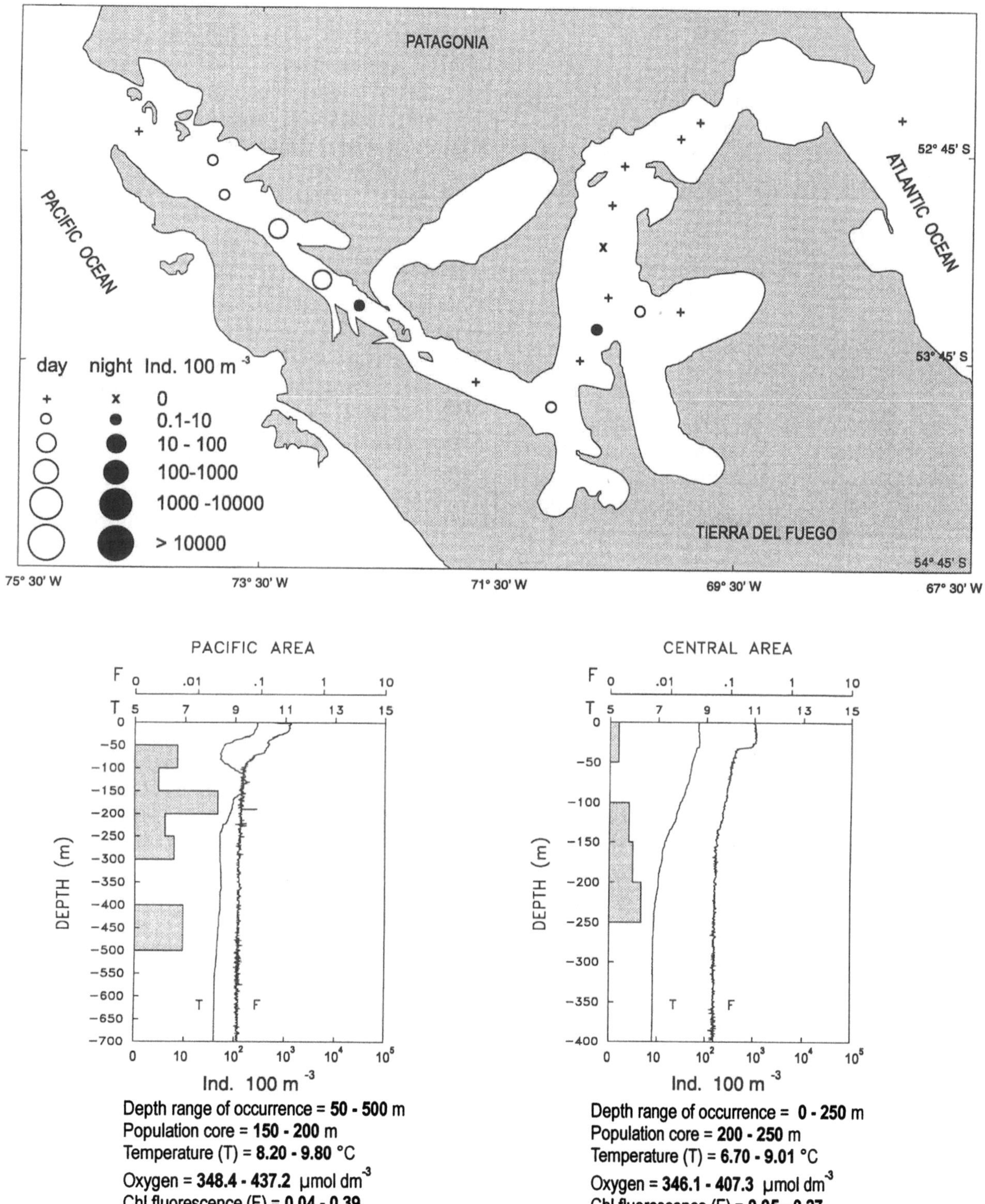

Fig. 3.8.4. *Candacia cheirura.* Distribution in the Straits of Magellan in February–March 1991

3.9 *Candacia norvegica* Boeck 1864

Candace norvegica Boeck, 1864: p. 235
Candacia norvegica Sars, 1903: p. 134, pls. 89, 90 – Wilson, 1932: p. 138, fig. 94 – Grice and Jones, 1960: p. 288, figs. 46–65 – Grice, 1963: p. 178, figs. 1, 27, 65.

Body length –
females: 2.90–3.14 mm (7 specimens)
males: 2.82–2.96 mm (10 specimens)

Female. Body robust with truncated forehead characteristic of genus. Last thoracic somite bears straight, pointed symmetrical processes. Proportional lengths of cephalothorax and abdomen 77:23; proportional lengths of abdominal somites and furca 48:21:15:16 = 100. In dorsal view, genital somite bears short bifid processes at midpoint of lateral margins; right process shorter than left. Above bifid processes, genital somite bears tufts of hairs on either side. Distal joint of P5 bears two broad, unequal terminal spines, two shorter spines on the outer margin and three setae on the inner margin.

Male. Same body shape as female but with asymmetrical posterior processes on last thoracic somite. Right corner lamellar in shape with depression along midline, turned inwards and reaching about midpoint of genital somite. Left corner pointed as in female. Proportional lengths of cephalothorax and abdomen 74:26; proportional lengths of abdominal somites and furca 28:19:19:11:8:15 = 100. In dorsal view, genital somite with a stout process on right side, distal end of which divided into 3 lobes. P5 uniramous, asymmetrical; left leg 3-jointed; 2 distal joints bear long, dense hairs along margin. Right leg ends in straight pair of tongs.

Remarks

Our female specimens bear bifid processes on both sides of the genital somite, as reported for the variety *tropica* by Sewell (1932). They thus differ from the simple spines reported by Wilson (1932) or bifid tip of the left spine by Grice and Jones (1960). These latter authors maintain that the size and shape of these processes are variable and of poor diagnostic value. The lamellar process on the posterior right corner of the male thorax has not been described or shown before in detail.

Distribution

C. norvegica has been reported for the North Atlantic (Boeck 1864; Sars 1903; Farran 1948), Indian (Sewell 1932) and North Pacific (Tanaka 1953; Grice and Jones 1960) Oceans, as well as off the Chilean coast (Fagetti 1962). In the present study, *C. norvegica* was only present from 100–700 m in the Pacific area of the Straits. This represents a first record for the Magellan region.

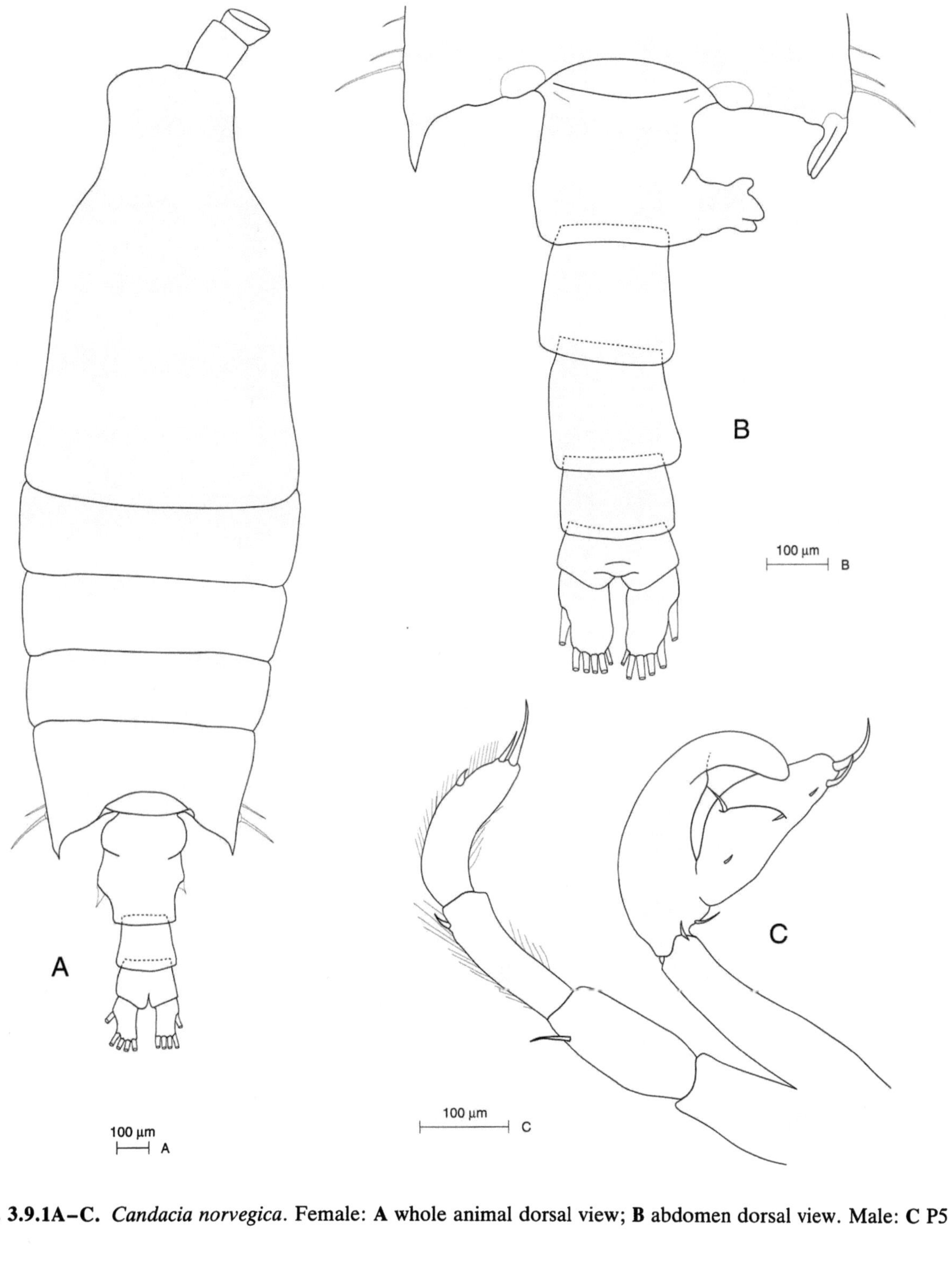

Fig. 3.9.1A–C. *Candacia norvegica*. Female: **A** whole animal dorsal view; **B** abdomen dorsal view. Male: **C** P5

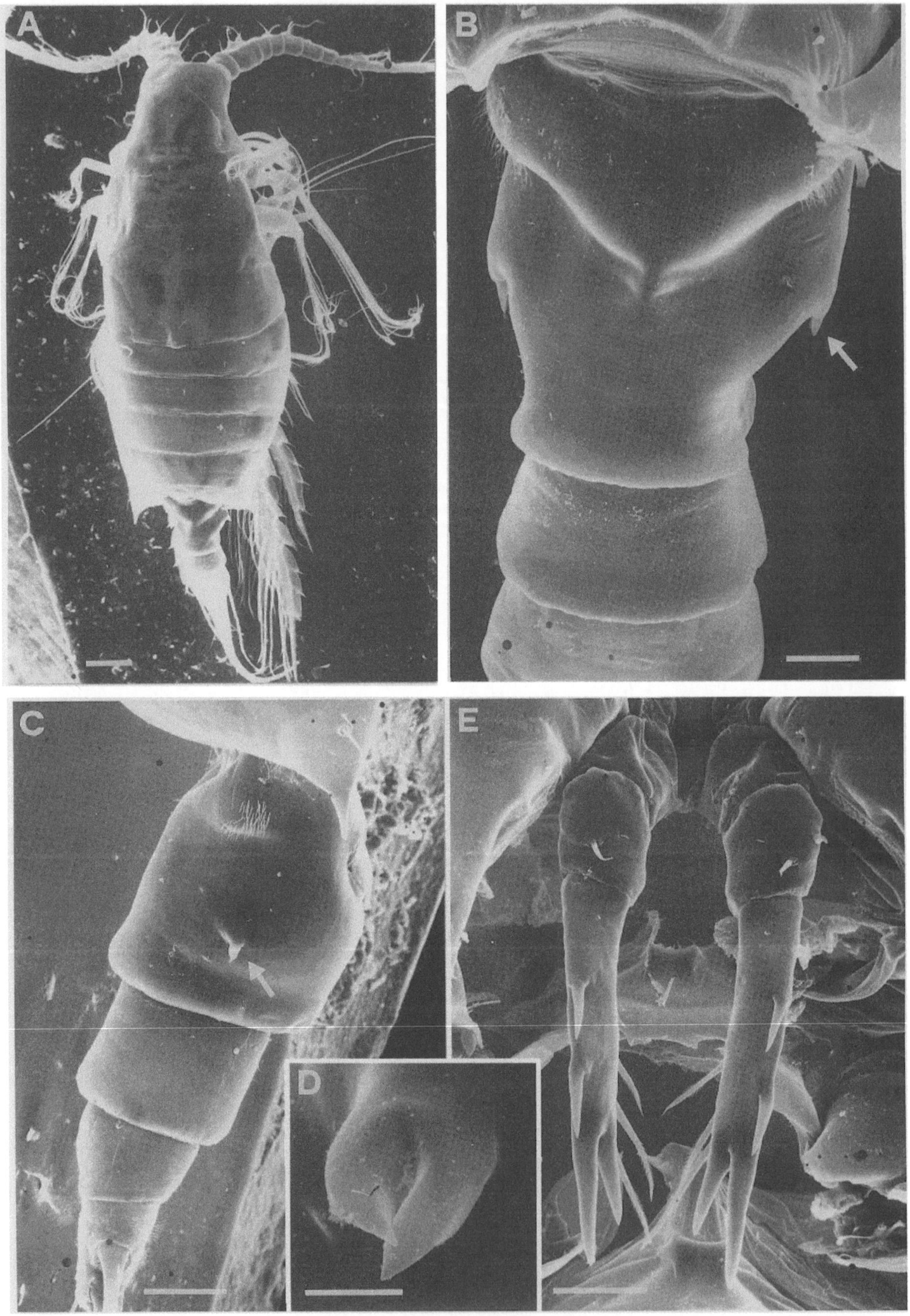

Fig. 3.9.2A–E. *Candacia norvegica*. Female: **A** whole animal dorsal view; **B** genital somite dorsal view; **C** abdomen lateral view; **D** detail of bifid process on genital somite (*arrows* in **B** and **C**); **E** P5. *Bars*: **A** 200 µm; **C** 100 µm; **B**, **E** 50 µm; **D** 10 µm

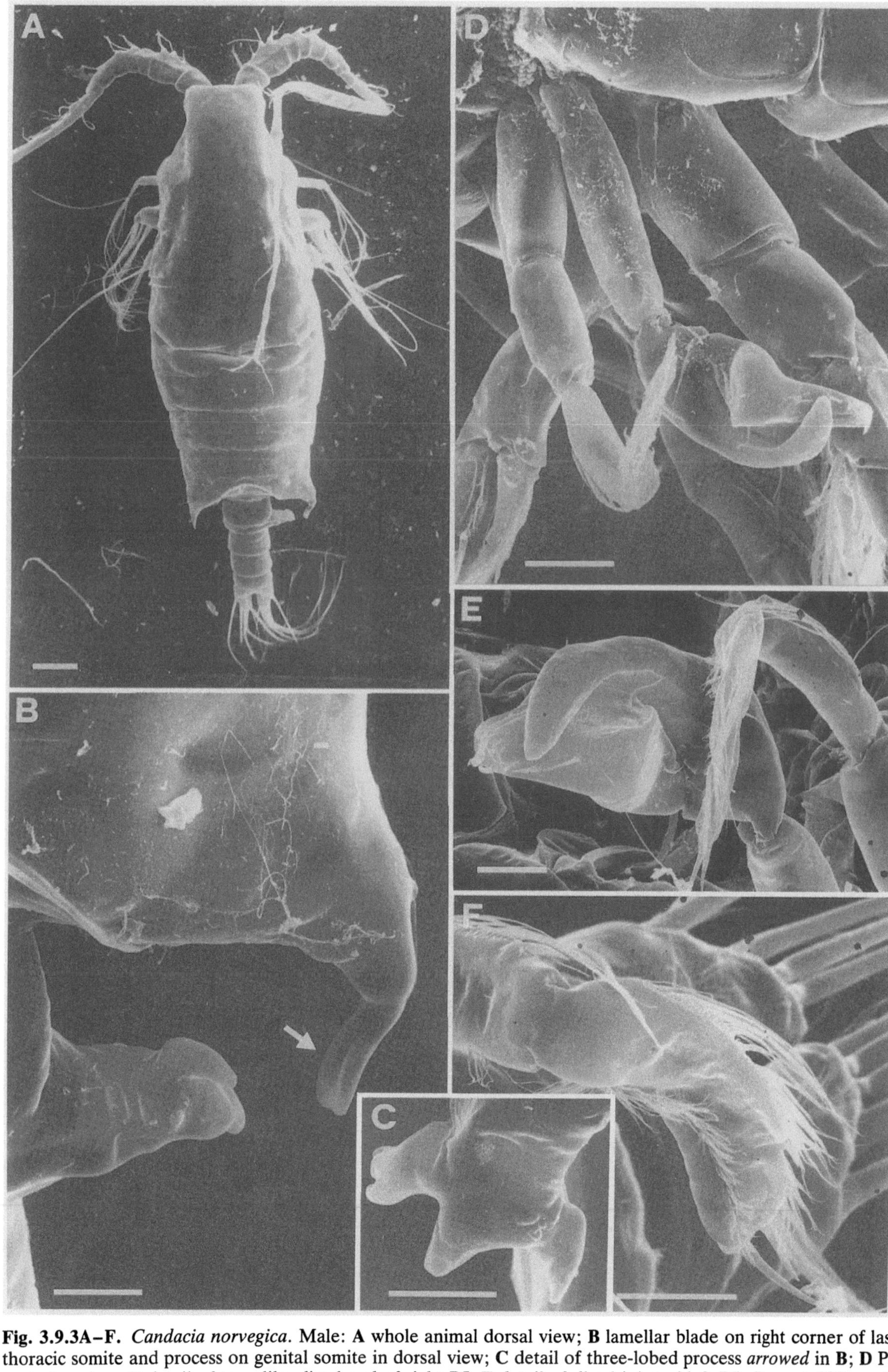

Fig. 3.9.3A–F. *Candacia norvegica.* Male: **A** whole animal dorsal view; **B** lamellar blade on right corner of last thoracic somite and process on genital somite in dorsal view; **C** detail of three-lobed process *arrowed* in **B**; **D** P5 posterior view; **E** detail of tong-like distal end of right P5; **F** detail of distal joint of left P5. *Bars*: **A** 200 µm; **D** 100 µm; **B**, **C**, **E**, **F** 50 µm

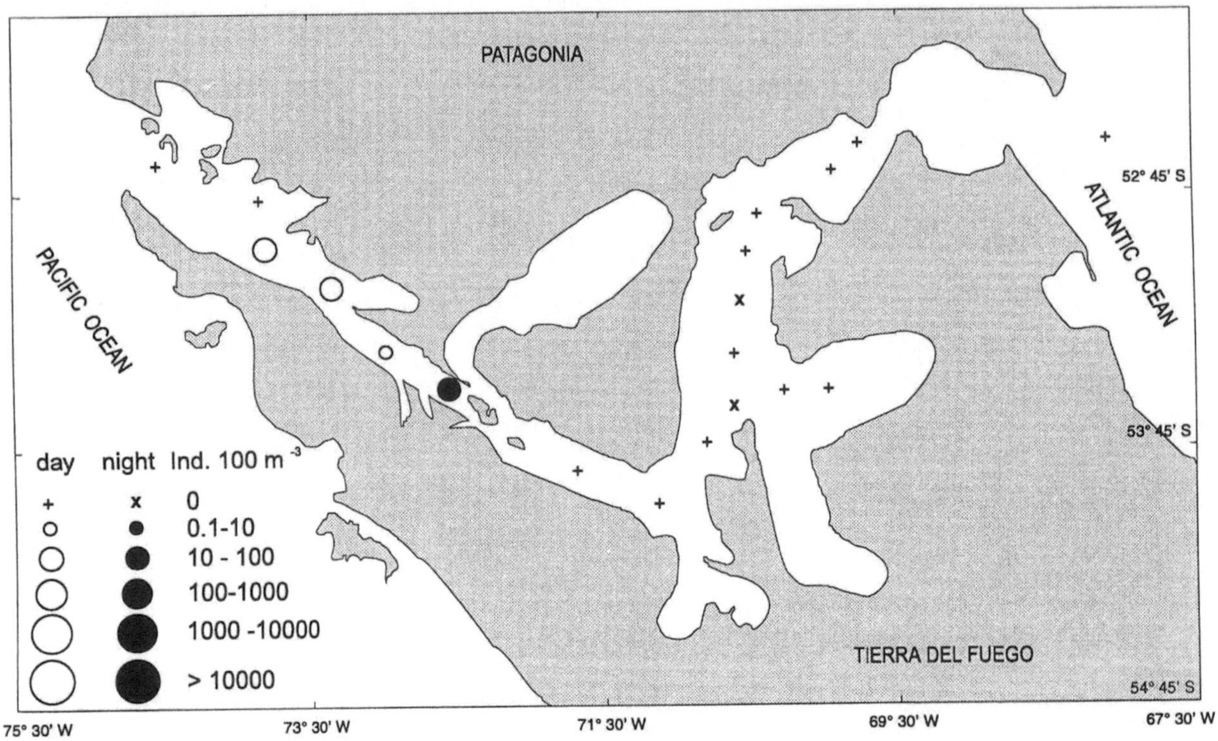

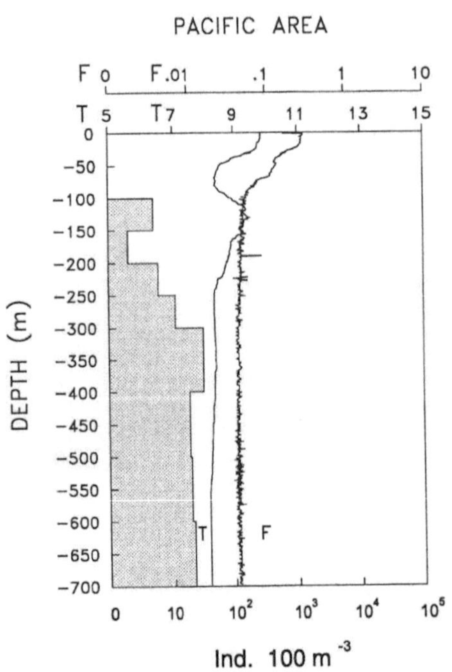

Depth range of occurrence = **100 - 700** m
Population core = **300 - 400** m
Temperature (T) = **8.20 - 9.80** °C
Oxygen = **348.4 - 437.2** μmol dm^{-3}
Chl fluorescence (F) = **0.04 - 0.39**

Fig. 3.9.4. *Candacia norvegica*. Distribution in the Straits of Magellan in February–March 1991

3.10 *Centropages brachiatua* Dana 1849

Pontella brachiata Dana, 1849: p. 27.
Calanopia brachiata Dana, 1853: pp. 1133, 1853, pl. 79 figs. 7a–b, 8a–b, 9a–g.
Centropages brachiatus Brady, 1883: p. 82, pl. 26 figs. 1–7 – Giesbrecht, 1892: p. 304, pl. 17 figs. 26, 37, pl. 18 fig. 7, pl. 38 figs. 2, 17 – Unterüberbacher, 1964: p. 28, pls. 29, 30 – Ramirez, 1966: p. 15, pl. 3 figs. 13–21 – Vidal, 1968: p. 35, pl. 12 figs. 1–6 – Ramirez, 1969: p. 75, pl. 14 figs. 105, 106, 110, 112, 113, 115 – Arcos, 1974: p. 221, figs. 9, 11, 14, 15; 1975: p. 20, pl. 7 figs. 58–64.

Body length –
females: 1.67–2.21 mm (7 specimens)
males: 1.76 – 1.86 mm (5 specimens)

Female. Body robust. Last thoracic somite with sinuous posterior margins; posterolateral corners produced into elongated, asymmetrical processes extending to mid-genital somite. Al as long as caudal rami and with spines on segments 1, 2 and 5. Proportional lengths of cephalothorax and abdomen 69:31; proportional lengths of abdominal somites and furca 30:32:13:25 = 100. Genital somite strongly asymmetrical with two spines on either side visible on different focal planes. Second abdominal somite bears cluster of hairs on dorsal side and short spine on ventral side. P5 biramous; Re2 with broad spiniform process on inner side. Terminal spine of last exopodal joint shorter than this joint and closely serrated along external margin.

Male. Posterior margins of last thoracic somite less sinuous than in females and with less pronounced asymmetrical pointed corners. Right A1 geniculated with spines on segments 1, 2, 5, 13, 15, 16 and segments 17–19 with finely serrated margins. Proportional lengths of cephalothorax and abdomen 70:30; proportional lengths of abdominal somites and furca 15:18:18:16:33 = 100. P5 biramous, asymmetrical. Right leg ending in strong chela; Re2 with long, thin, naked spine. Ril and Ri3 of both right and left legs ornated with clusters of hairs.

Remarks

Our specimens differ from those drawn by Brady (1883) in the male left P5. Ri3 bears 6 setae compared to 7 reported by Brady. Re2 bears 3 short, strong spines, also reported by Ramirez (1969) and Arcos (1975), compared to Brady's figures that show 4 apical spines.

Distribution

C. brachiatus is a cold-living species which occurs more frequently in the Southern rather than Northern hemisphere. The presence of this species in Southern Oceans has been reported by numerous authors including Björnberg (1963) for the coast of Brazil and Ramirez (1969, 1970, 1971, 1977) off Argentina. It has been reported as one of the dominant species along the Chilean coast (Vidal 1968; Marin and Antezana 1985; Peterson et al. 1988). In other cases, it seems to be sporadic in occurrence, as in the Bahia de Conception (Arcos 1975). Mazzocchi and Ianora (1991) found it at only seven stations situated on the Pacific side of the Straits of Magellan. In the present study, *C. brachiatus* was sporadically present in surface waters throughout the Straits.

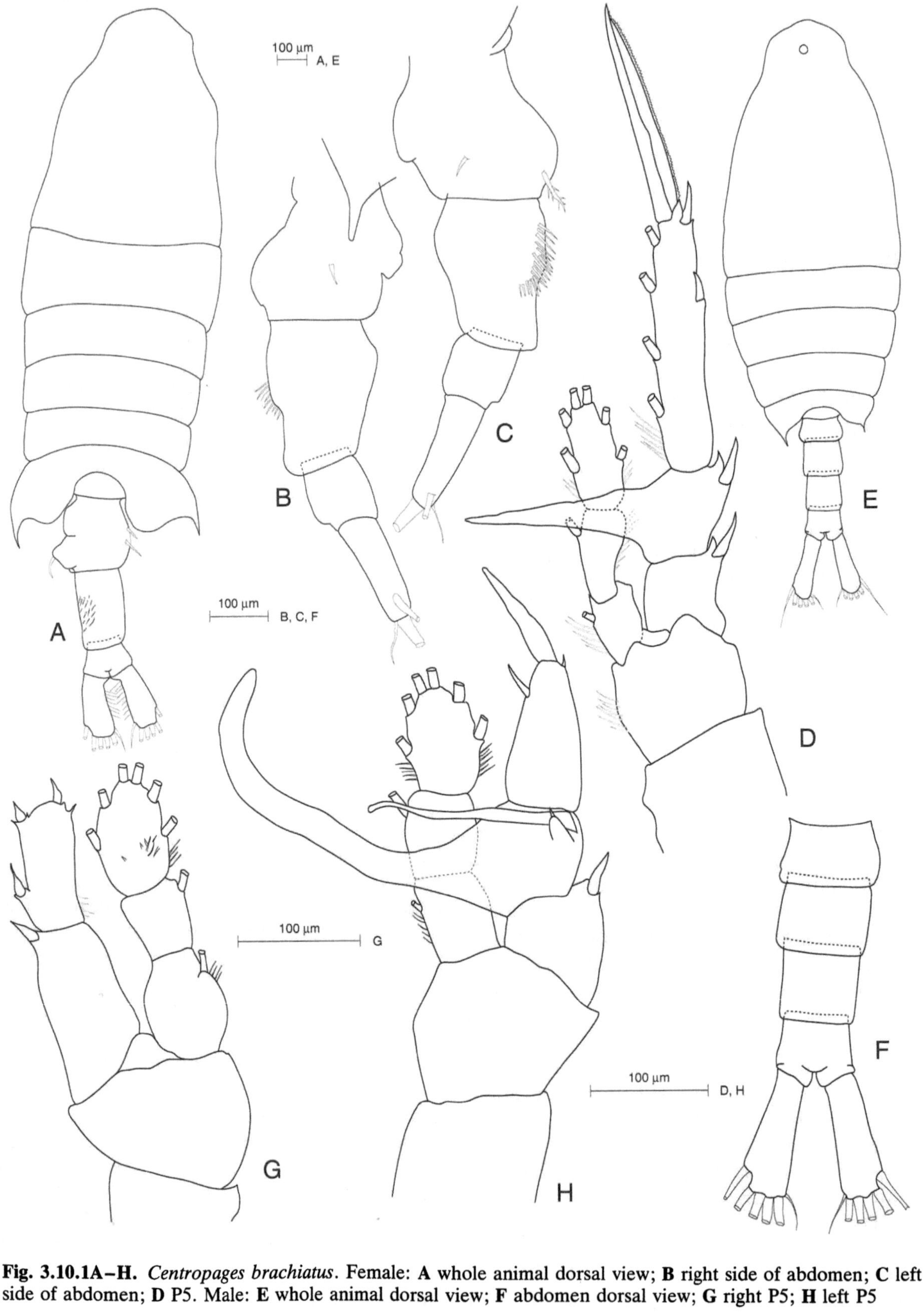

Fig. 3.10.1A–H. *Centropages brachiatus*. Female: **A** whole animal dorsal view; **B** right side of abdomen; **C** left side of abdomen; **D** P5. Male: **E** whole animal dorsal view; **F** abdomen dorsal view; **G** right P5; **H** left P5

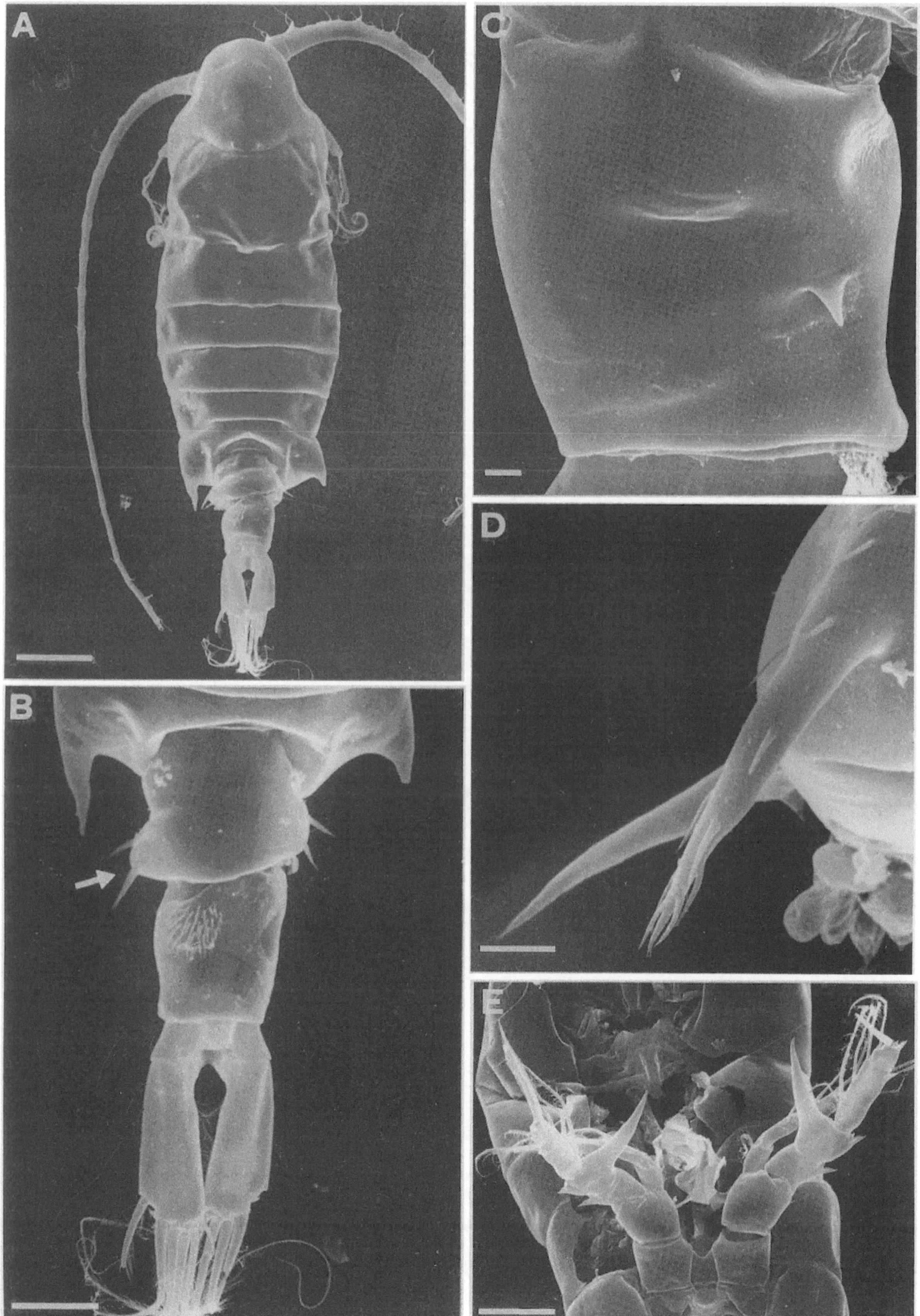

Fig. 3.10.2A–E. *Centropages brachiatus*. Female: **A** whole animal dorsal view; **B** abdomen dorsal view; **C** second abdominal somite ventral view; **D** ramified spine on genital somite (*arrow* in **B**); **E** P5 posterior view. *Bars*: **A** 200 μm; **B**, **E** 100 μm; **C**, **D** 10 μm

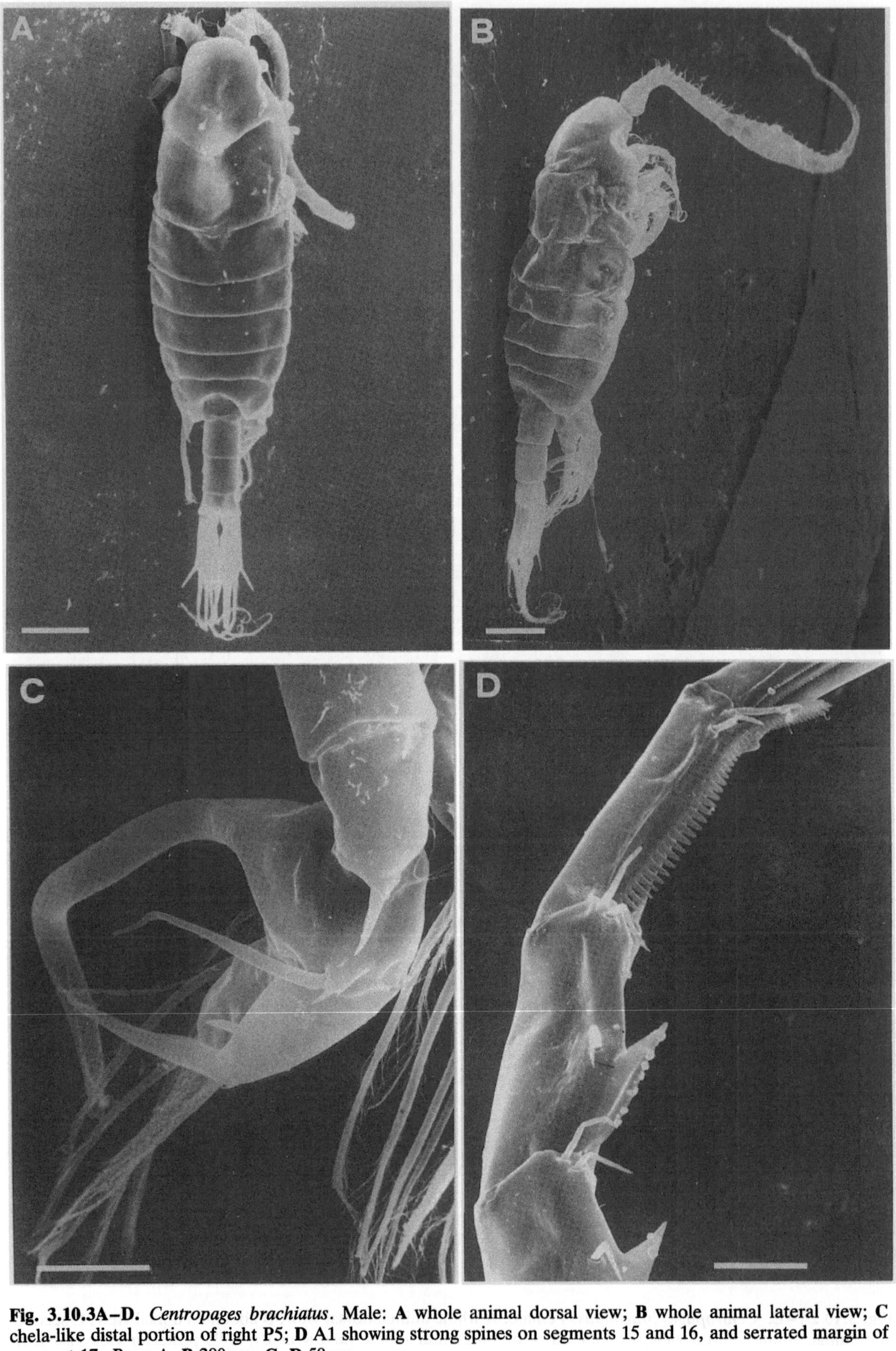

Fig. 3.10.3A–D. *Centropages brachiatus*. Male: **A** whole animal dorsal view; **B** whole animal lateral view; **C** chela-like distal portion of right P5; **D** A1 showing strong spines on segments 15 and 16, and serrated margin of segment 17. *Bars*: **A**, **B** 200 µm; **C**, **D** 50 µm

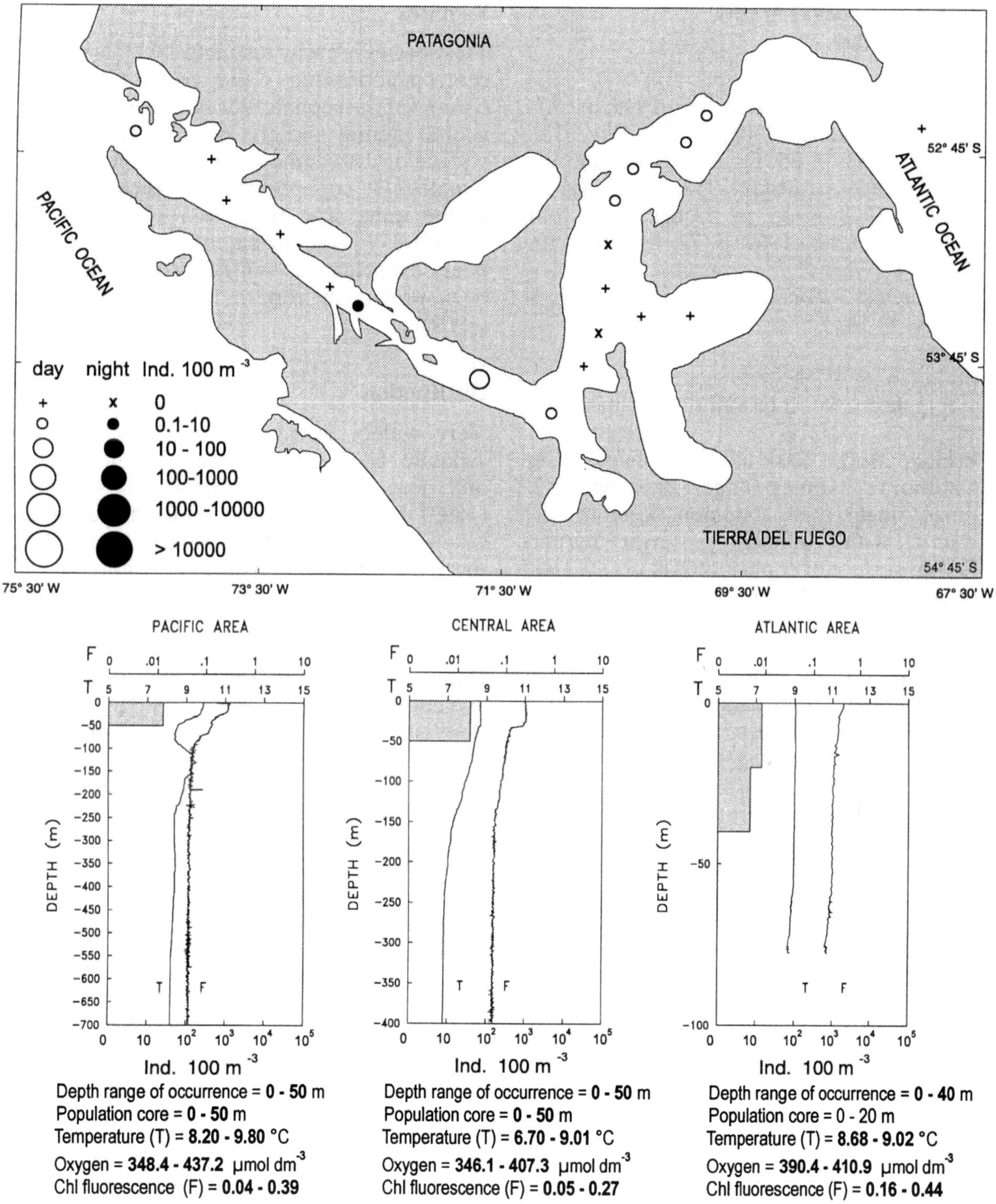

Fig. 3.10.4. *Centropages brachiatus*. Distribution in the Straits of Magellan in February–March 1991

3.11 *Centropages bradyi* Wheeler 1899

Centropages bradyi Wheeler, 1899: p. 174, fig. 12 – Esterly, 1905: p. 172, fig. 32 – Wilson, 1932: p. 86, fig. 59 – Rose, 1933: p. 187, fig. 216 – Brodskii, 1950: p. 318, fig. 221 – Vidal, 1968: p. 35, pl. 13 figs. 6–11 (only male) – Ramirez, 1969: p. 74, figs. 107–109, 111, 114.
Centropages violaceus Brady, 1883: p. 83, pl. 27 figs. 1–14.

Body length –
females: 1.96–2.06 mm (3 specimens)

Female. Body flask-like in shape. Cephalothorax 2 times longer than wide, 2.7 times longer than abdomen. Last thoracic somite with rounded posterior corners. Dorsal surface of cephalothorax ornated with dense spinules. Postero-dorsal margin of first somite ornated with conspicuous protuberance. Abdomen 3-segmented, symmetrical in dorsal view. Proportional lengths of abdominal somites and furca 24:12:30:34 = 100. Genital somite symmetrical in dorsal view, extending in part over second abdominal somite on ventral side. Anal somite with strong prominence on ventral side. Caudal rami robust and divergent, with strong tubercle on ventral surface at distal margin between two outer setae. P5 biramous; Re2 with a robust spiniform process on inner side and finely serrated margin on external side. Endopod not protruding beyond apex of spiniform process of exopod.

Remarks

This species is easily distinguished from other centropagiides due to the rounded posterior corners of the cephalothorax, the symmetrical genital somite and the pegs on the ventral surface of the caudal rami. The specimens (3 females, 1 copepodite) we recorded are in the same size range as reported in the literature. In our specimens, the anal somite bears a prominent ventral process that has been previously reported only by Ramirez (1969).

Distribution

Very widely distributed in the temperate Atlantic, Mediterranean Sea, Pacific Ocean, and near the coast of Australia. Wheeler (1899) described this species from specimens taken in the Gulf Stream off the East coast of the United States. Farran (1929) found it off New Zealand in surface and subsurface waters. Brodskii (1950) recorded *C. bradyi* near the Kurile Islands in the NW Pacific. Ramirez (1969) reported this species in Argentinian waters from 100 to 1000 m depths. Fagetti (1962), Vidal (1968) and Marin and Antezana (1985) reported it in Chilean waters. In the present study, *C. bradyi* was found from 0–200 m at two stations at the Pacific entrance of the Straits and from 0–50 m at a station in front of Baia Inutil. This represents a first record in the Magellan region.

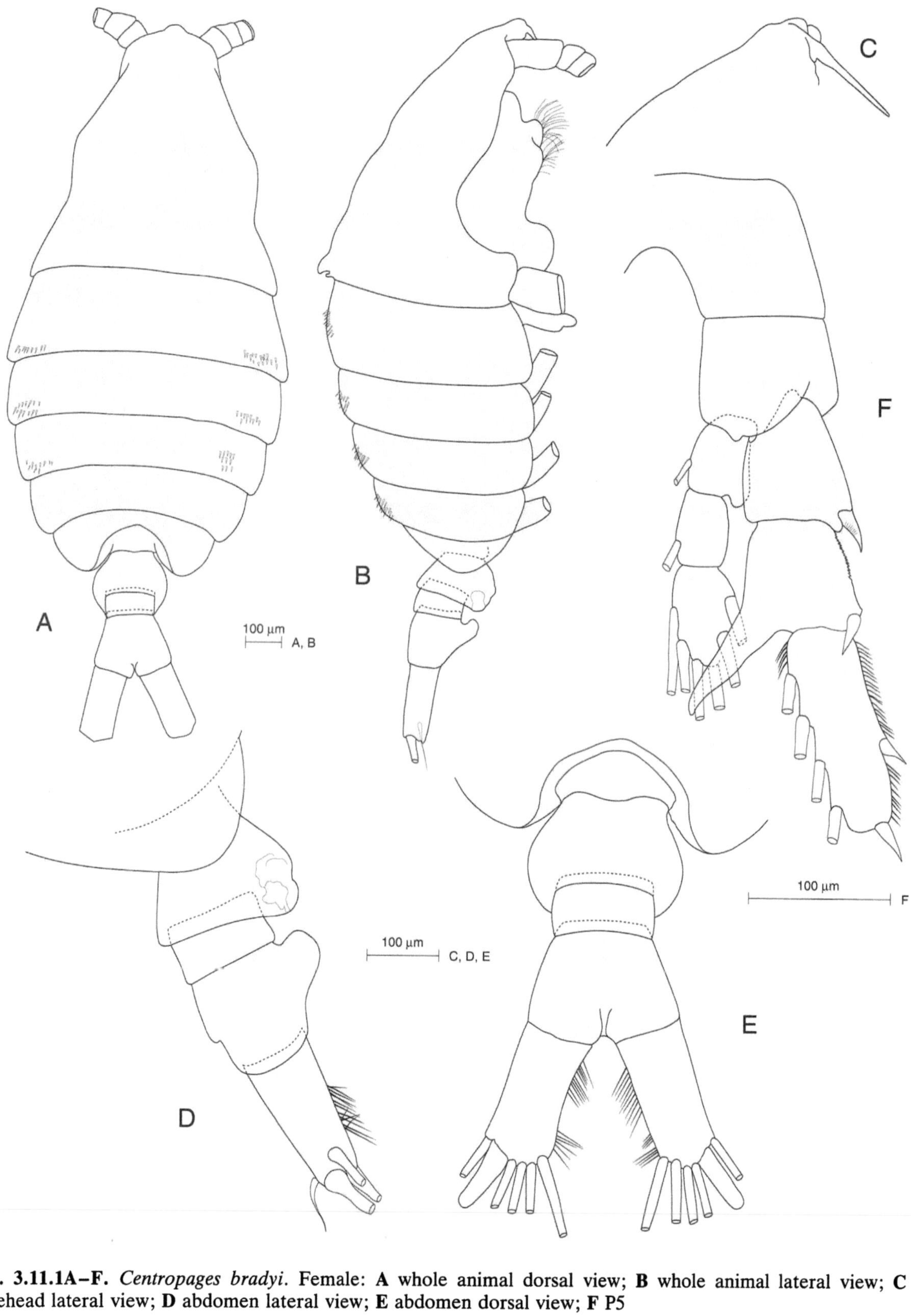

Fig. 3.11.1A–F. *Centropages bradyi.* Female: **A** whole animal dorsal view; **B** whole animal lateral view; **C** forehead lateral view; **D** abdomen lateral view; **E** abdomen dorsal view; **F** P5

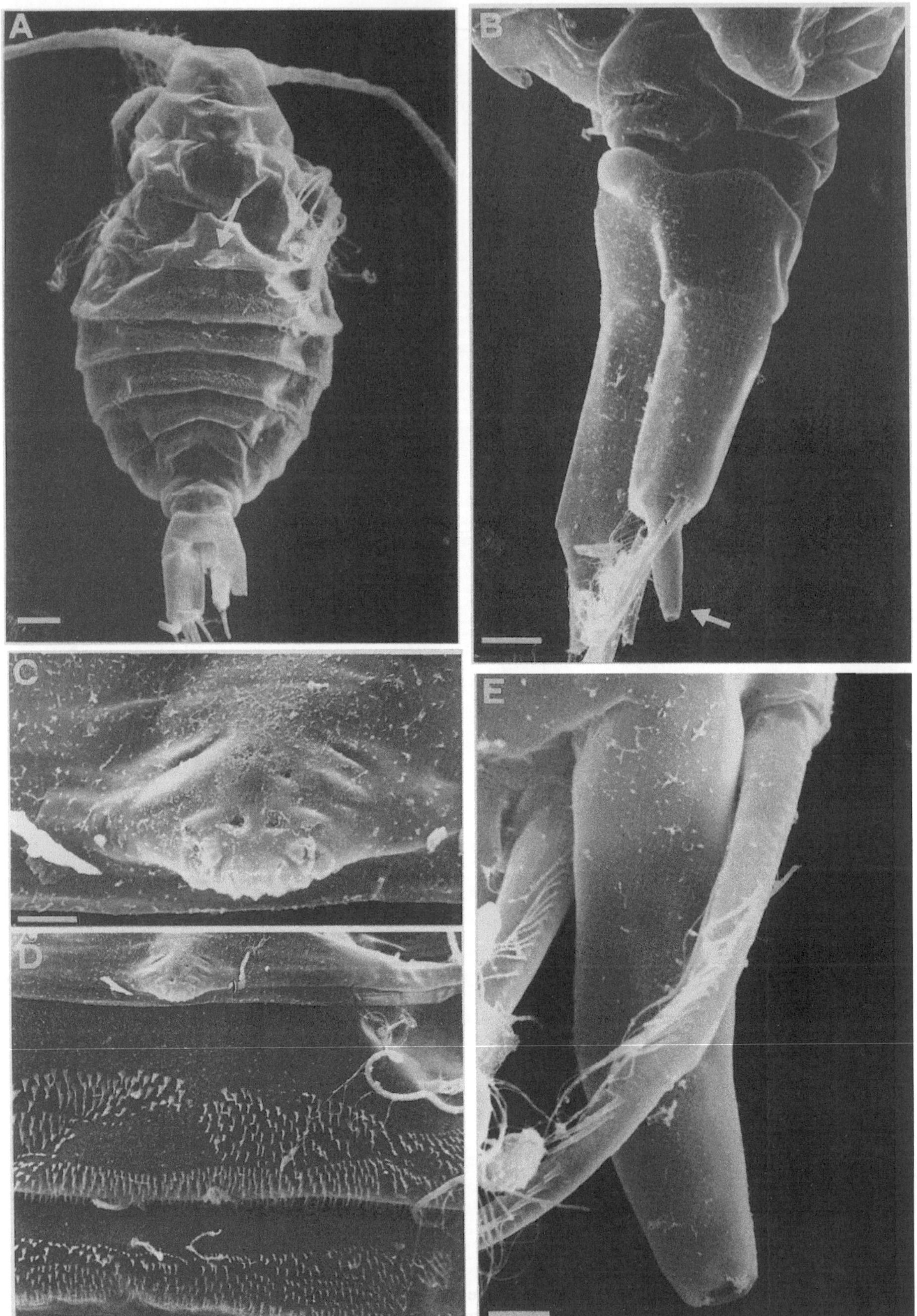

Fig. 3.11.2A–E. *Centropages bradyi*. Female: **A** whole animal dorsal view; **B** abdomen ventral view with prominent bump on ventral surface; **C**, **D** protuberance on midpoint of postero-dorsal margin (*arrow* in **A**); **D** detail of dense spinulation covering 2nd to 5th thoracic somites; **E** detail of caudal tubercule (*arrow* in **B**). *Bars*: **A** 100 µm; **B**, **D** 50 µm; **C**, **E** 10 µm

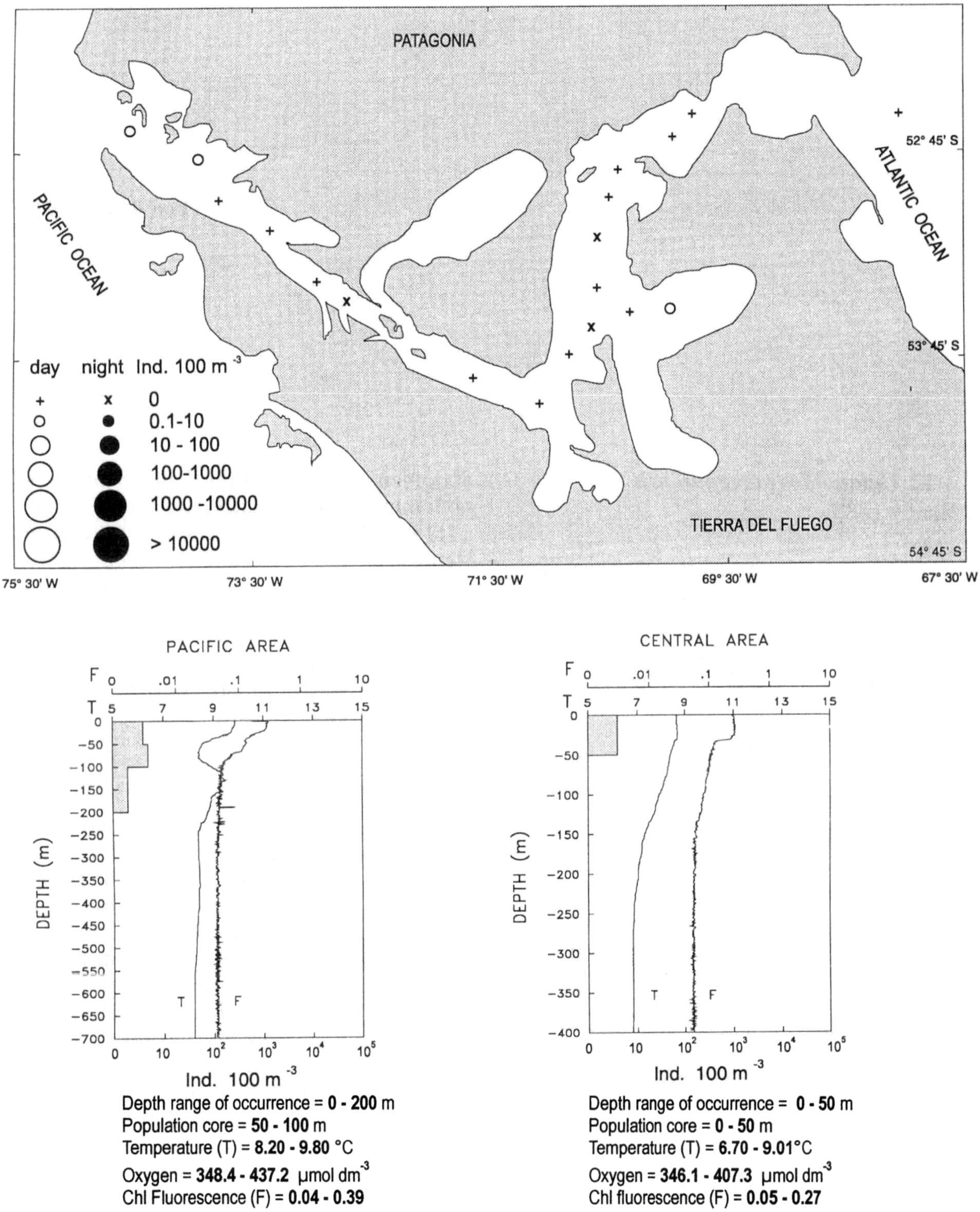

Fig. 3.11.3. *Centropages bradyi.* Distribution in the Straits of Magellan in February–March 1991

3.12 *Calanoides patagoniensis* Brady 1883

Calanoides patagoniensis Brady, 1883: p. 75, pl. 23 figs. 1–10 – Arcos, 1974: p. 219, figs. 7, 8–10; 1975: p. 9, pl. 1 figs. 6–8.
Calanus patagoniensis Giesbrecht, 1892: p. 91, pl. 6 figs. 8, 10, 17, pl. 8 fig. 29.

Body length –
females: 2.08–2.37 mm (10 specimens)
males: 2.02–2.41 mm (11 specimens)

Female. Body slender and elongated. Long bifid rostrum. Postero-lateral margins of last thoracic somite somewhat elongated. Proportional lengths of cephalothorax and abdomen 76:24; proportional lengths of abdominal somites and furca 33:16:14:16:21 = 100. In lateral view, genital somite bears wide, flap-like structure that almost completely covers genital aperture. This plate observed in an upright position or folded over the genital pores. Often, mucous secretions visible in the gonopore area. P5 B1 lacks denticulation along inner margin characteristic of *Calanus* species. Endopod of P5 with 6 setae.

Male. Same body shape as female but with more elongated and slender abdomen. Proportional lengths of cephalothorax and abdomen 72:28; proportional lengths of abdominal somites and furca 15:25:17:14:14:15 = 100. P5 asymmetrical, very long and slender, extending beyond abdomen. B1 with naked inner border. Left P5 longer than right and with reduced, one-jointed endopod; last joint of exopod ornated with thick tufts of hairs. Right P5 with longer 3-jointed endopod. Exopods of left and right P5 ending with long spine-like seta.

Distribution

This antiboreal species has also been reported for the Arabian Sea by Sewell (1947). Farran (1929) recorded its occurrence in the tropical Atlantic Ocean, off Rio de Janeiro. It has been frequently sampled off the west coast of South America (Brady 1883) and inshore off the Chilean coast in sub-Antarctic waters (Fagetti 1962; Björnberg 1973; Arcos 1975; Marin and Antezana 1985; Hirakawa 1989). Arcos (1974) recorded it as frequent in September-October in the Straits of Magellan, whereas Mazzocchi and Ianora (1991) reported low abundances for this species in November when the species was present mainly at the three most Pacific stations. Very little is known about the vertical distribution of this species. In the present study, *C. patagoniensis* was frequently sampled in subsurface waters throughout the Straits of Magellan.

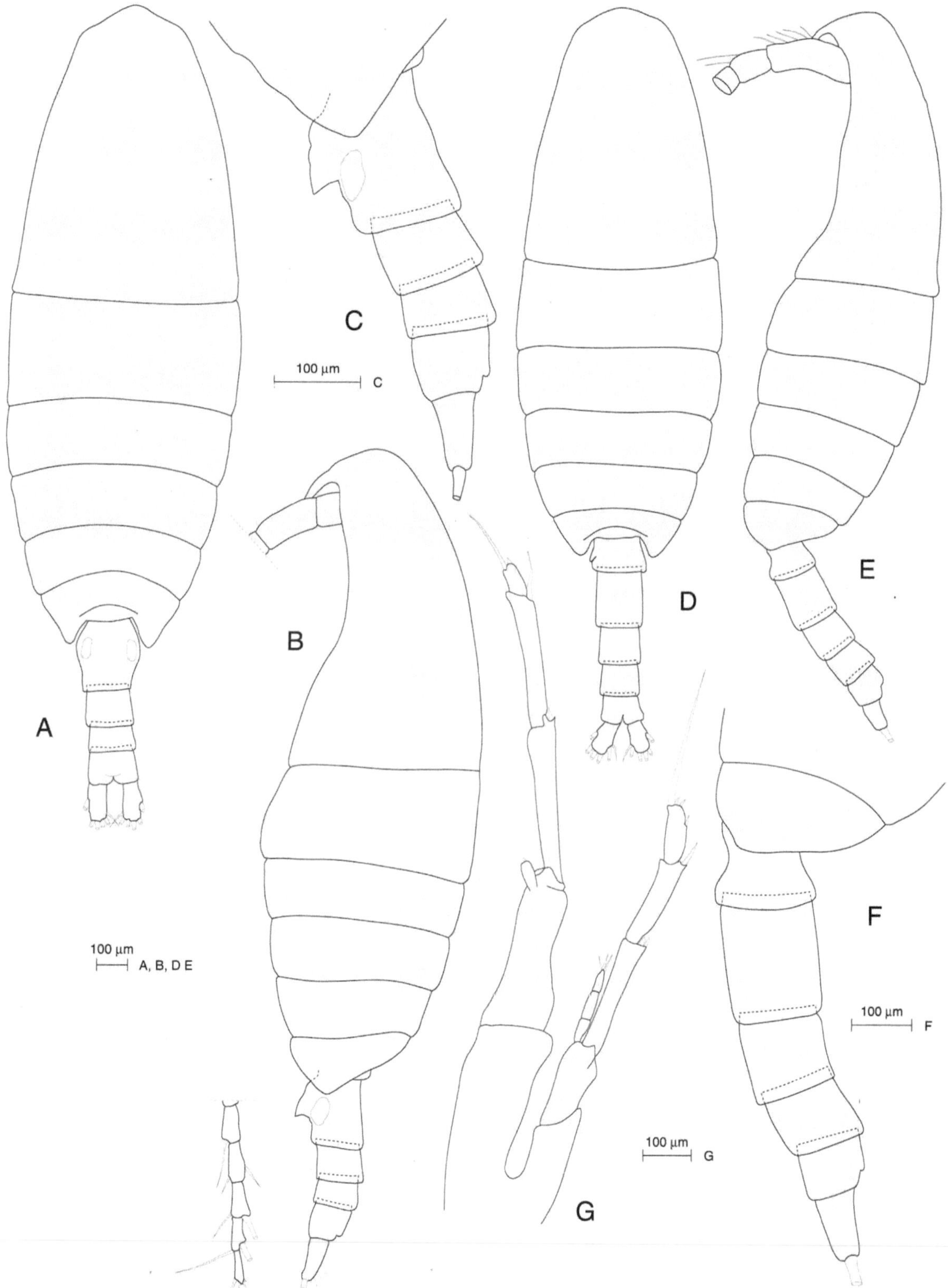

Fig. 3.12.1A–G. *Calanoides patagoniensis*. Female: **A** whole animal dorsal view; **B**, whole animal lateral view; **C** abdomen lateral view. Male: **D** whole animal dorsal view; **E** whole animal lateral view; **F** abdomen lateral view; **G** P5

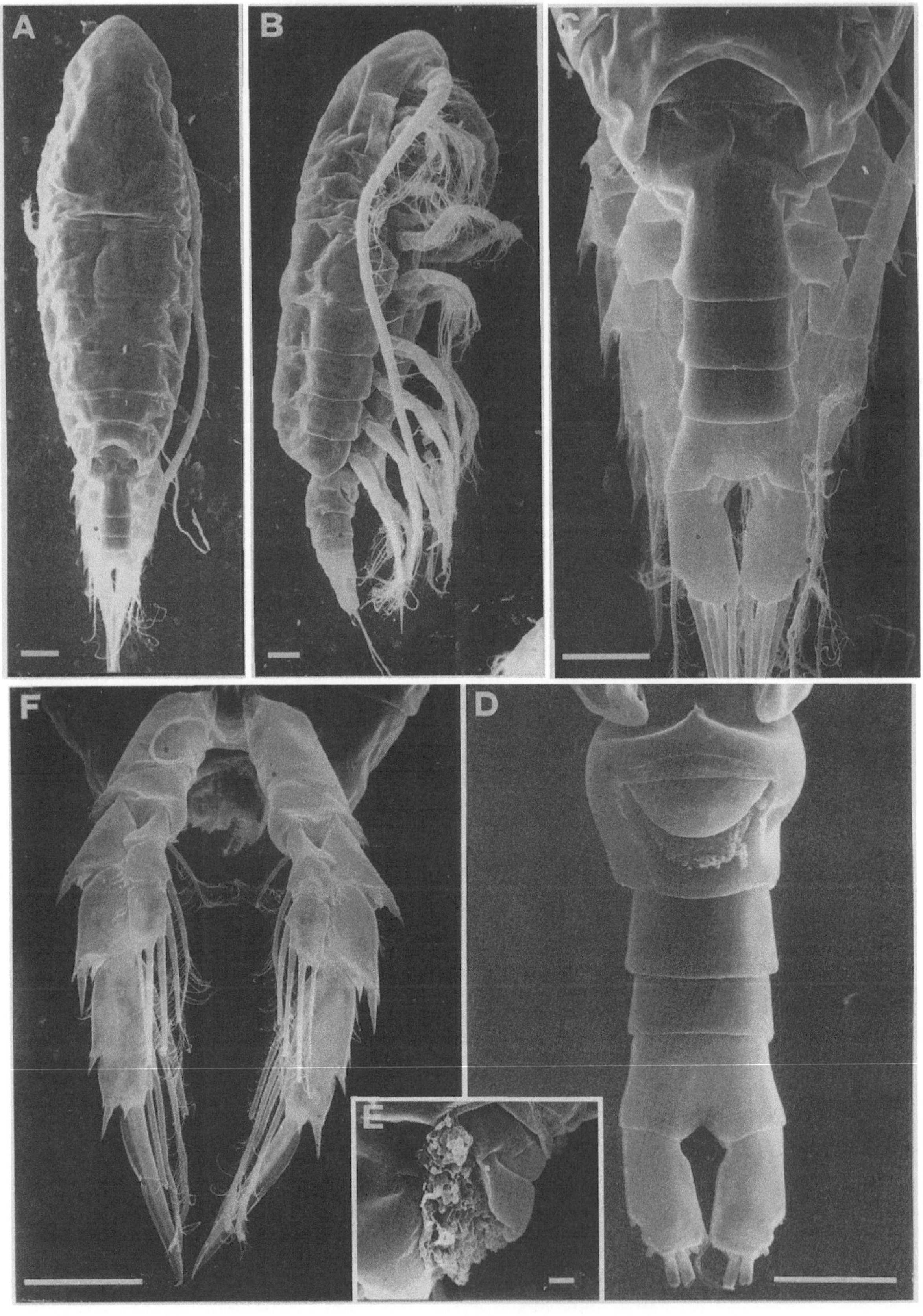

Fig. 3.12.2A–F. *Calanoides patagoniensis*. Female: **A** whole animal dorsal view; **B** whole animal lateral view; **C** abdomen dorsal view; **D** abdomen ventral view; **E** detail of plate covering genital aperture in lateral view; **F** P5. *Bars*: **A**, **B**, **C**, **D**, **E**, **F** 100 μm

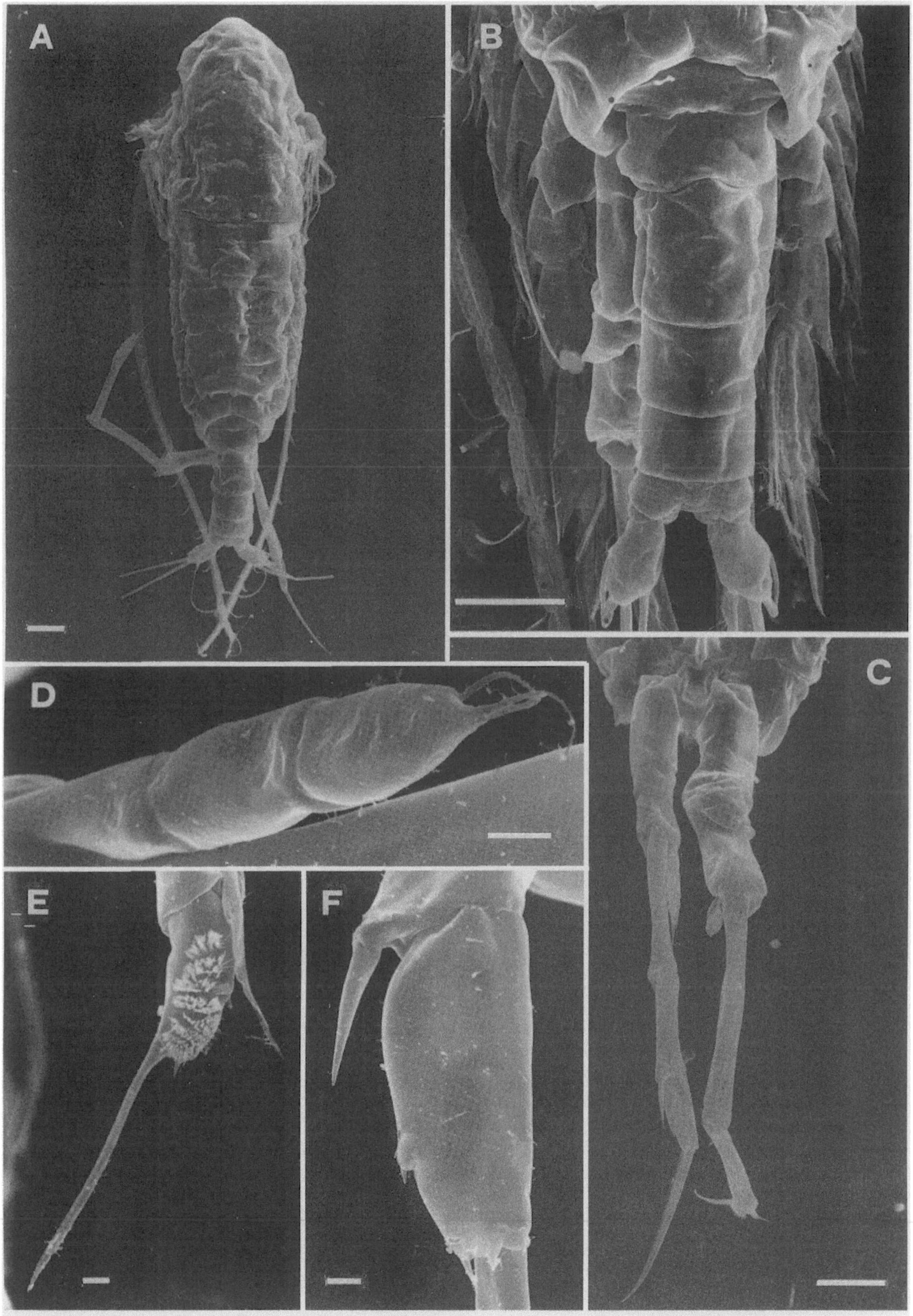

Fig. 3.12.3A–F. *Calanoides patagogeniensis*. Male: **A** whole animal dorsal view; **B** abdomen dorsal view; **C** P5; **D** detail of three-jointed endopod of right P5; **E** detail of last joint of exopod of left P5; **F** detail of last joint of exopod of right P5. *Bars*: **A**, **B**, **C** 100 µm; **D**, **E**, **F** 10 µm

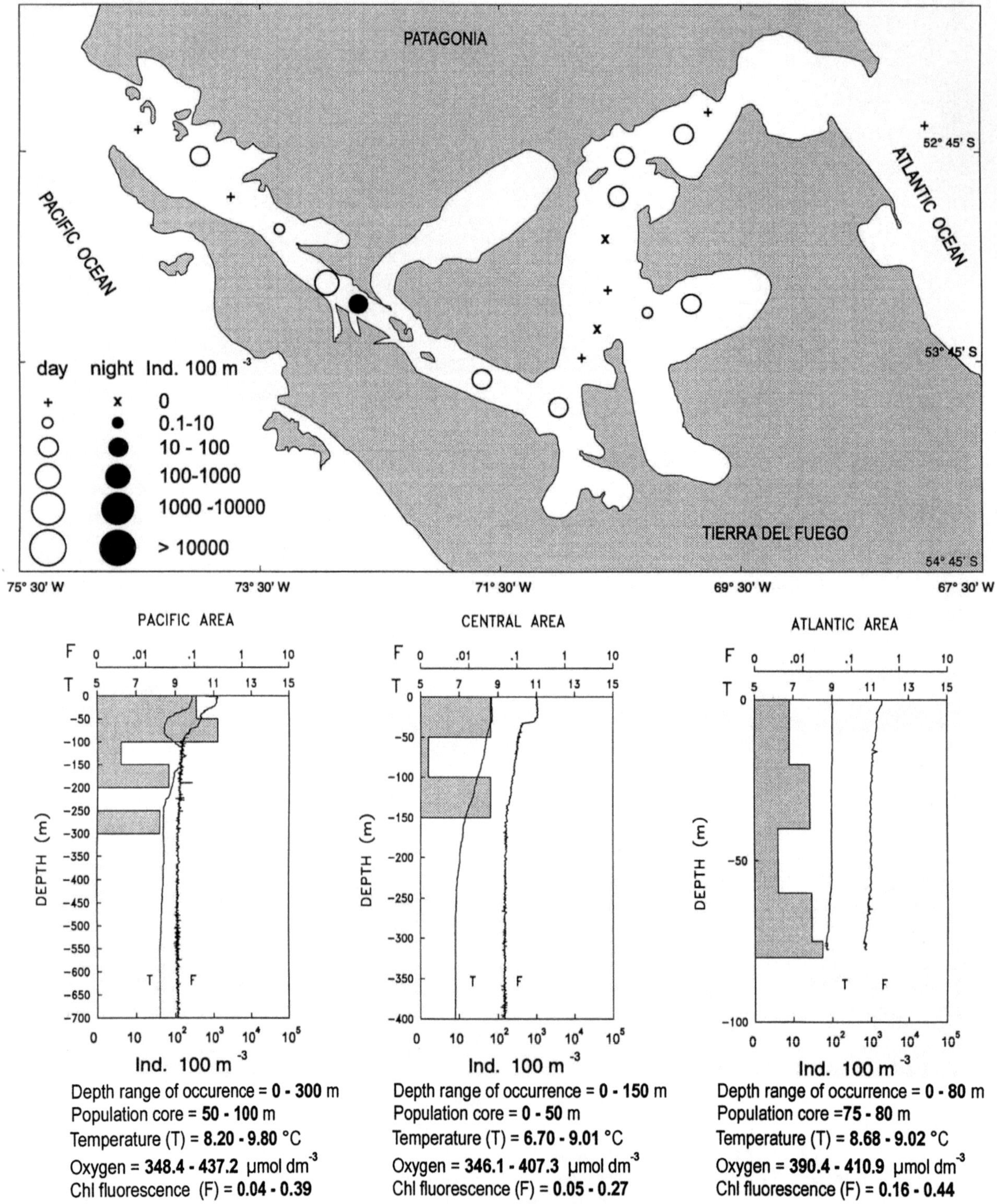

Fig. 3.12.4. *Calanoides patagoniensis*. Distribution in the Straits of Magellan in February–March 1991

3.13 *Calanus australis* Brodskii 1959

Calanus australis var. *atlanticus* Brodskii, 1959: p. 1939, pl. 1 figs. 11, 12, pl. 2 fig. 11, pl. 3 fig. 14, pl. 4 figs. 8, 9.
Calanus australis Vidal, 1968: p. 13, pl. 1 figs. 1–10 – Grice and Hülsemann, 1967: p. 20, fig. 3 – Arcos, 1975: p. 7, pl. 1 figs. 1, 2.

Body length –
females: 2.47–3.04 mm (30 specimens)
male: 2.78 mm (1 specimen)

Female. Body slender and gradually tapering anteriorly; head slightly rounded in dorsal view. Cephalothorax more than 3 times longer than wide. Last thoracic somite with rounded postero-lateral corners reaching 1/3 length of genital somite. Proportional lengths of cephalothorax and abdomen 78:22; proportional lengths of abdominal somites and furca 36:19:13:16:16 = 100. P5 B1 ornated with 14–20 slightly curved, triangular teeth. Number of teeth along margins of left and right P5 rarely the same.

Male. Same body shape as female. Proportional lengths of cephalothorax and abdomen 77:23; proportional lengths of abdominal somites and furca 12:30:17:12:11:18 = 100. P5 B1 with asymmetrical left and right sides ornated with 12–18 teeth. Endopods of both right and left P5 3-jointed. Left endopod short and stout, reaching 1/3 length of Re2 of same leg. Last joint of left exopod about 1/2 length of preceding joint and ending with two spines of different length; this joint ornated with long hairs and bearing one small spine on external margin. Exopod of right P5 about 2/3 the length of the left.

Remarks

Brodskii (1959) describes two morphotypes for this species, *C. australis* var. *atlanticus* and *C. australis* var. *pacificus*, the former occurring in the Atlantic Ocean off Argentina and the latter sampled in New Zealand waters and along the Pacific coast off Chile. Our specimens resemble the former in that they are slender and have a length/width ratio of the cephalothorax that varies from 1:3.1 to 1:3.3 in accordance to the key reported by the same Author.

Distribution

Brodskii (1959) suggested that this species occurs in temperate latitudes in the Southern hemisphere where it has been reported by several Authors (Grice and Hülsemann 1967 and others). It has been collected repeatedly off the Argentine (Ramirez 1969, 1970, 1971) and Chilean (Fagetti 1962; Vidal 1968; Arcos 1975; Marin and Antezana 1985) coasts. The species seems to be more abundant in northern Chilean and Argentine shelf waters and Vidal (1968) and Ramirez (1970) report it as one of the most common species in their collections. Not much information is available on the vertical distribution of this species but Arcos (1975) maintains that it undergoes wide vertical migrations. In the present study, *C. australis* was uniformly distributed throughout the Straits of Magellan, with maximum abundances in the upper 200 m.

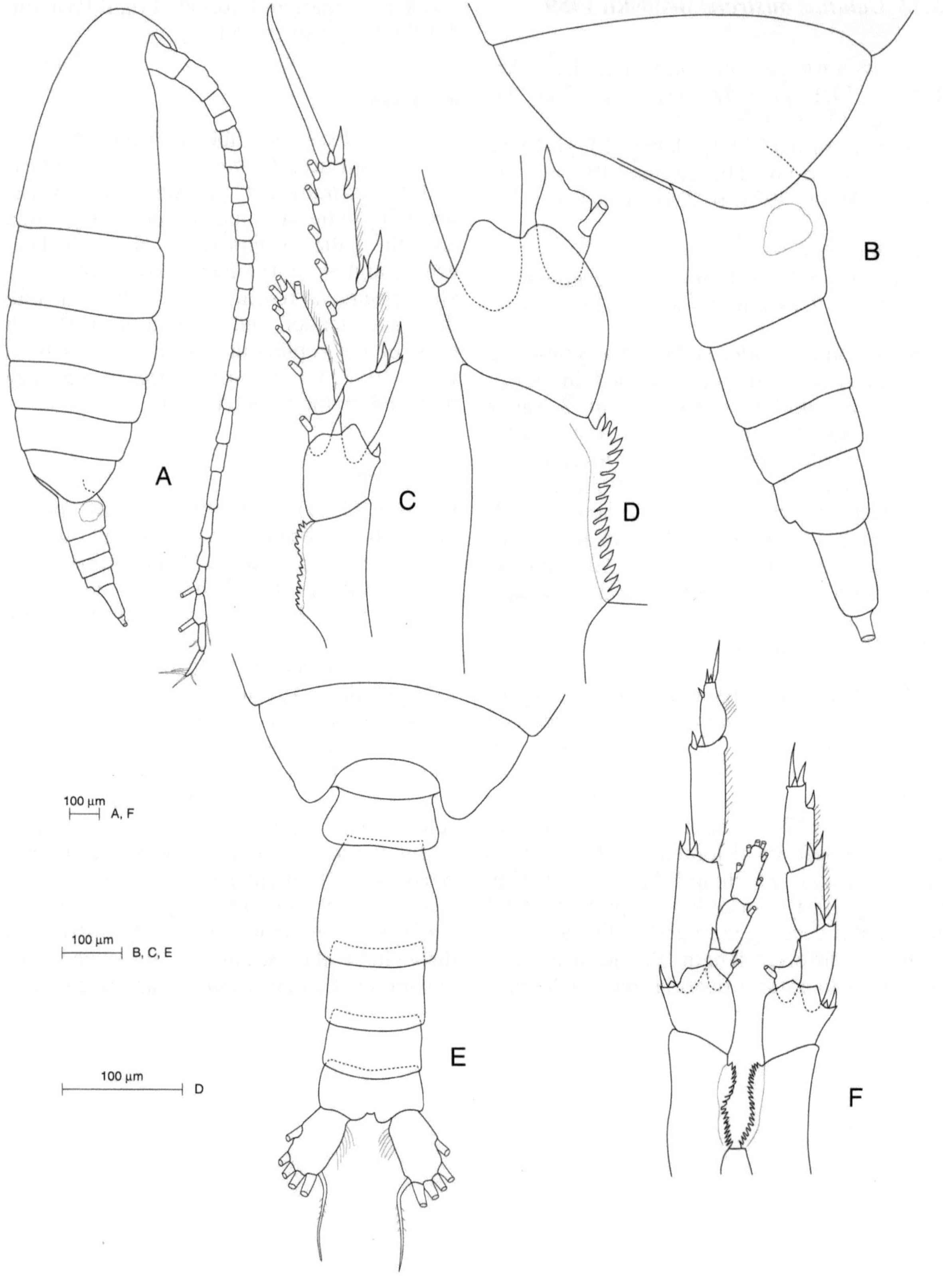

Fig. 3.13.1A–F. *Calanus australis*. Female: **A** whole animal lateral view; **B** abdomen lateral view; **C** P5; **D** detail of P5 B1. Male: **E** abdomen dorsal view; **F** P5

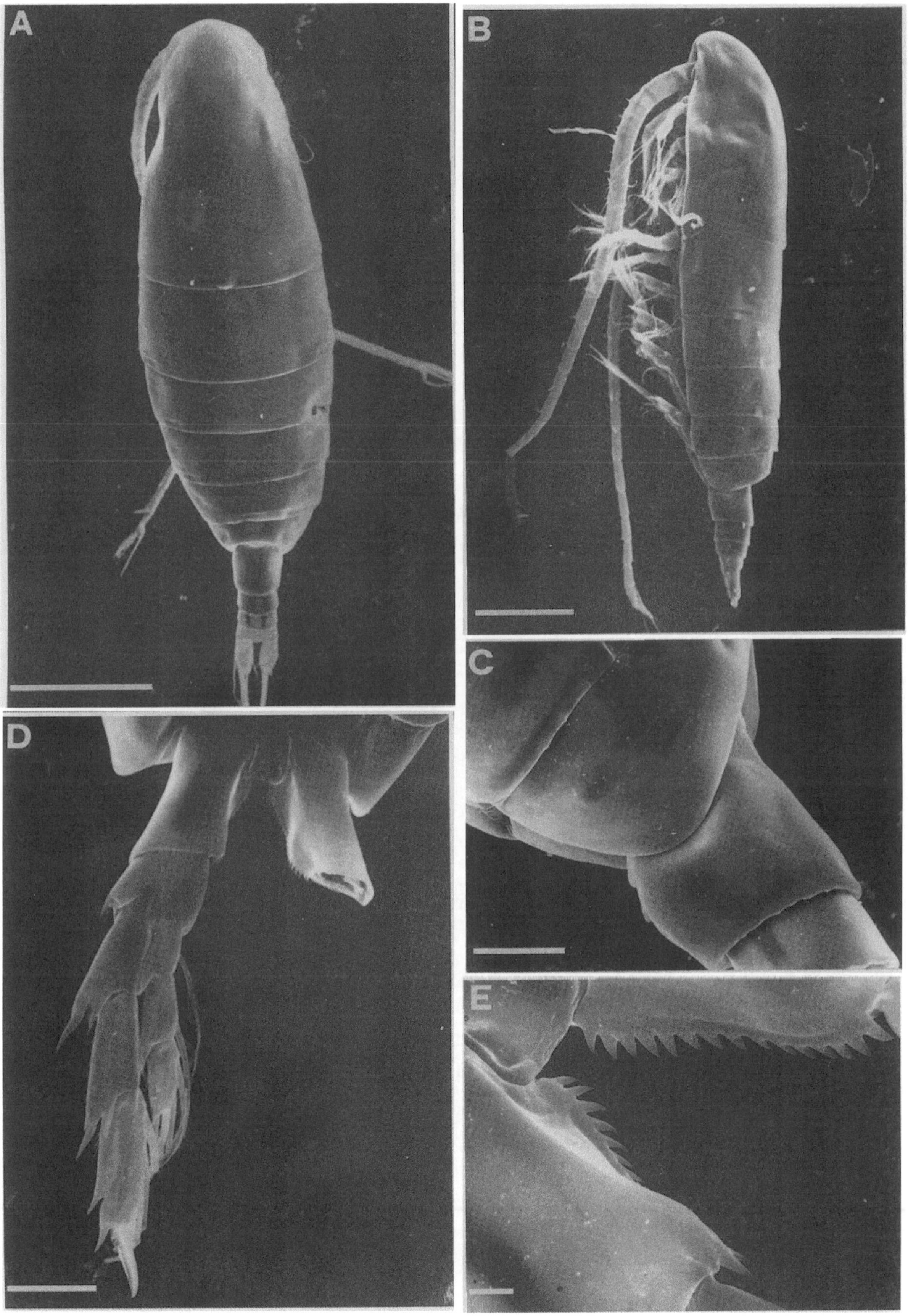

Fig. 3.13.2A–E. *Calanus australis*. Female: **A** whole animal dorsal view; **B** whole animal lateral view; **C** rounded left postero-lateral corner of last thoracic somite; **D** P5; **E** detail of P5 B1. *Bars*: **A, B** 500 μm; **C, D** 100 μm; **E** 10 μm

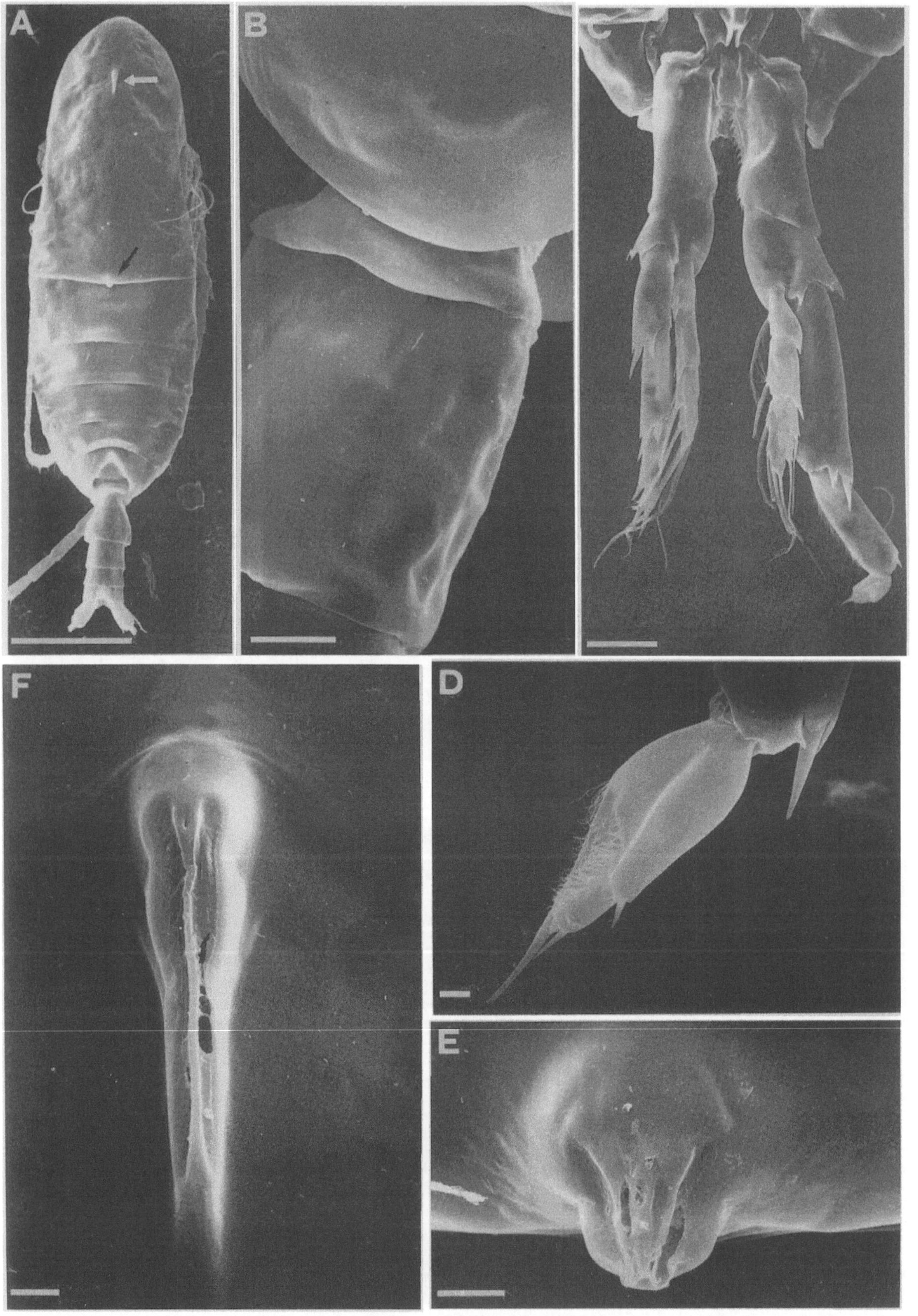

Fig. 3.13.3A–F. *Calanus australis*. Male: **A** whole animal dorsal view; **B** postero-lateral right corner of last thoracic somite; **C** P5; **D** last joint of P5 left exopod; **E** integumental organ on midpoint of postero-dorsal margin of cephalon (*black arrow* in **A**); **F** cephalic integumental organ situated in mid-head region (*white arrow* in **A**). *Bars*: **A** 500 μm; **B** 50 μm **C** 100 μm; **D**, **E**, **F** 10 μm

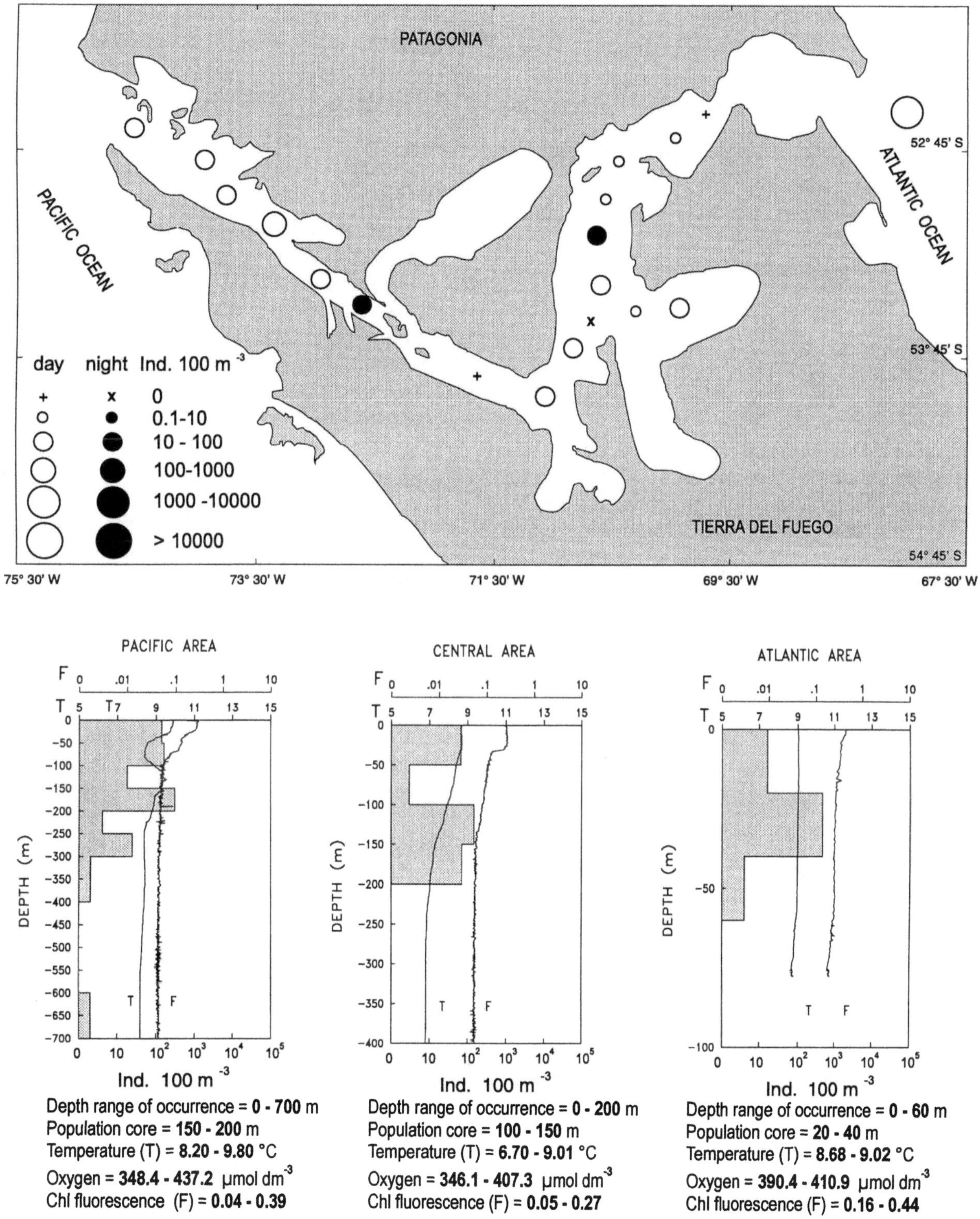

Fig. 3.13.4. *Calanus australis*. Distribution in the Straits of Magellan in February–March 1991

3.14 *Calanus simillimus* Giesbrecht 1902

Calanus simillimus Giesbrecht, 1902: p. 16, pl. 1 fig. 9 – van Breemen, 1908: p. 10, fig. 5 – Wolfenden, 1908: p. 10, pl. 1 figs. 5, 6 – Mackintosh, 1934: p. 85, figs. 3c, c^1 – Vervoort, 1951: p. 11, figs. 3–14 – Tanaka, 1960a: p. 12, pl. 2 figs. 3–10.
Calanus propinquus Giebrecht, 1892: p. 91, pl. 7 figs. 31, 34, pl. 8 figs. 14, 22, 24, 32.

Body length –
females: 2.93–3.97 mm (31 specimens)
males: 2.89–3.20 mm (10 specimens)

Female. Body slender, with a distinctly obtuse head in dorsal view. Proportional lengths of cephalothorax and abdomen 75:25. Last thoracic somite with triangular-shaped margins which extend half-way across genital somite; postero-lateral corners pointed, with points slightly produced. Proportional lengths of abdominal somites and furca 36:17:14:13:20 = 100. Genital somite longer than wide. Each caudal ramus carries 5 thin setae and a curved short internal seta bearing fine hairs on inner margin. P5 symmetrical. Internal margins of B1 carry about 18 triangular teeth of equal size and small cluster of 3–4 teeth isolated from the others.

Male. Same body shape as female with similar proportional lengths between cephalothorax and abdomen 74:26. Proportional lengths of abdominal somites and furca 12:31:17:13:13:14 = 100. Caudal rami diverging and carrying 5 thin setae and curved inner seta. P5 asymmetrical. B1 with asymmetrical left and right sides, this latter shorter than the left. Both furnished with triangular teeth on inner margin (20–22 on left B1). Left exopod twice the length of exopod of right P5. Endopod of left leg reaching base of Re2. Last joint of left exopod with 2 small lateral spines and 1 very long apical spine about 1.5 times length of joint. Inner margin of last joint ornated with short bristles and small cluster of longer hairs.

Remarks

Our specimens are similar to those described by Vervoort (1951) except for the toothing of the margins of the P5 B1 in both females and males. The teeth are more numerous in our specimens. Also, the last joint of the left P5 of the male differs by the presence of an ornamentation along the inner margin that is not reported in Vervoort's figure.

Distribution

C. simillimus ranges widely over northern Antarctic and sub-Antarctic oceanic areas. According to Mackintosh (1934) and Vervoort (1951), this species is a normal inhabitant of sub-Antarctic waters. It may also occur near the Convergence and is at times carried south of the polar front in certain areas (Ramirez and Dinofrio 1976). Specimens have been reported for Chilean fjords by many Authors (Arcos 1976; Marin and Antezana 1985; Hirakawa and Zama 1985; Hirakawa 1989). Mazzocchi and Ianora (1991) reported it as the most common *Calanus* species in previous collections from the Straits of Magellan, where it was present, mainly as copepodite stages, only in the Pacific sector. Hardy and Gunther (1935) reported that the vertical distribution of *C. simillimus* ranges up to 1000 m throughout the area between South Georgia and the Falkland Islands. They also noted that this species exhibits marked and rapid diurnal vertical migrations. In the present study, *C. simillimus* was distributed throughout the Straits of Magellan. Maximum abundances were recorded in the Central area, where it was found at all sampled depths.

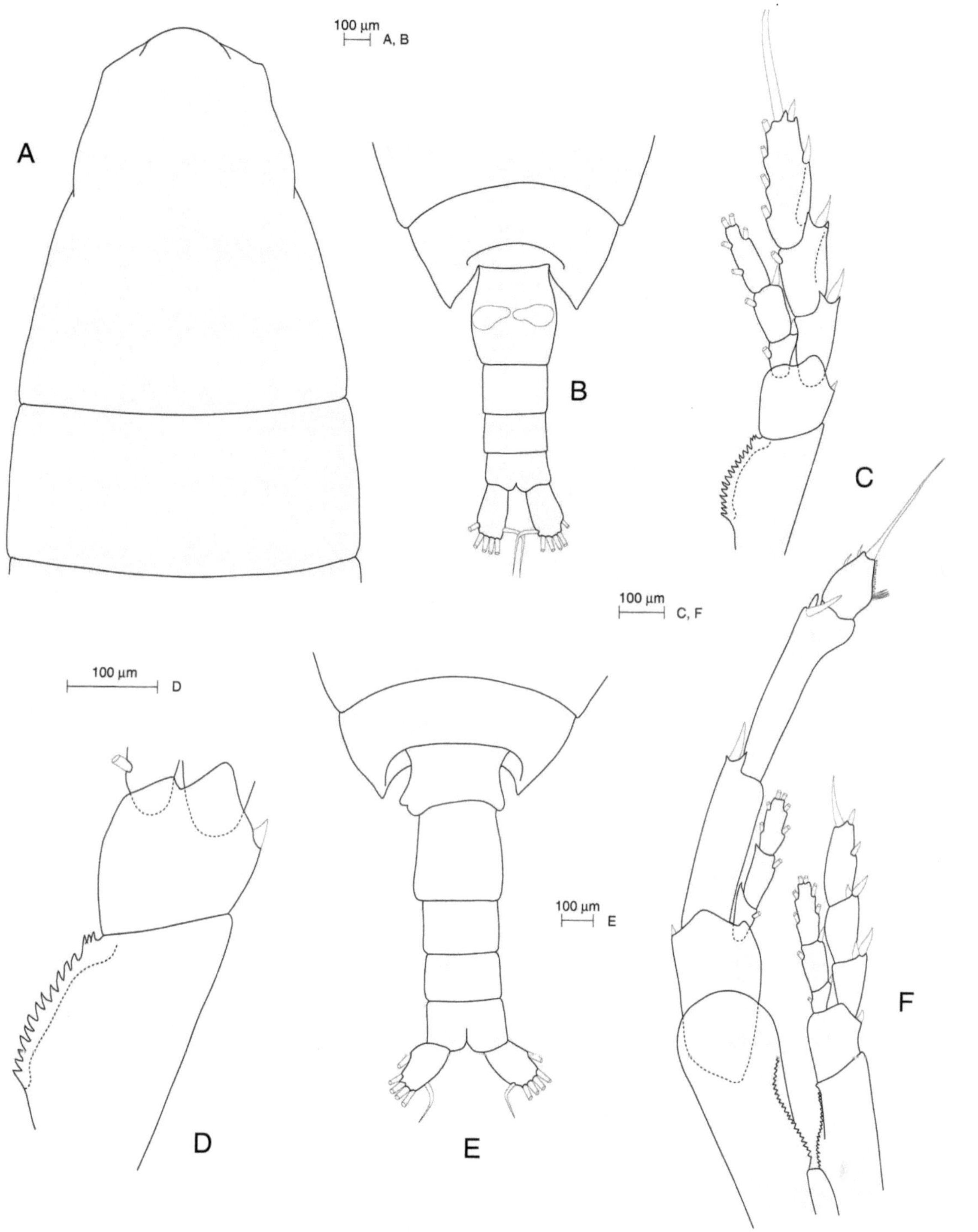

Fig. 3.14.1A–F. *Calanus simillimus*. Female: **A** cephalosome dorsal view; **B** abdomen dorsal view; **C** P5; **D** detail of P5 B1. Male: **E** abdomen dorsal view; **F** P5

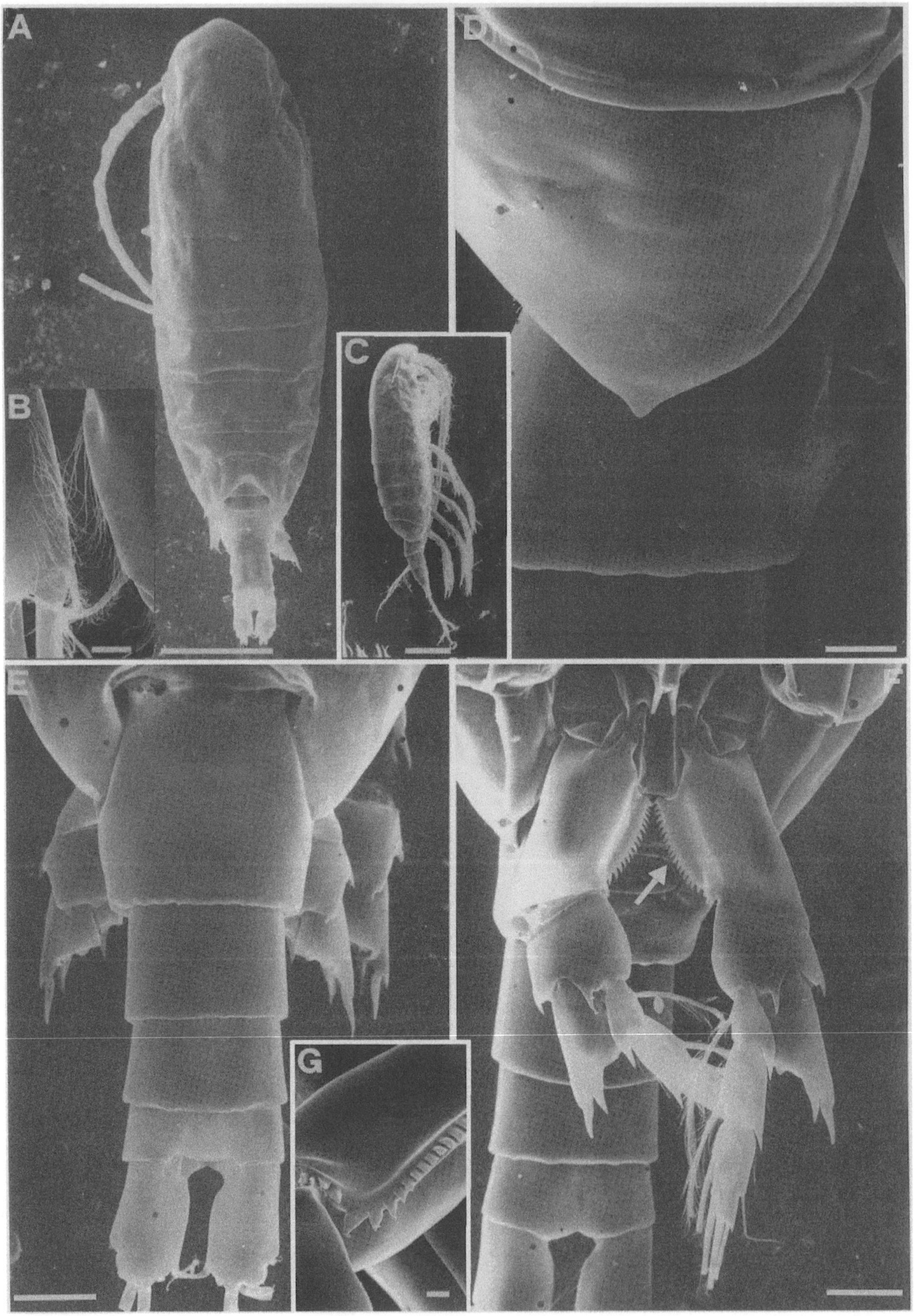

Fig. 3.14.2A–G. *Calanus simillimus*. Female: **A** whole animal dorsal view; **C** whole animal lateral view; **B** detail of fine hairs on inner margins of caudal rami; **D** pointed postero-lateral right corner of last thoracic somite; **E** abdomen dorsal view; **F** P5; **G** detail of internal margin of P5 B1 bearing triangular teeth (*arrow* in **F**). *Bars*: **A**, **C** 500 µm; **B**, **D** 50 µm; **E**, **F** 100 µm; **G** 10 µm

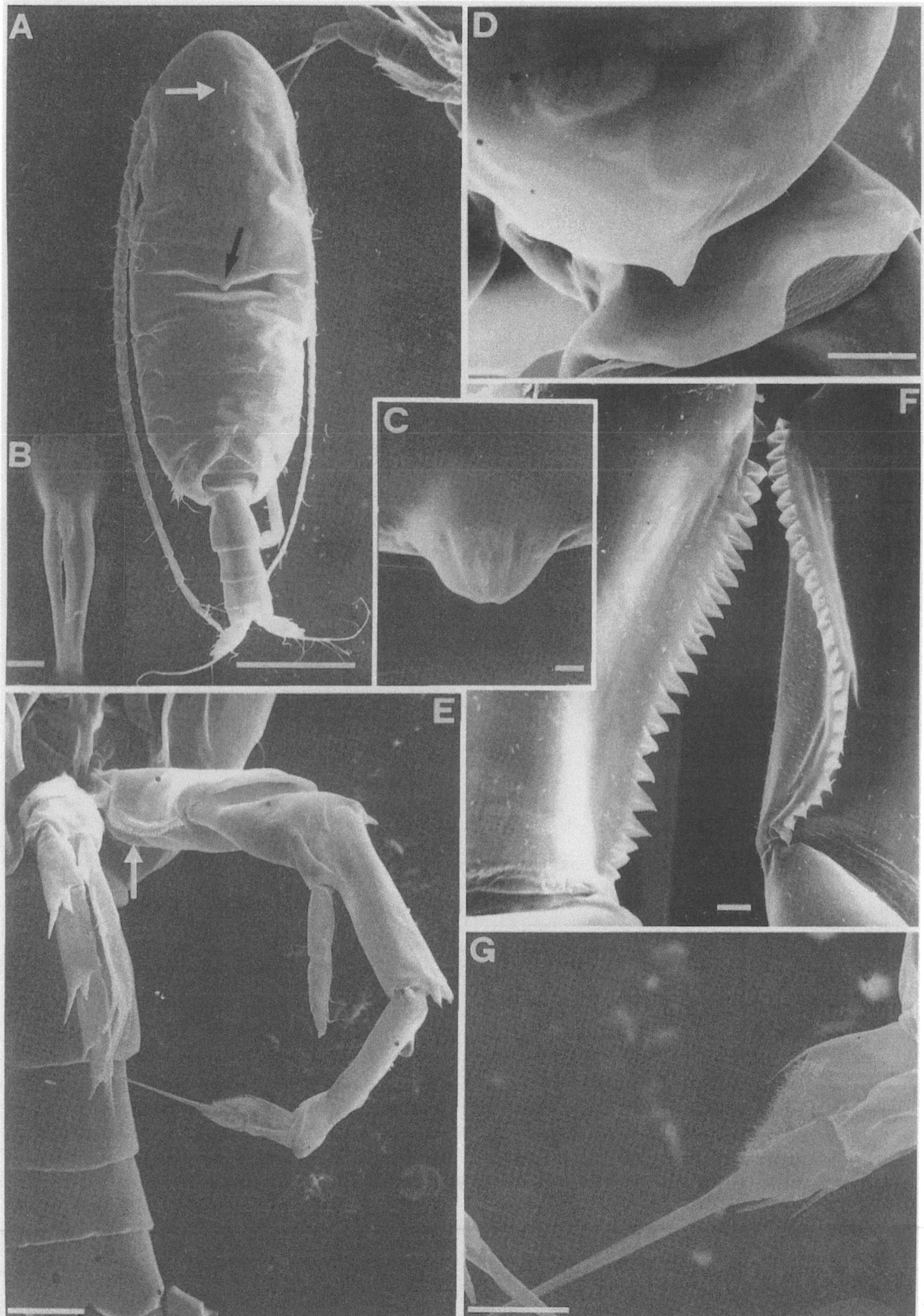

Fig. 3.14.3A–G. *Calanus simillimus*. Male: **A** whole animal dorsal view; **B** cephalic integumental organ (*white arrow* in **A**); **C** integumental protuberance at midpoint of postero-dorsal margin of cephalon (*black arrow* in **A**); **D** pointed postero-lateral left corner of last thoracic somite; **E** P5; **F** inner margins of P5 B1 bearing triangular teeth (*arrow* in **E**); **G** detail of last somite of P5 left exopod. *Bars*: **A** 500 µm; **B**, **C**, **F** 10 µm; **D**, **G** 50 µm; **E** 100 µm

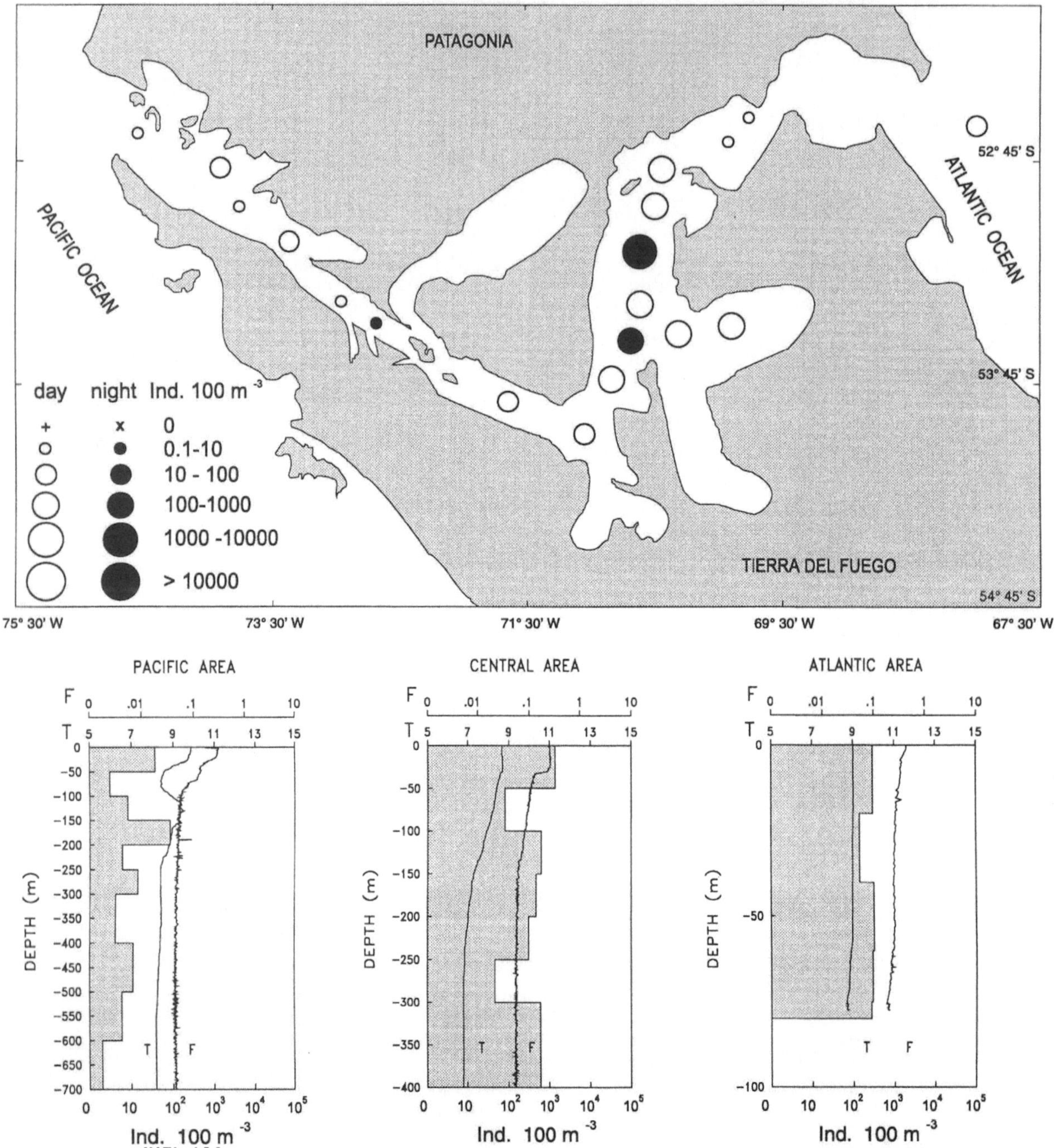

PACIFIC AREA	CENTRAL AREA	ATLANTIC AREA
Depth range of occurrence = **0 - 700** m	Depth range of occurrence = **0 - 400** m	Depth range of occurrence = **0 - 80** m
Population core = **150 - 200** m	Population core = **0 - 50** m	Population core =**0 - 20** m
Temperature (T) = **8.20 - 9.80** °C	Temperature (T) = **6.70 - 9.01** °C	Temperature (T) = **8.68 - 9.02** °C
Oxygen = **348.4 - 437.2** µmol dm^{-3}	Oxygen = **346.1 - 407.3** µmol dm^{-3}	Oxygen = **390.4 - 410.9** µmol dm^{-3}
Chl fluorescence (F) = **0.04 - 0.39**	Chl fluorescence (F) = **0.05 - 0.27**	Chl fluorescence (F) = **0.16 - 0.44**

Fig. 3.14.4. *Calanus simillimus*. Distribution in the Straits of Magellan in February–March 1991

3.15 *Neocalanus tonsus* Brady 1883

Calanus tonsus Brady, 1883: p. 34, pl. 4 figs. 8, 9 – Wolfenden, 1908: p. 11, pl. 1 figs. 7, 8 – Farran, 1929: p. 216, fig. 2 – Brodskii, 1948: p. 32, pl. 3 figs. 1–7, pl. 4 figs. 1, 2; 1950: p. 92, fig. 22 (*C. tonsus* f. *plumchrus* and f. *typica*) – Tanaka, 1956a: p. 49, figs. 1–4 – Vervoort, 1957: p. 27, figs. 3–6 – Jillett, 1968: p. 27, fig. 3.
Neocalanus tonsus Bradford, 1988: p. 79.

Body length –
female: 3.92 mm (1 specimen)

Female. Cephalothorax elongated and tapering towards the head with poorly defined protuberance. Cephalothorax of our only specimen 2.8 times longer than wide. Posterior corners of last thoracic somite produced backward into short round processes. Proportional lengths of cephalothorax and abdomen 79:21; proportional lengths of abdominal somites and furca 42:22:11:11:14 = 100. P5 B1 with smooth inner margins without teeth; B2 with 4 spines on distal inner surface.

Remarks

In our samples, we recorded only copepodite specimens and one adult female. Unfortunately, this latter was heavily damaged, with a P5 having only B1. According to Brodskii (1950), there are two forms of this species: f. *typica* and f. *plumchrus*, the latter being synonymous with *Calanus plumchrus* Marukawa. The only difference between these forms is the presence of elongated, very hairy caudal bristles in the CV of f. *plumchrus*. Brodskii (1950) stated that possibly f. *plumchrus* represents a seasonal form.

Distribution

N. tonsus is an abundant oceanic, autiboreal copepod distributed mainly in the sub-Antarctic, from New Zealand across the Southern Pacific, but not entering the true Antarctic region (Jillett 1968). It is abundant over the edges of the continental shelf and slope, somewhat less plentiful over oceanic depths, but avoids neritic conditions inshore. This was the dominant species found by Marin and Antezana (1985) in Chilean fjords, where it was present only as copepodite stages (mainly CV) comprising 32–89% of the copepod stock. *N. tonsus* is also reported for the Atlantic, north of the Patagonian region, in external slope waters subjected to sub-Antarctic currents (Ramirez 1971). The depth of occurrence of this species varies seasonally; Jillett (1968) reports it as abundant in the upper 150 m layer from midspring to midsummer (October to January) but it was virtually absent from surface waters from late summer (February) until spring (September), when it again became abundant. Ohman (1987) maintains that reproductive copepods of this species occur in mesopelagic depths (500–1000 m) in austral winter and epipelagic depths (0–150 m) in spring. In the present study, *N. tonsus* was found only from 150–200 m at a Pacific station in the Straits of Magellan. It was also sporadically present throughout the water column at two other stations in the Central area with maximum abundances from 300–400 m.

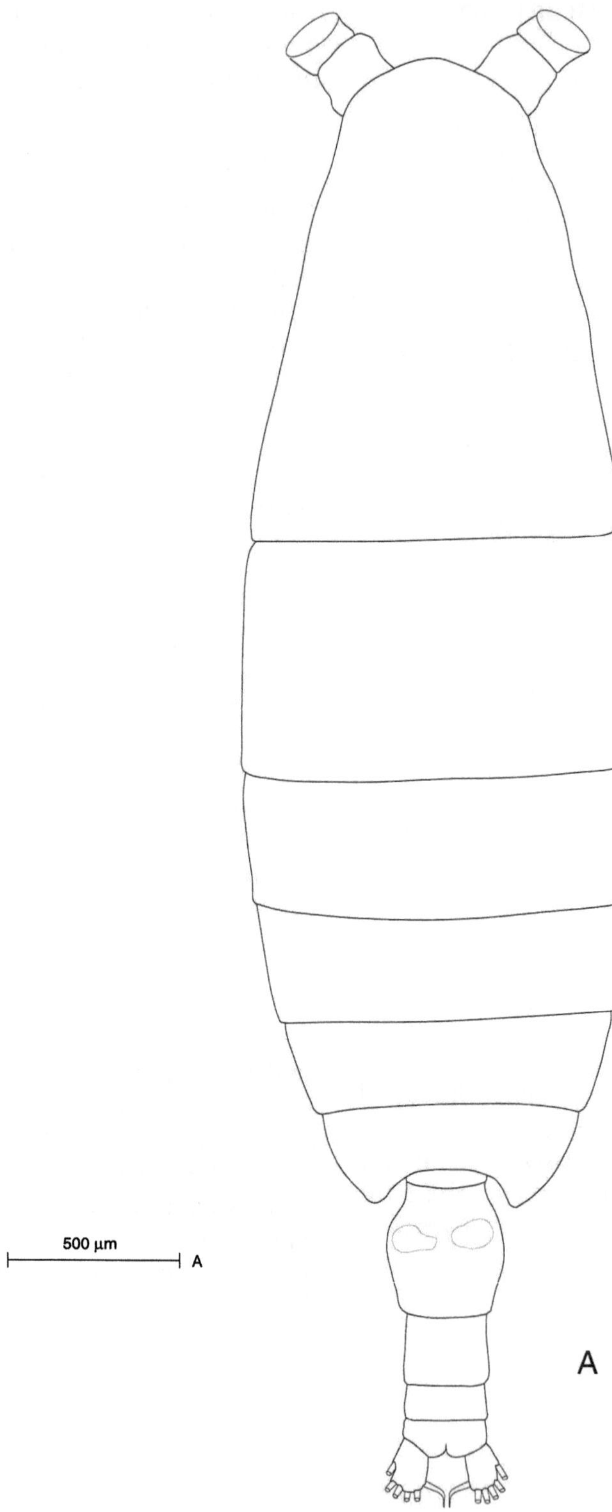

Fig. 3.15.1. *Neocalanus tonsus*. Female: whole animal dorsal view

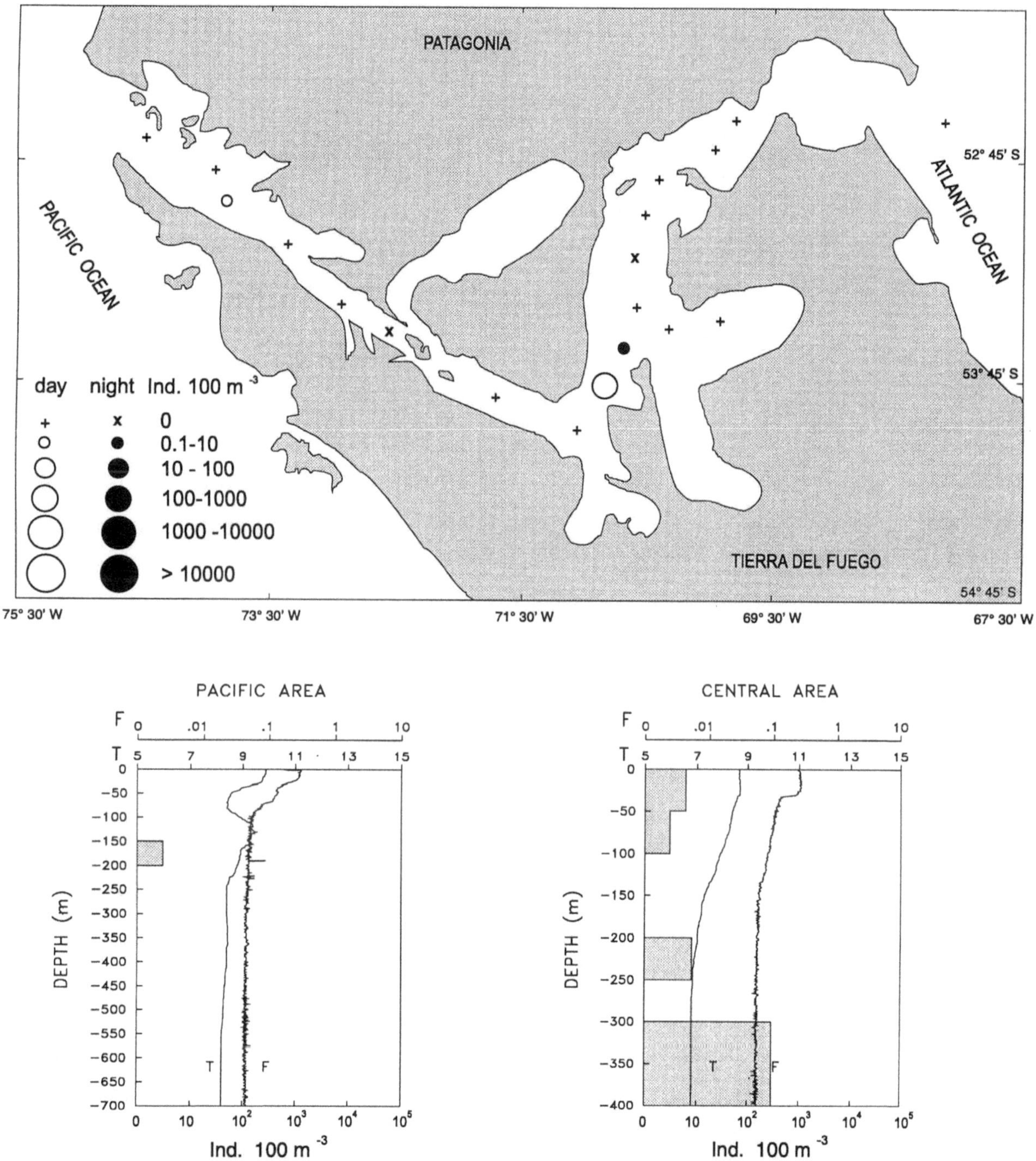

Depth range of occurrence = **150 - 200** m
Population core = **150 - 200** m
Temperature (T) = **8.20 - 9.80** °C
Oxygen = **348.4 - 437.2** µmol dm^{-3}
Chl fluorescence (F) = **0.04 - 0.39**

Depth range of occurrence = **0 - 400** m
Population core = **300 - 400** m
Temperature (T) = **6.70 - 9.01**°C
Oxygen = **346.1 - 407.3** µmol dm^{-3}
Chl fluorescence (F) = **0.05 - 0.27**

Fig. 3.15.2. *Neocalanus tonsus*. Distribution in the Straits of Magellan in February–March 1991

3.16 *Calocalanus pavoninus* Farran 1936

Calocalanus pavoninus Farran, 1936: p. 83, fig. 3 – Tanaka 1956b: p. 380, fig. 9 – Corral, 1972: p. 4, fig. 3a–d.

Body length –
females: 0.71, 0.76 mm (2 specimens)

Female. Cephalothorax 2.7 times longer than wide, 5 times longer than abdomen. Abdomen 2-segmented; genital somite 1.6 times wider than long. Caudal rami only slightly longer than wide. P5 symmetrical, 4-jointed; last joint terminating in long, strong plumose spine with row of spinules at base.

Remarks

Farran (1936) described the female of *C. pavoninus* with a 3-jointed P5. He stated that this species can be distinguished from *C. contractus*, which it resembles in form and size, by the single stout seta at the end of the P5. We collected only two specimens, with 4-jointed P5s as reported by Tanaka (1956b) and Corral (1972).

Distribution

Calocalanus pavoninus is mainly an oceanic and epiplanktonic species. It has been reported from the Gulf of Mexico and Caribbean Sea, Canary Islands, Great Barrier Reef Region, Izu Region (Japan), and the Mediterranean Sea (Corral 1972). To our knowledge, this species has never been reported in Chilean or Argentinian waters. In the present study, it was present in very low numbers from 0–50 m at two stations in the Pacific area of the Straits of Magellan and at one station off Cape Froward.

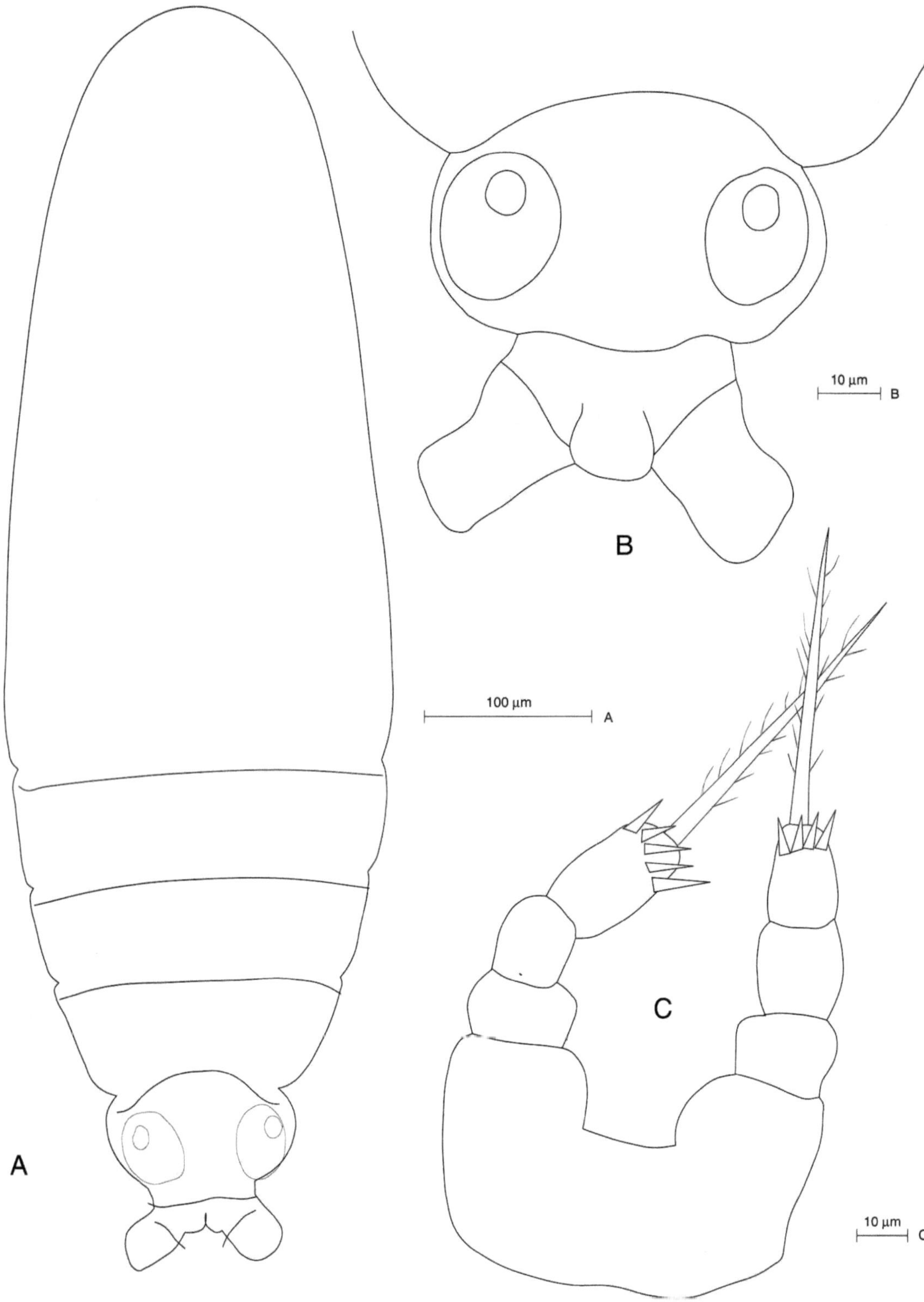

Fig. 3.16.1A–C. *Calocalanus pavoninus*. Female: **A** whole animal dorsal view; **B** abdomen dorsal view; **C** P5

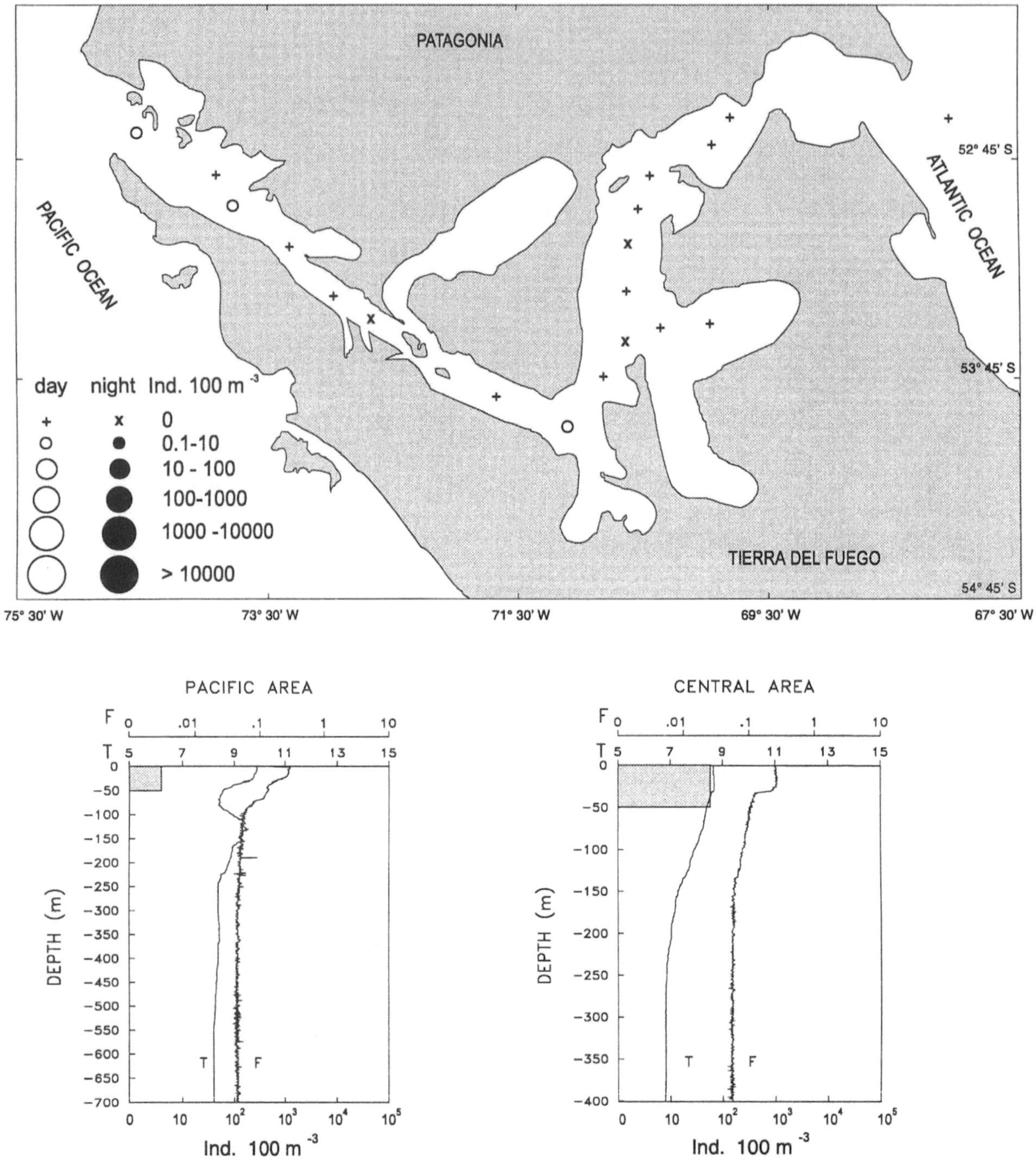

PACIFIC AREA	CENTRAL AREA
Depth range of occurrence = **0 - 50** m	Depth range of occurrence = **0 - 50** m
Population core = **0 - 50** m	Population core = **0 - 50** m
Temperature (T) = **8.20 - 9.80** °C	Temperature (T) =**6.70 - 9.01** °C
Oxygen = **348.4 - 437.2** μmol dm^{-3}	Oxygen = **346.1 - 407.3** μmol dm^{-3}
Chl fluorescence (F) = **0.04 - 0.39**	Chl fluorescence (F) = **0.05 - 0.27**

Fig. 3.16.2. *Calocalanus pavoninus*. Distribution in the Straits of Magellan in February–March 1991

3.17 *Paracalanus parvus* Claus 1863

Calanus parvus Claus, 1863: p. 173, pl. 26 figs. 10–14, pl. 27 figs. 1–4.
Paracalanus parvus Giesbrecht, 1892: p. 48, 165, pl. 1 fig. 5, pl. 6 figs. 28–30, pl. 9 figs. 5, 9, 11, 17, 23, 25, 27, 31–34 – Sars, 1903: p. 17, pls. 8, 9 – Brodskii, 1948: p. 41, pl. 7 figs. 9, 10, pl. 8 figs. 1–6; 1950: p. 108, fig. 33 – Farran and Vervoort, 1951a: p. 3, fig. 1 – Marques, 1959: p. 208, pl. 1 fig. d – Vervoort, 1962: p. 394, figs. 1, 2 – Ramirez, 1966: p. 9, pl. 4 figs. 13–19 – Vidal, 1968: p. 22, pl. 5 figs. 10–14 – Bowman, 1971: p. 28, fig. 23b–p – Arcos, 1975: p. 12, pl. 3 figs. 17–21.

Body length –
females: 1.02–1.18 mm (10 specimens)
males: 1.08 mm (2 specimens)

Female. Body stout with wide, smoothly rounded head. Cephalothorax 2.6 times longer than wide. Rostrum with two short filaments. A1 not reaching caudal rami, with slightly thickened base. Proportional lengths of cephalothorax and abdomen 73:27; proportional lengths of abdominal somites and furca 27:17:14:24:18 = 100. P2–P4 B1 with very variable ornamentation (spines, spinules and hairs). P2–P4 Ri2 with numerous spines. P2, P3 Re2 with a clutch of 2–3 spines. These spines are absent in P4. P2–P4 Re3 with serrated external margin but smooth outer distal edge. P5 small, symmetrical, uniramous and 2-jointed. Terminal joint long with two spines, inner one almost 3 times longer than other and with marked serrated margin.

Male. Body shape similar to female. Cephalothorax 2.4 times longer than wide. A1 with very thick base. Proportional lengths of cephalothorax and abdomen 68:32; proportional lengths of abdominal somites and furca 9:25:19:17:16:14 = 100. P5 asymmetrical and uniramous. Right leg 2-jointed and left 5-jointed. Both terminal joints of left and right P5 with 2 short spines; in right joint one spine much shorter than other, in left joint one spine about half the length of the other. This latter joint bears a longitudinal row of fine hairs.

Remarks

Some authors have observed a very small spine on the margin of the long spine of the terminal joint of the female P5 (Giesbrecht 1892; Sars 1903). This spine was absent in our females, as also reported by other authors (Brodskii 1950; Farran and Vervoort 1951a; Vervoort 1962; Bowman 1971).

Distribution

P. parvus is a neritic species with a remarkably wide distribution. Though plentiful inshore, it can also occur in surface oceanic waters. Rose (1933) and Wilson (1932) describe its wide occurrence, extending to all three major oceans, the Mediterranean and Black Seas and including warm as well as temperate waters under greatly varying ecological conditions. Most authors report that it does not occur at very high latitudes either in the Northern or Southern hemispheres; but it has been recorded as plentiful inshore for the Chilean coast by Arcos (1975), Marin and Antezana (1985) and Peterson et al. (1988), and for the Argentine coast by Ramirez (1969, 1970, 1971). In the Mediterranean Sea, it is typically found in surface waters throughout the year, with a depth of occurrence ranging from 0–100 m (Hure 1955; Scotto di Carlo et al. 1984). In the present study, *P. parvus* was found from 0–50 m throughout the Straits of Magellan.

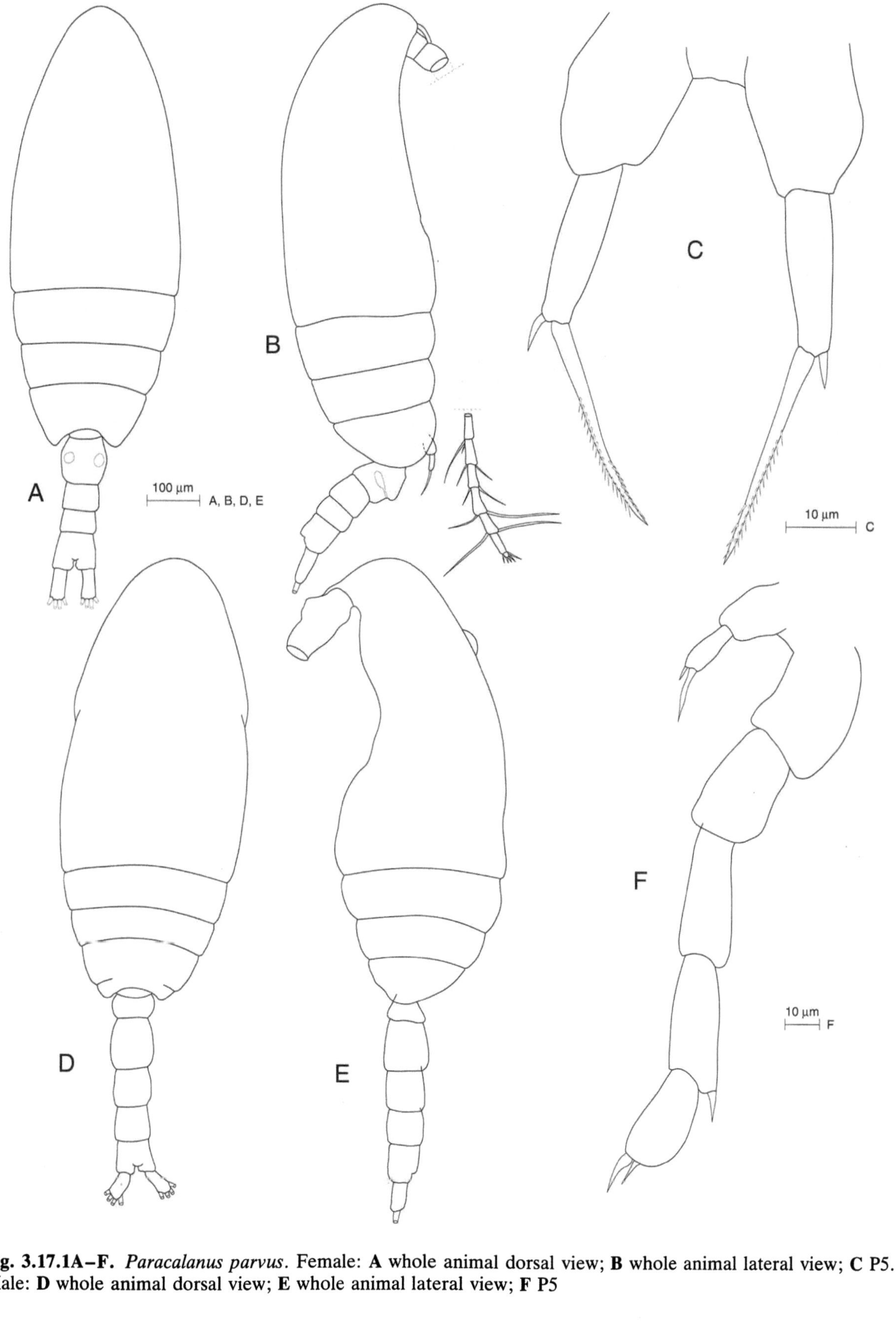

Fig. 3.17.1A–F. *Paracalanus parvus*. Female: **A** whole animal dorsal view; **B** whole animal lateral view; **C** P5. Male: **D** whole animal dorsal view; **E** whole animal lateral view; **F** P5

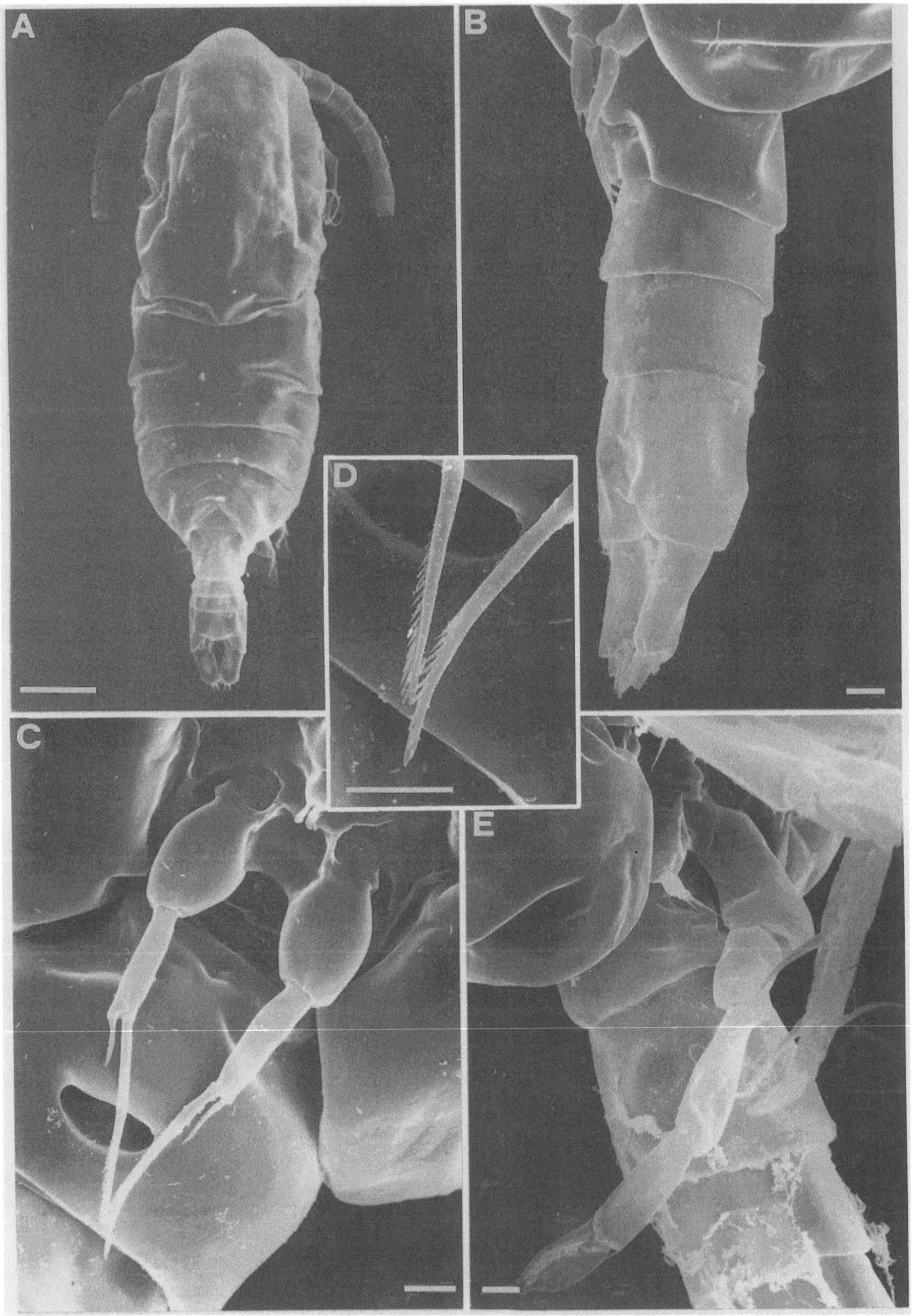

Fig. 3.17.2A–E. *Paracalanus parvus*. Female: **A** whole animal dorsal view; **B** abdomen lateral view; **C** P5; **D** detail of terminal part of P5 apical spines. Male: **E** P5. *Bars*: **A** 100 μm; **B**, **C**, **D**, **E** 10 μm

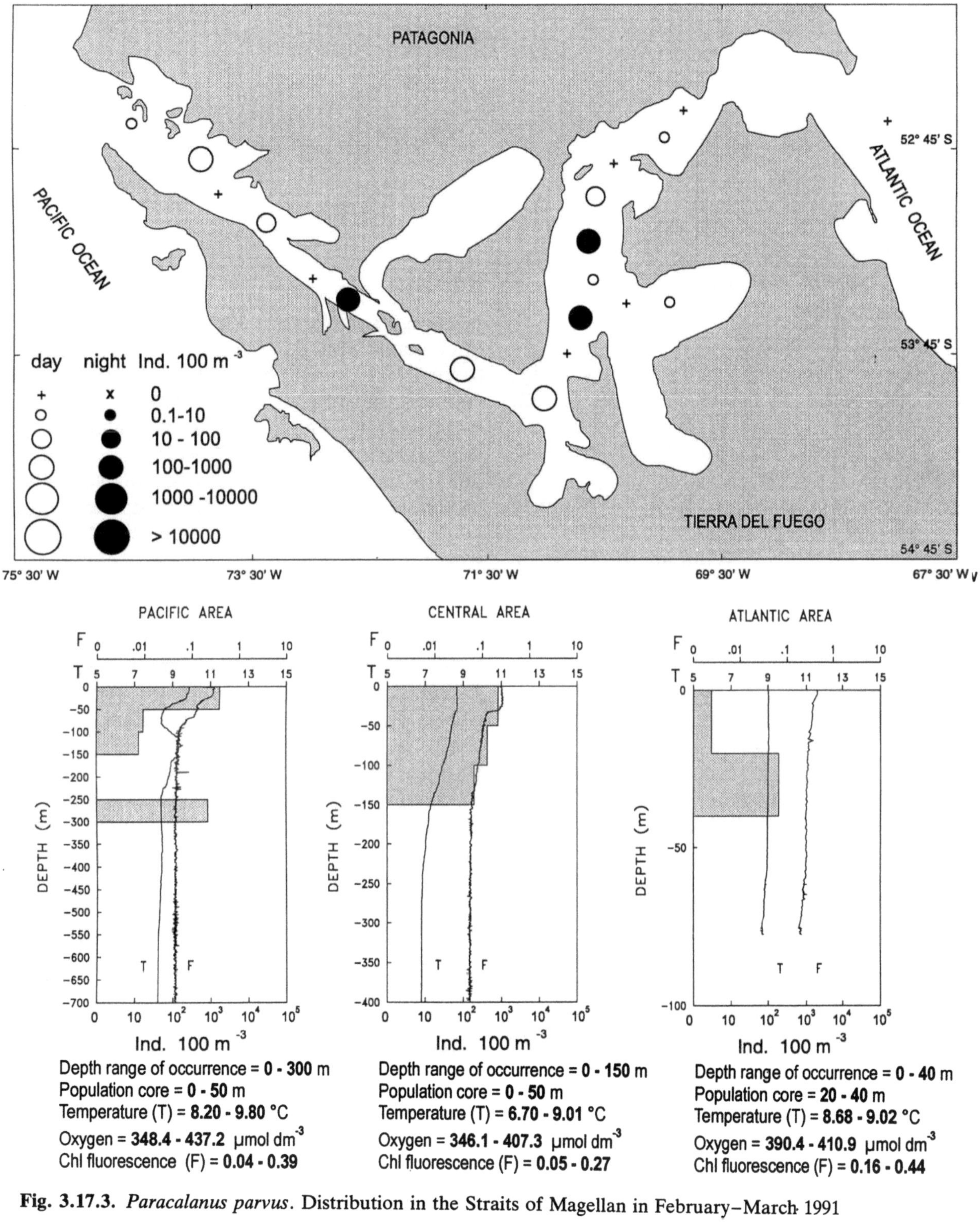

Fig. 3.17.3. *Paracalanus parvus*. Distribution in the Straits of Magellan in February–March 1991

3.18 *Eucalanus elongatus hyalinus* Giesbrecht 1892

Eucalanus elongatus hyalinus Giesbrecht, 1892: p. 131, pl. 11 figs. 2, 7, 12, 20, 25, 32, 36, pl. 35 figs. 1, 2, 13, 23, 24 – Johnson, 1938: p. 170, figs. 3, 17, 19, 21, 24 – Brodskii, 1950: p. 103, fig. 29.
Eucalanus spinifer Scott, 1894: p. 29, pl. 1 figs. 15–23.
Eucalanus atlanticus Wolfenden, 1904: p. 111, pl. 9 figs. 3, 4.
Eucalanus elongatus Ramirez, 1969: p. 45, pl. 7 figs. 43, 44.

Body length –
females: 6.90–7.35 mm (4 specimens)

Female. Body very transparent and limp. Rostrum robust, elongated, and ending in fine filaments. Cephalothorax about 3 times longer than wide. Head triangular; last thoracic somite with pointed and slightly elongated posterior corners. Proportional lengths of cephalothorax and abdomen 86 : 12. Last thoracic somite ornated with small spinules on dorsal posterior margin. Abdomen 4-segmented; anal somite and furca fused. Proportional lengths of abdominal somites and furca 48 : 15 : 9 : 28 = 100. Genital somite longer than wide. Left caudal ramus larger than right; left second inner seta much stronger and longer than right. Anal somite with tufts of hairs and sparse spinules on ventral side and spinules along ridge of anal plate. P1 with 3-jointed exopod, distal joint with long terminal spine and one outer, shorter spine; endopod 2-jointed; proximal joint with tuft of hairs on inner edge. P5 absent.

Remarks

Although *Eucalanus* is a morphologically distinct genus and its species are widespread, there is still much confusion regarding the validity and rank of its nominal species and subspecies (Fleminger 1973). Difficulties in the identification of this species are largely due to the absence of distinctive, sexually modified appendages. Our specimens (4 adult females and a few copepodites) agree with the descriptions for *E. elongatus hyalinus* by Giesbrecht (1892). This species was discussed by Vervoort (1946, 1963) and reported by Brodskii (1950), although this latter author considered only the two species *E. elongatus* and *E. hyalinus* in a successive paper, distinguishing them on the basis of the shape of the posterior corners of the last thoracic somite (Brodskii et al. 1983). Fleminger (1973) widely discussed the genus *Eucalanus* and commented on the *elongatus* complex. He considered as useful characters the seminal receptacles and integumental organs (sensilla and glandular pore). We did not use this latter character since we had only a limited number of specimens and these were heavily damaged.

Distribution

E. elongatus hyalinus has been recorded in the North Atlantic, Indian Ocean and Mediterranean Sea. It seems to have a fairly wide distribution in the northern Pacific (Johnson 1938). However, according to Brodskii (1950), it is difficult to determine the range of this species in the Pacific owing to its confusion with *E. bungii*. This species is considered mainly, though not exclusively, Atlantic, being distributed over nearly the whole Atlantic Ocean from the sub-Antarctic region in the south to the seas around Iceland in the north (Vervoort 1963). It was reported from sub-Antarctic localities by Vervoort (1957). Fagetti (1962) and Ramirez (1969), respectively reported it in Chilean and Argentine waters. *E. elongatus hyalinus* is considered a bathypelagic oceanic species and has rarely been found in surface layers; it varies in abundance in deep hauls (Roe 1972). In the present study, the species was found from 150–500 m at two stations in the Pacific area of the Straits of Magellan.

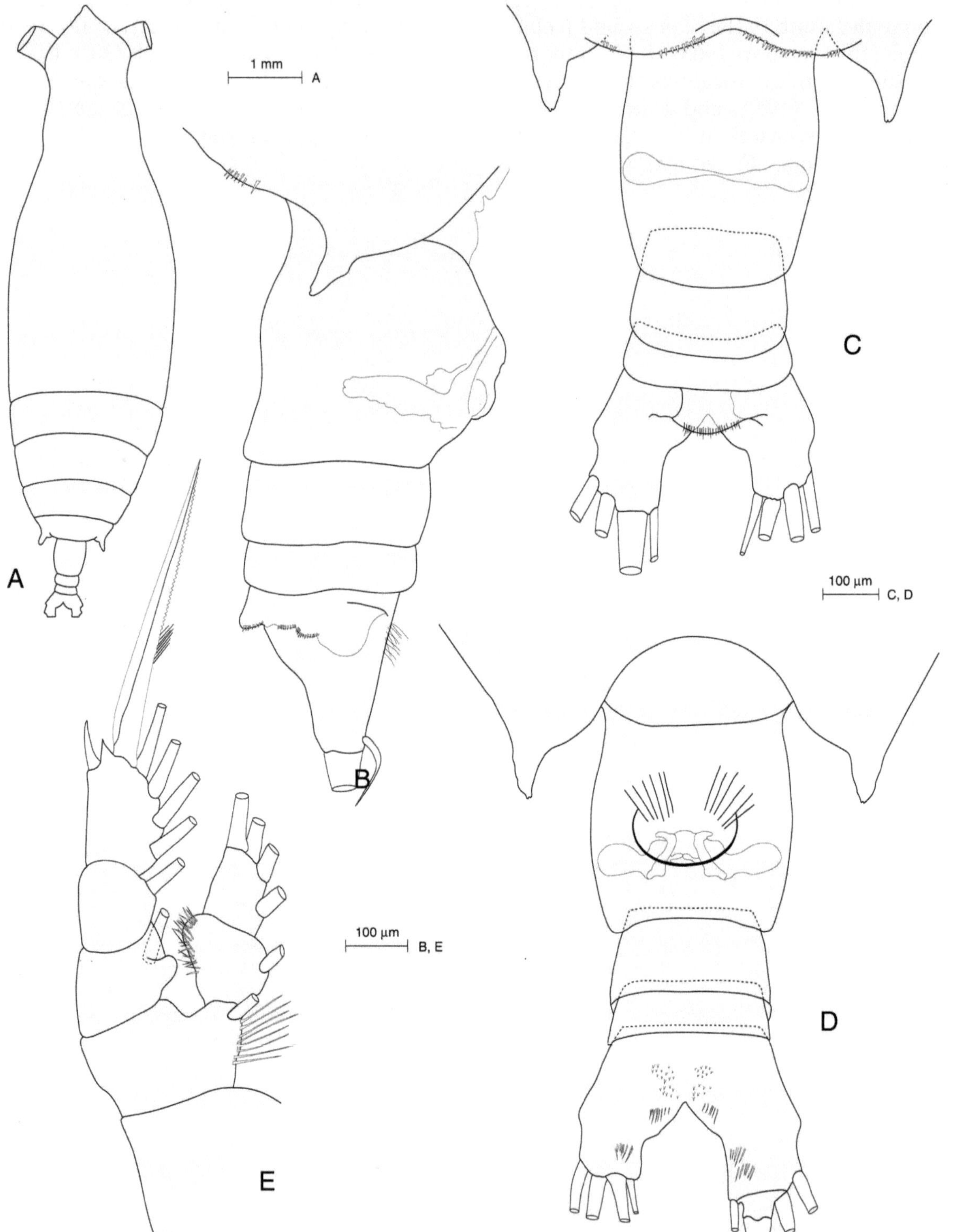

Fig. 3.18.1A–E. *Eucalanus elongatus hyalinus*. Female: **A** whole animal dorsal view; **B** abdomen lateral view; **C** abdomen dorsal view; **D** abdomen ventral view; **E** P1

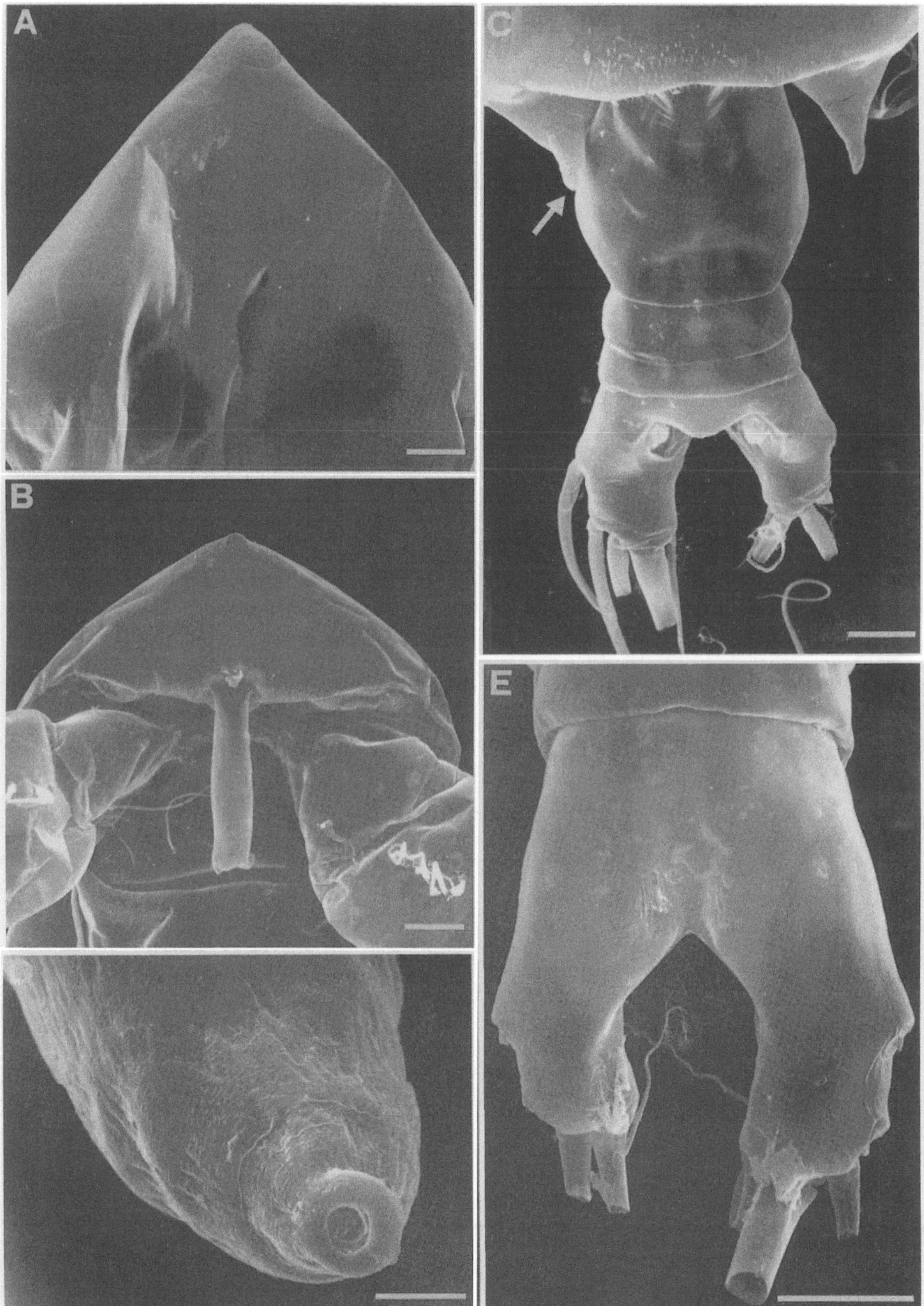

Fig. 3.18.2A–E. *Eucalanus elongatus hyalinus*. Female: **A** forehead dorsal view; **B** forehead ventral view with elongated rostrum; **C** abdomen dorsal view; **D** terminal points of last thoracic somite ending in conspicuous pores (*arrow* in **C**); **E** asymmetrical caudal rami ventral view; *Bars*: **A**, **B**, **C**, **E** 100 μm; **D** 5 μm

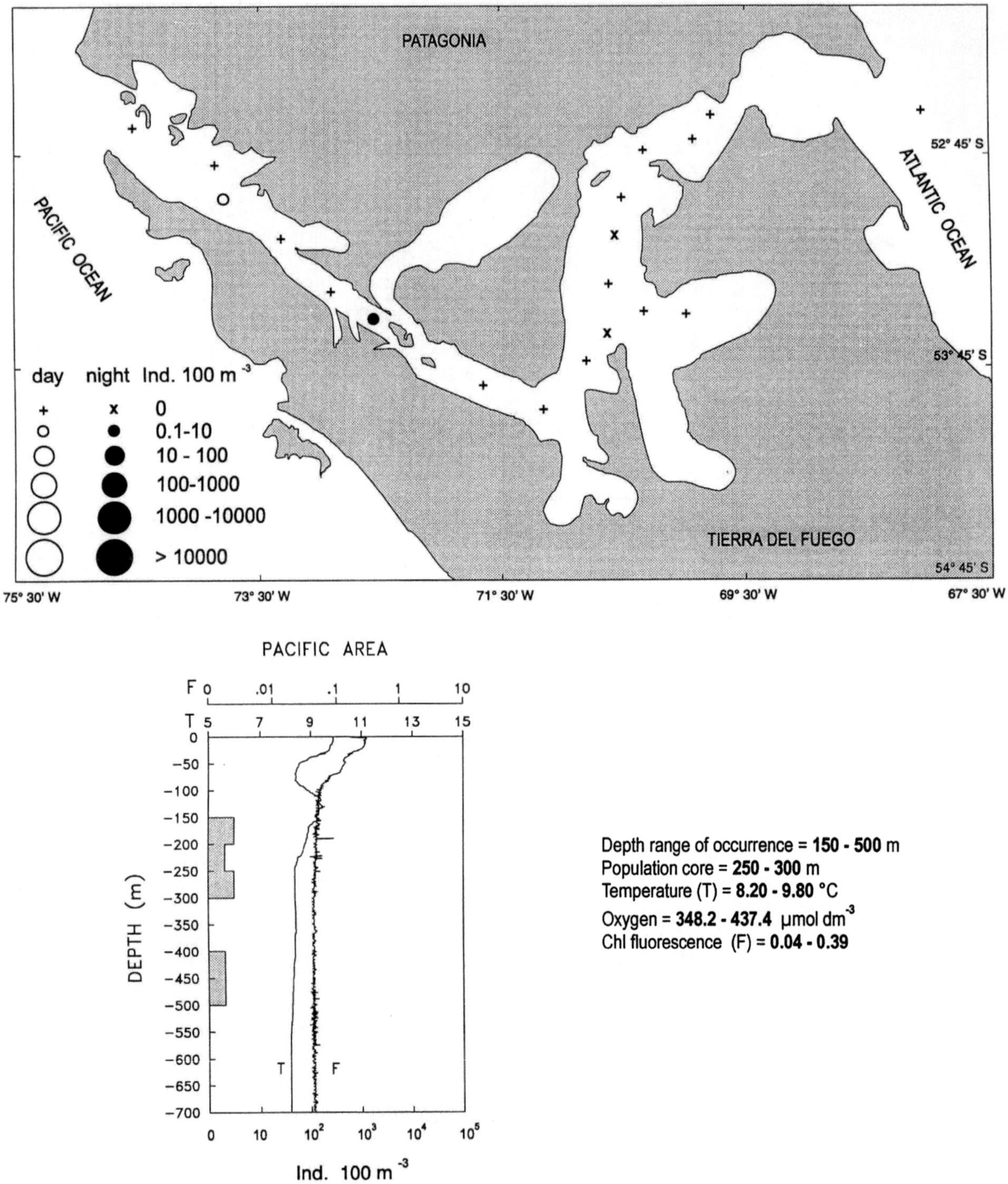

Fig. 3.18.3. *Eucalanus elongatus hyalinus*. Distribution in the Straits of Magellan in February–March 1991

3.19 *Eucalanus longiceps* Matthews 1925

Eucalanus longiceps Matthews, 1925: p. 127, pl. 9 figs. 1–10 – Ramirez, 1969: p. 43, figs. 35, 36.
Eucalanus acus Farran, 1929: p. 218, fig. 3.

Body length –
females: 4.21–5.10 mm (3 specimens)

Female. Cephalothorax elongate and slender, slightly enlarged at base of first somite. Head tapered into acute conical point. Postero-lateral corners of last thoracic somite rounded. Proportional lengths of cephalothorax and abdomen 89:11; proportional lengths of abdominal somites and furca 54:13:20:13 = 100. Abdomen 3-segmented. Genital somite laterally swollen in dorsal view, with pronounced circular genital pores in proximity of the lateral margins. Caudal rami very short, more or less as long as wide.

Remarks

E. longiceps was originarily described by Matthews (1925) and then erroneously redescribed by Farran (1929) as *E. acus* n.sp. This mistake was noted and corrected for by Sewell (1947). *E. longiceps* is very easily distinguished from all other eucalanid species due to its very characteristic elongated conical head. In our samples, we found very few adult females and a number of juvenile stages but no males.

Distribution

The species has frequently been reported in cold waters of the Southern hemisphere. Farran (1929) found *E. acus* (synonymus) from the surface to 80 m south of New Zealand. Hardy and Gunther (1935) reported it from intermediate cold Antarctic waters around South Georgia. Senô et al. (1963) reported this species as characteristic of sub-Antarctic waters. Also Ramirez (1969, 1970) recorded it in sub-Antarctic waters from the surface to 500–600 m in the SW Atlantic Ocean. Atkinson and Peck (1988) reported *E. longiceps* as a strong migrant species concentrated at depths greater than 500 m in the oceanic area around South Georgia. Marin and Antezana (1985) reported it for the first time in Chilean fjords. In the present study, the species was mainly sampled in the Pacific area and at a single station off Cape Froward. It was distributed from 100–500 m with a maximum abundance from 250–300 m.

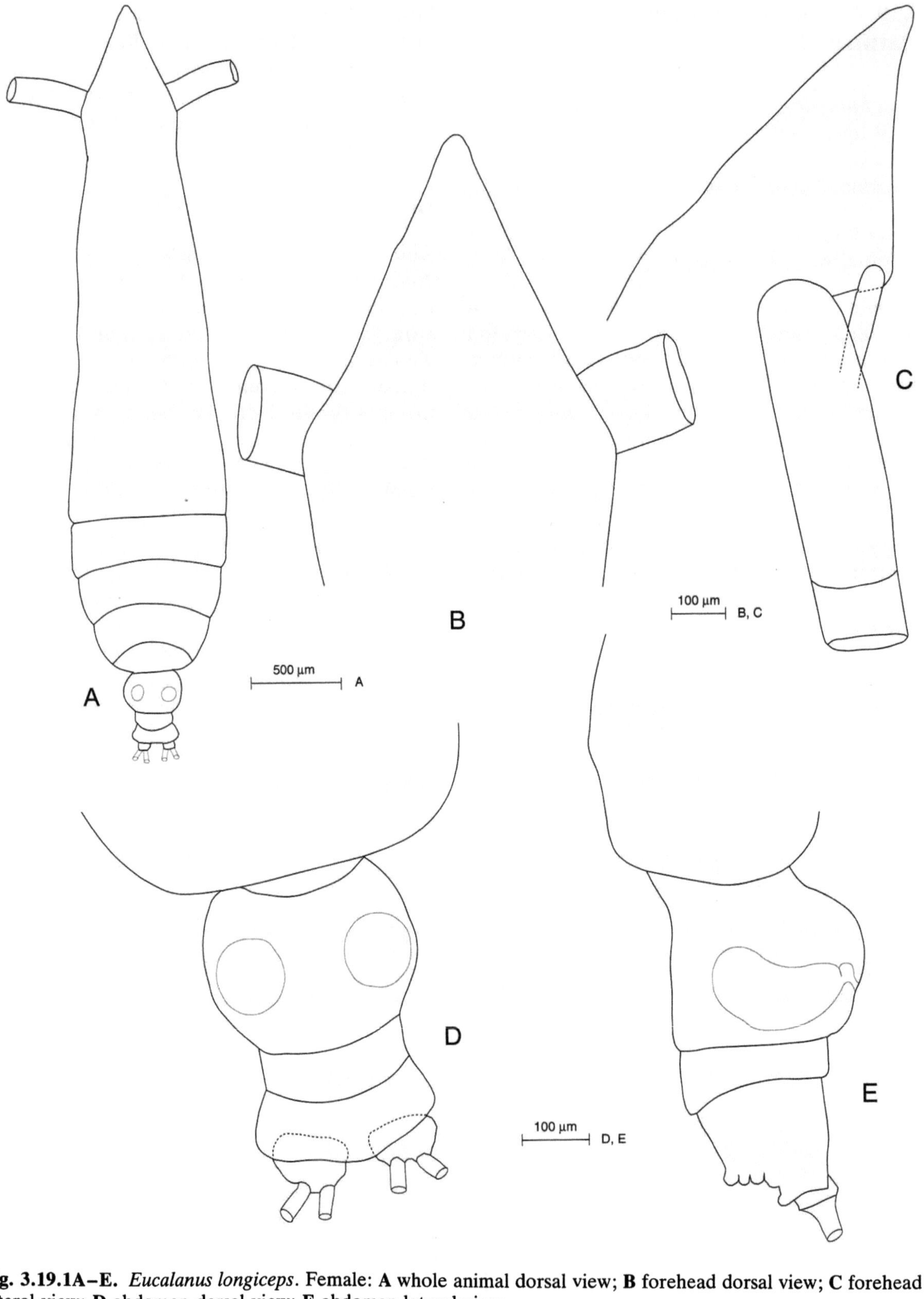

Fig. 3.19.1A–E. *Eucalanus longiceps*. Female: **A** whole animal dorsal view; **B** forehead dorsal view; **C** forehead lateral view; **D** abdomen dorsal view; **E** abdomen lateral view

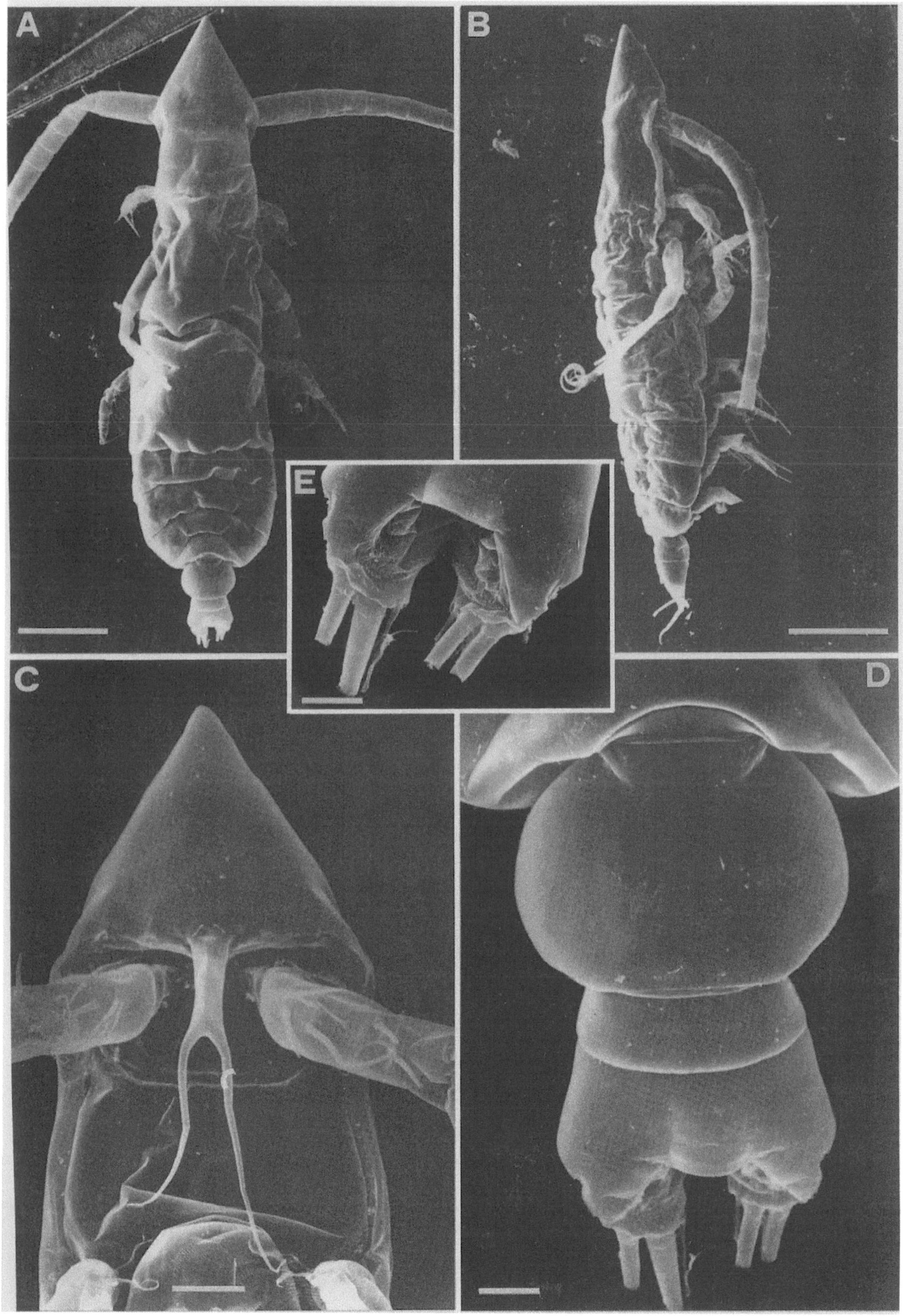

Fig. 3.19.2A–E. *Eucalanus longiceps*. Female: **A** whole animal dorsal view; **B** whole animal lateral view; **C** forehead and rostral filaments ventral view; **D** abdomen dorsal view; **E** furca. *Bars*: **A**, **B** 500 μm; **C** 100 μm; **D**, **E** 50 μm

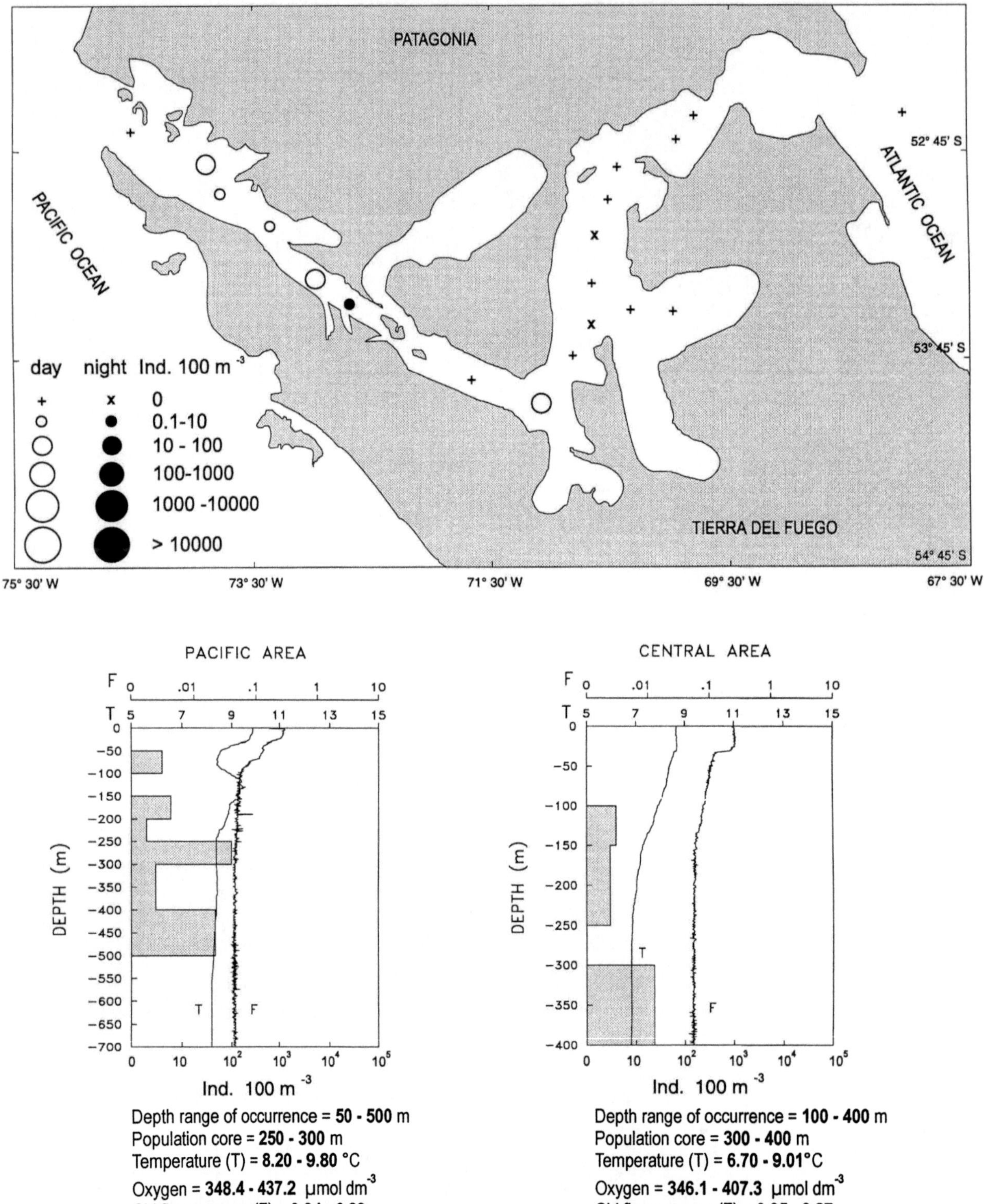

Fig. 3.19.3. *Eucalanus longiceps*. Distribution in the Straits of Magellan in February–March 1991

3.20 *Rhincalanus nasutus* Giesbrecht 1888

Rhincalanus nasutus Giesbrecht, 1888: p. 334; 1892: p. 152, pl. 3 fig. 6, pl. 9 figs. 6, 14, pl. 12 figs. 9–12, 14, 16, pl. 35 figs. 46, 49–Sars, 1903: p. 15, pls. 6, 7 – Esterly, 1905: p. 136, fig. 10 – Wilson, 1932: p. 34, fig. 18 – Sewell, 1947: p. 59, fig. 19 – Brodskii, 1950: p. 105, fig. 32 – Farran and Vervoort, 1951b: p. 3, fig. 4 – Marques, 1956: p. 9, pl. 1 figs. a, b – Fish, 1962: p. 9, fig. 4 – Grice and Hart, 1962: pl. 3.

Body length –
females: 5.63–6.08 mm (5 specimens)

Female. Body slender and elongated; beak-like projection on forehead bearing long, slender bifid rostrum. Rostral filaments not visible in dorsal view. Cephalothorax 3.5 times longer than wide. Proportional lengths of cephalothorax and abdomen 84:16. Head separated from thorax; thorax 5-segmented. Second-fourth thoracic somites bear two small lateral spines. Third and fourth thoracic somites with two dorsal spines on posterior margins. The 3-segmented abdomen is due to fusion of the anal somite with the furca. Proportional lengths of abdominal somites and furca 38:11:51: = 100. Two large spines visible on dorsal side of genital somite. Caudal rami slightly asymmetrical bearing 2 tufts of hairs on ventral side of anal region. P5 uniramous, symmetrical, with one seta on the second joint and three on third.

Remarks

In our specimens, the head is clearly separated from the first thoracic somite. By contrast, other authors report that the head is fused with the first thoracic somite. We found only 1 male specimen in our samples. Since it was heavily damaged, it was not considered for this study.

Distribution

Giesbrecht (1888) described this species from specimens also collected from Punta Arenas, Chile. *R. nasutus* is widely distributed throughout cold and temperate waters of the Northern and Southern hemispheres. In the south, it overlaps with the Antarctic species *R. gigas* at its extreme southern limit. The species has been reported both for the Chilean coast (Arcos 1975; Marin and Antezana 1985; Peterson et al. 1988) and Argentinian offshore waters (Ramirez 1969, 1970, 1971). It was previously reported for the Straits of Magellan by Mazzocchi and Ianora (1991), who recorded mainly juvenile specimens only at Pacific stations. The species has a wide distribution range from the surface to 1000–1500 m in the Indian Ocean, but to lesser depths in the Atlantic and Pacific Oceans (Sewell 1947). In the Mediterranean Sea, it has been recorded from 0–800 m (Scotto di Carlo et al. 1984). In the present study, *R. nasutus* was present throughout the water column in the Pacific and Central areas of the Straits of Magellan. It was also sampled in low numbers in the Atlantic area up to the entrance of the Primera Angostura.

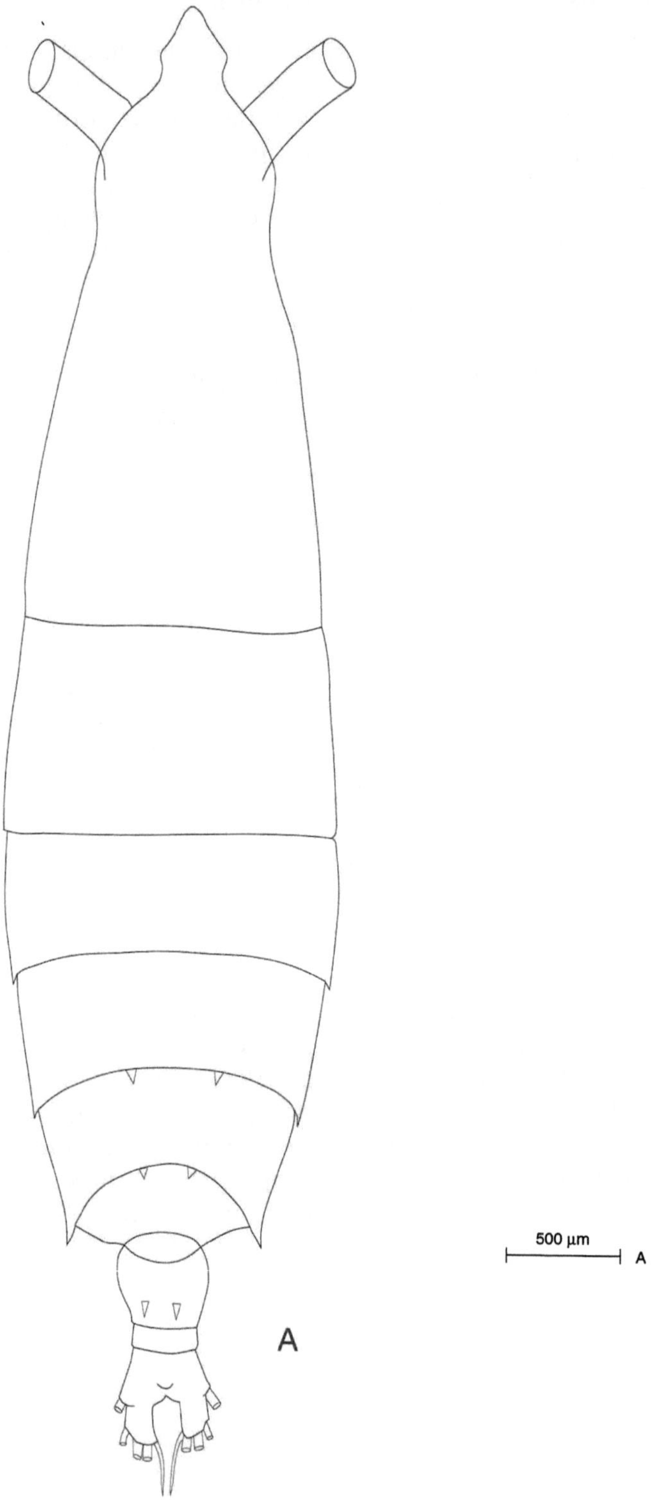

Fig. 3.20.1. *Rhincalanus nasutus*. Female: whole animal dorsal view

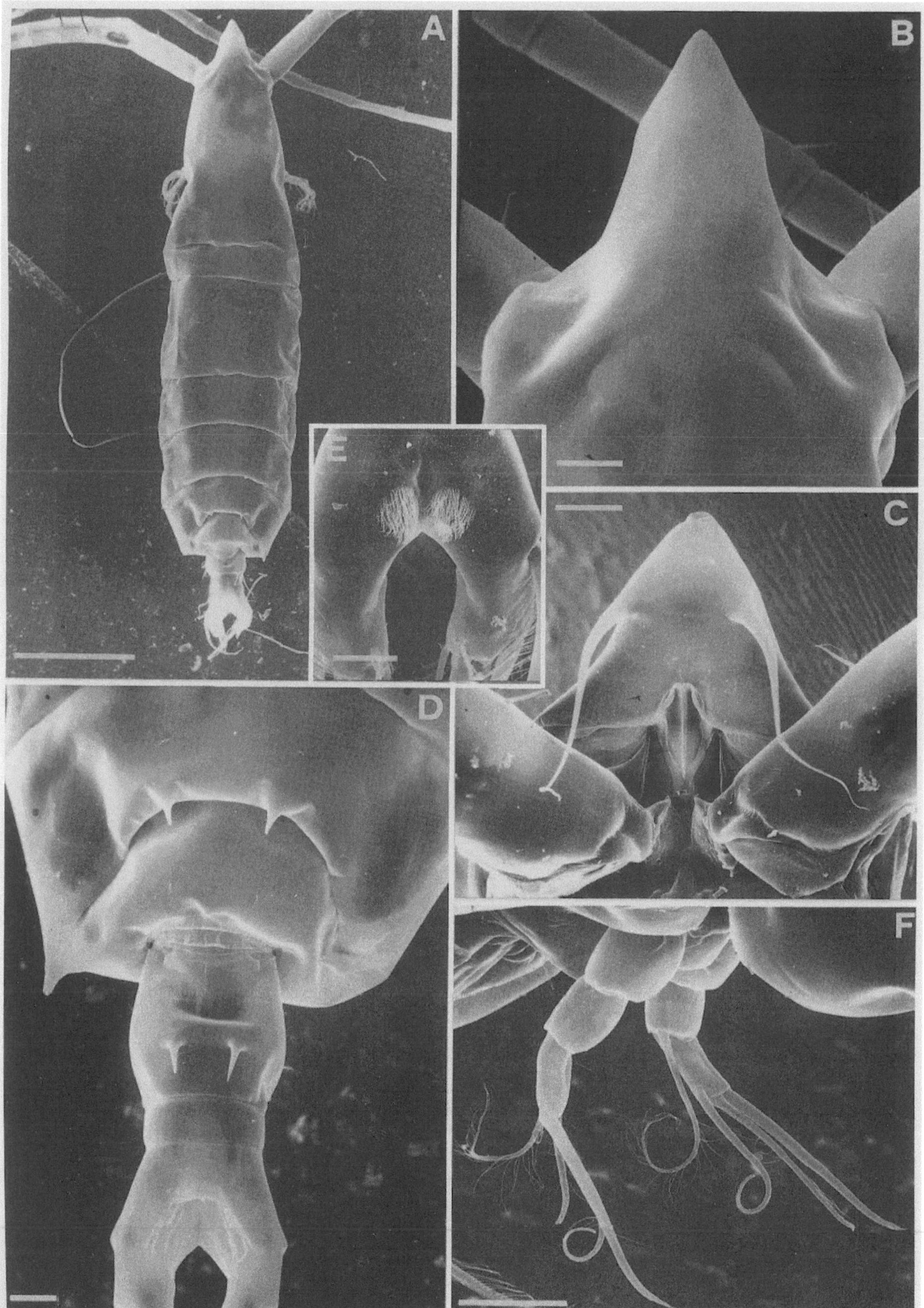

Fig. 3.20.2A–F. *Rhincalanus nasutus*. Female: **A** whole animal dorsal view; **B** forehead in dorsal view extending into beak-like projection; **C** rostral filaments ventral view; **D** 4th thoracic somite and abdomen bearing two pronounced spines in dorsal view; **E** anal region bearing tufts of hairs on ventral surface; **F** P5. *Bars*: **A** 1 mm; **B**, **C**, **D**, **E**, **F** 100 µm

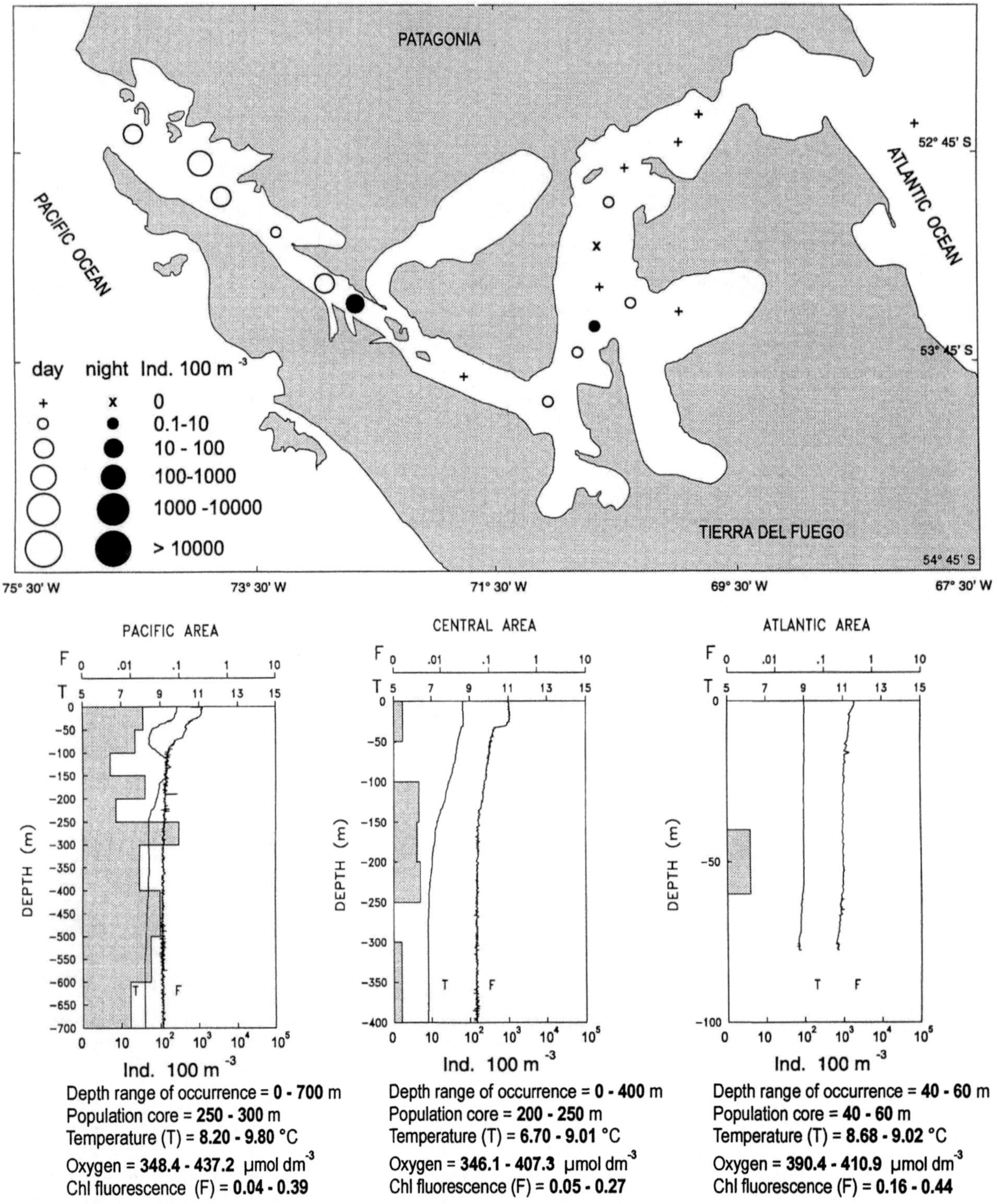

Fig. 3.20.3. *Rhincalanus nasutus*. Distribution in the Straits of Magellan in February–March 1991

3.21 *Aetideus arcuatus* Vervoort 1949

Snelliaetideus arcuatus Vervoort, 1949: p. 4, fig. 1–Park, 1978: p. 108, figs. 4, 5.
Aetideus arcuatus Roe, 1975: p. 304, fig. 5a–m.

Body length –
females: 1.76–1.96 mm (18 specimens)
male: 1.55 (1 specimen)

Female. Body slender. Cephalothorax 2.7 times longer than wide. Head terminates ventrally in pronounced rostrum with two strong pointed rami. Margins of last thoracic somite rounded. Proportional lengths of cephalothorax and abdomen 77:23; proportional lengths of abdominal somites and furca 36:15:12:14:23 = 100. Genital somite swollen in dorsal view with large comma-shaped seminal receptacle in side view. Genital aperture ovoid in ventral view. Caudal rami more than 2 times longer than wide. P4 bears a few small spines and hairs at base of seta on inner margin of B1. P5 absent.

Male. Body shape similar to female with same proportions between cephalothorax and abdomen. Forehead without rostrum. Posterior corners of cephalothorax rounded. Abdomen 4-segmented. Proportional lengths of abdominal somites and furca 23:18:19:21:19 = 100. First abdominal somite asymmetrical with small protuberance on left side, in dorsal view. P4 B1 lacks internal spines. P5 present only on left side, uniramous, 5-jointed; last joint very short, ending in a cluster of 5 spines and 1 lateral spine surrounded by a row of small teeth; inner margin adorned by setae.

Remarks

Vervoort (1949) described *Snelliaetideus arcuatus* as a new genus erected to accommodate this species. Bradford (1971a) noted that the only difference between *Snelliaetideus* Vervoort, 1949 and *Aetideus* Brady, 1883 females was the absence of postero-lateral thoracic points. Roe (1975), after discovering the male, concluded that *Snelliaetideus* was a junior synonym for *Aetideus*, a conclusion which is now generally accepted (Razouls 1993). Our female specimens, though larger in size, are similar in morphology to those originally described by Vervoort. The males concord with the description given by Roe (1975) except for the last joint of the P5, which bears a lateral spine and setae not reported by Roe.

Distribution

The species was recorded for the first time in the Indo-Pacific Ocean during the Snellius expedition in the Malay Archipelago (Vervoort 1949). Since then, it has also been recorded in the Caribbean Sea (Park 1970), the NE Atlantic (Roe 1975), and the Antarctic Convergence in the Pacific Ocean (Park 1978). Its depth of occurrence has been reported in one haul from 610–790 m (Roe 1975), whereas Park (1978) collected the species at two stations from 0–1230 m. To the best of our knowledge, this species has never been recorded in Chilean and Argentine waters. In the present study, *A. arcuatus* was present from 250–500 m at only three stations in the Pacific area of the Straits of Magellan.

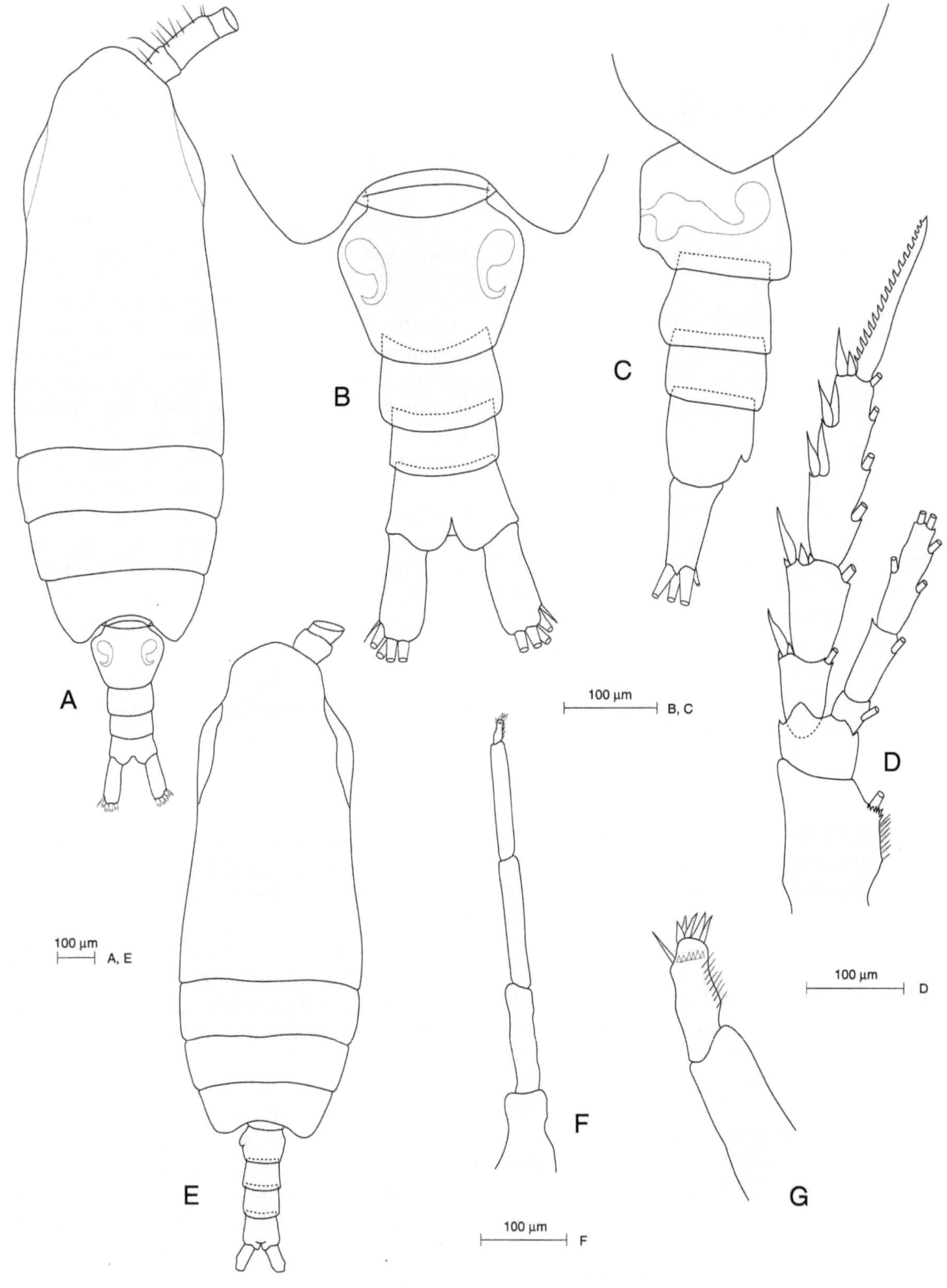

Fig. 3.21.1A–G. *Aetideus arcuatus*. Female: **A** whole animal dorsal view; **B** abdomen dorsal view; **C** abdomen lateral view; **D** P4. Male: **E** whole animal dorsal view; **F** P5; **G** last joint of P5

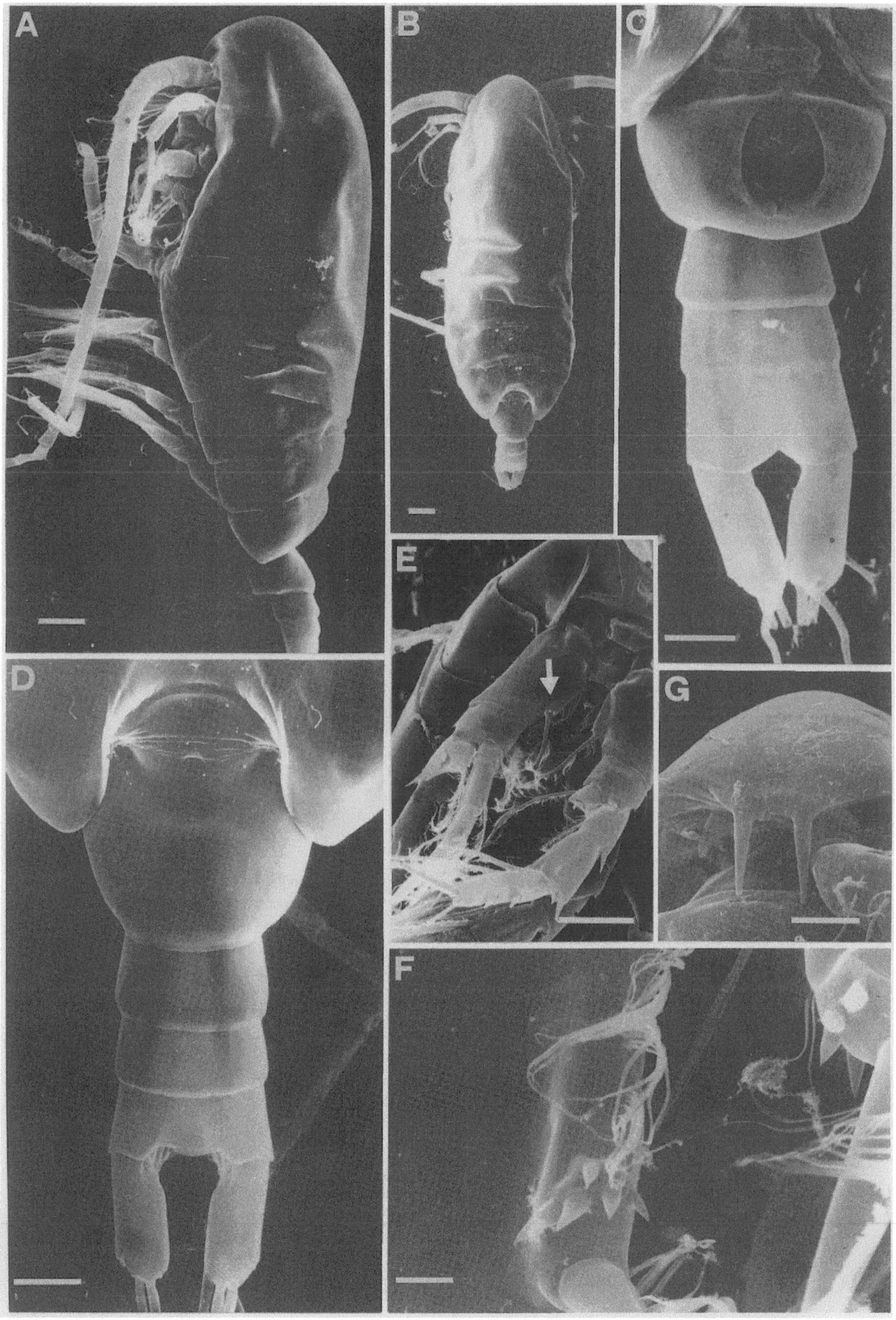

Fig. 3.21.2A–G. *Aetideus arcuatus*. Female: **A** whole animal lateral view; **B** whole animal dorsal view; **C** abdomen ventral view; **D** abdomen dorsal view; **E** P4 posterior view; **F** detail of spines and hairs on P4 B1 (*arrow* in **F**); **G** head ventral view with rostrum. *Bars*: **A**, **B**, **E** 100 µm; **C**, **D**, **G** 50 µm; **F** 10 µm

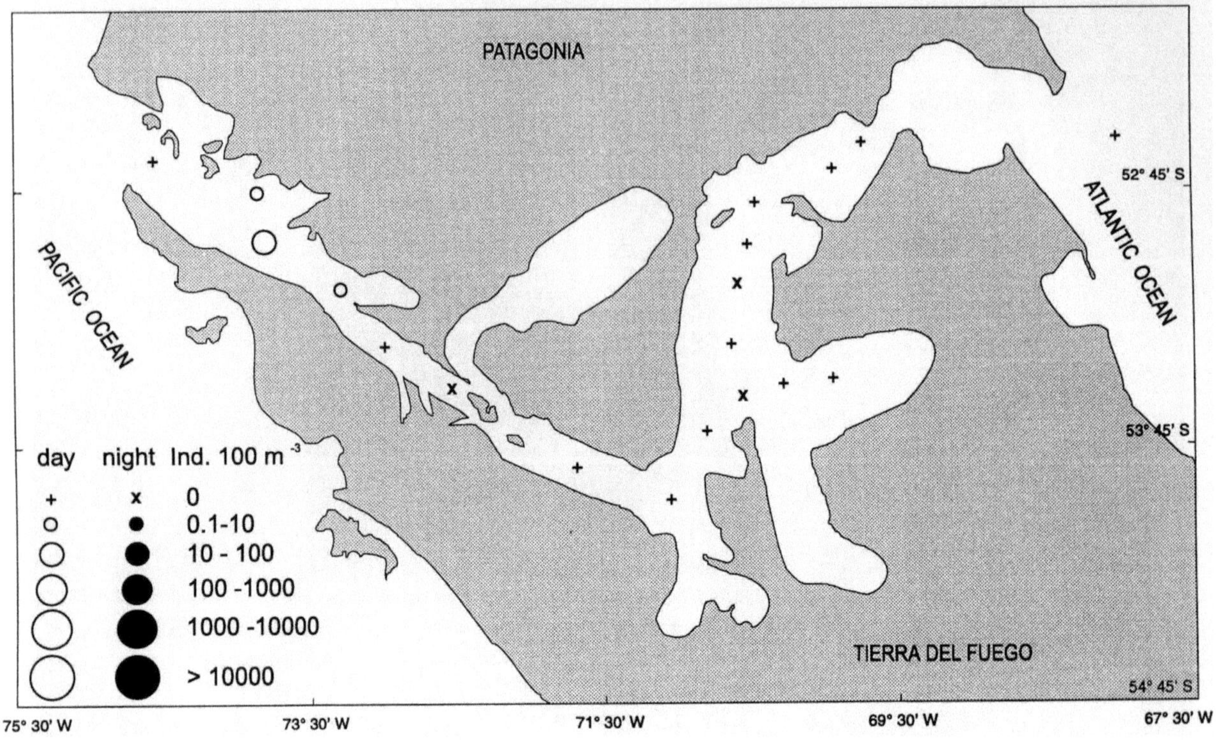

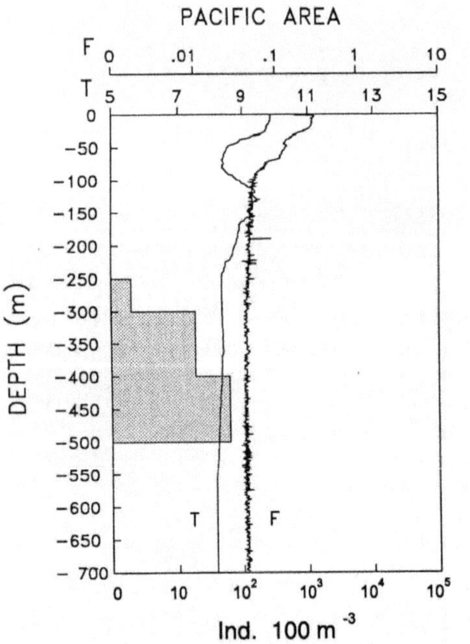

Depth range of occurrence = **250 - 500** m
Population core = **400 - 500** m
Temperature (T) = **8.20 - 9.80** °C
Oxygen = **348.4 - 437.2** μmol dm^{-3}
Chl fluorescence (F) = **0.04 - 0.39**

Fig. 3.21.3. *Aetideus arcuatus*. Distribution in the Straits of Magellan in February–March 1991

3.22 *Aetideus australis* Vervoort 1957

Euaetideus australis Vervoort, 1957: p. 46, figs. 16–19, 20a – Park, 1978: p. 105, figs. 2, 3.
Aetideus australis Bradford, 1971a: p. 19, figs. 5, 6a, 7 – Bradford and Jillett, 1980: p. 14, fig. 6.

Body length –
females: 1.76–1.96 mm (12 specimens)
males: 1.55–1.57 mm (3 specimens)

Female. Cephalothorax 2.5 times longer than wide. Proportional lengths of cephalothorax and abdomen 77:23. Head resembling pointed arch in dorsal view, tapering ventrally into large bifurcated rostrum with strong base and large U-shaped excavation between points. Posterior corners of last thoracic somite produced distally into large spiniform processes that become more subtle at extremities and reaching half length of second abdominal somite. Proportional lengths of abdominal somites and furca 32:14:10:16:28: = 100. Genital somite slightly wider than long in dorsal view, with widest part at mid-length. Very conspicuous seminal receptacles both in dorsal and lateral views, with distal end expanded into large bursa. Caudal rami about 3 times longer than wide, with long fragile hairs on inner margins. P4 B1 with row of strong, wide spinules at base of internal seta.

Male. Cephalothorax more slender than female, 2.7 times longer than wide. Proportional lengths of cephalothorax and abdomen 74:26. Rostrum absent. Posterior corners of last thoracic somite produced distally into spiniform processes extending beyond distal margin of first abdominal somite. Proportional lengths of abdominal somites and furca 18:22:18:22:18:2 = 100. Anal somite extremely reduced, hidden between fourth abdominal somite and furca. Caudal rami 3.2 times longer than wide. P5 single, uniramous, 5-jointed; third joint the longest. Fourth joint with few short bristles on distal surface; last joint very short, with longitudinal row of dense bristles that also ornate tip.

Remarks

A. australis was described by Vervoort (1957) as *Euaetideus a.* This species is very similar to *A. armatus* but, also according to Bradford and Jillett (1980) and Park (1978: *Euaetideus australis*), *A. australis* is characterized and distinguished by its longer posterior thoracic points and, in the female, by the genital somite, which is widest at the middle in dorsal view (in *A. armatus* it is widest anterior to the middle). This latter difference is very evident when Mediterranean specimens of *A. australis* and *A. armatus* are compared. The male can be readily distinguished by the relatively long caudal rami, which are about 3 times longer than wide; according to Tanaka (1957), the caudal rami are about 2 times longer than wide in *A. armatus*. In our specimens we did not observe (up to 7×10^4 magnification at SEM) the chitinous thickenings on the anterior integument reported by Bradford (1971a).

Distribution

A. australis was originally reported from Antarctic and sub-Antarctic waters, in the Indian Ocean and Australian sector of the Pacific (Vervoort 1957). It was later found in Antarctic and sub-Antarctic areas of the Pacific Ocean by Park (1978). According to Bradford (1971a), *A. australis* was sampled in the SW Pacific (0–1000 m depth) and in the SE Atlantic (0–300 m depth). It has been recorded once north of the Subtropical Convergence off Southern Africa (Bradford 1971a) and once south of the Antarctic Convergence (Vervoort 1957). In Chilean waters, only *A. armatus* (very similar to *A. australis*) is reported (Ramirez 1969; Arcos 1975; Ramirez and Dinofrio 1976). In the present study, *A. australis* was present throughout the water column at all stations in the Pacific area of the Straits of Magellan. A few individuals were also sampled from 200–250 m off Cape Froward.

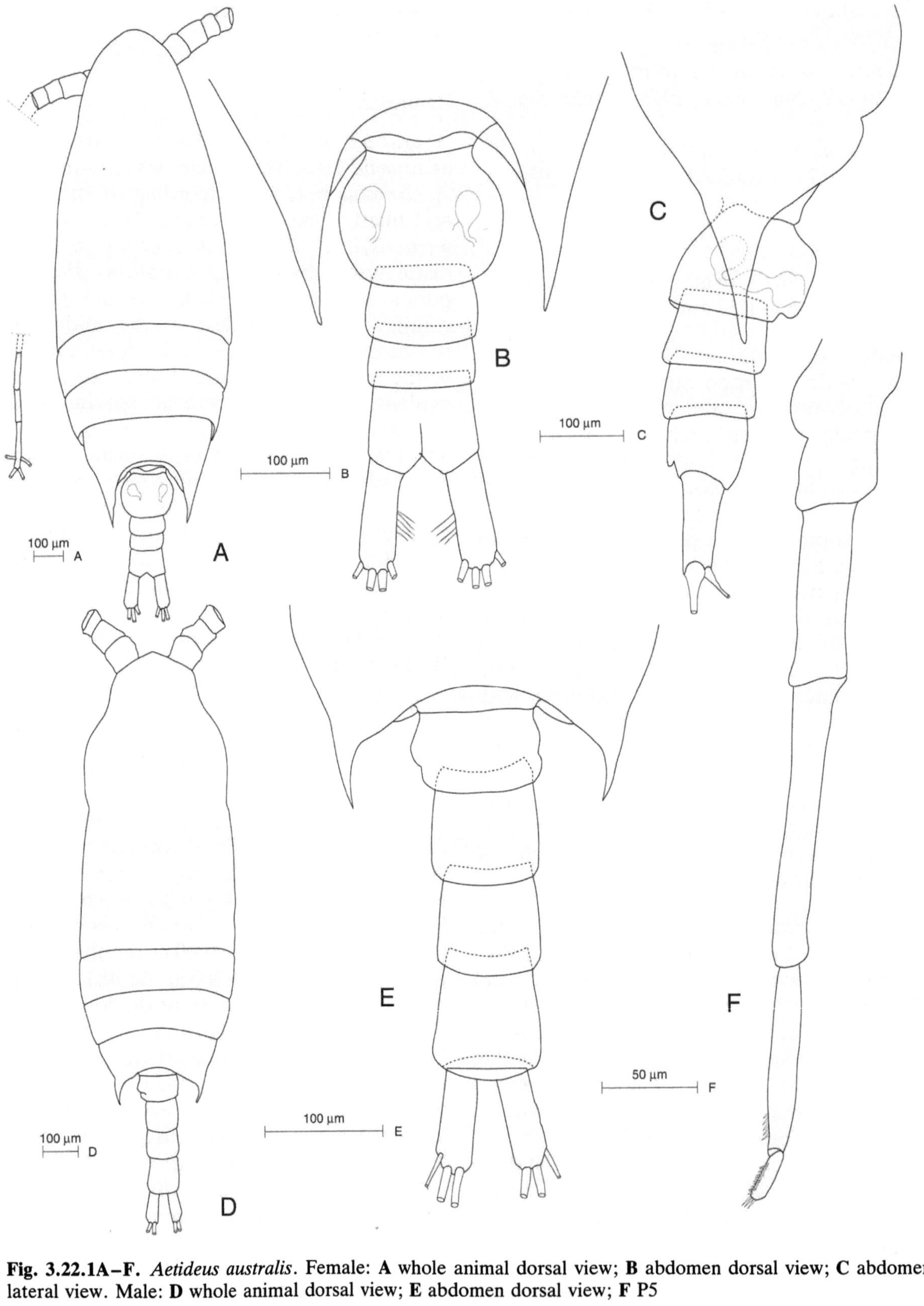

Fig. 3.22.1A–F. *Aetideus australis*. Female: **A** whole animal dorsal view; **B** abdomen dorsal view; **C** abdomen lateral view. Male: **D** whole animal dorsal view; **E** abdomen dorsal view; **F** P5

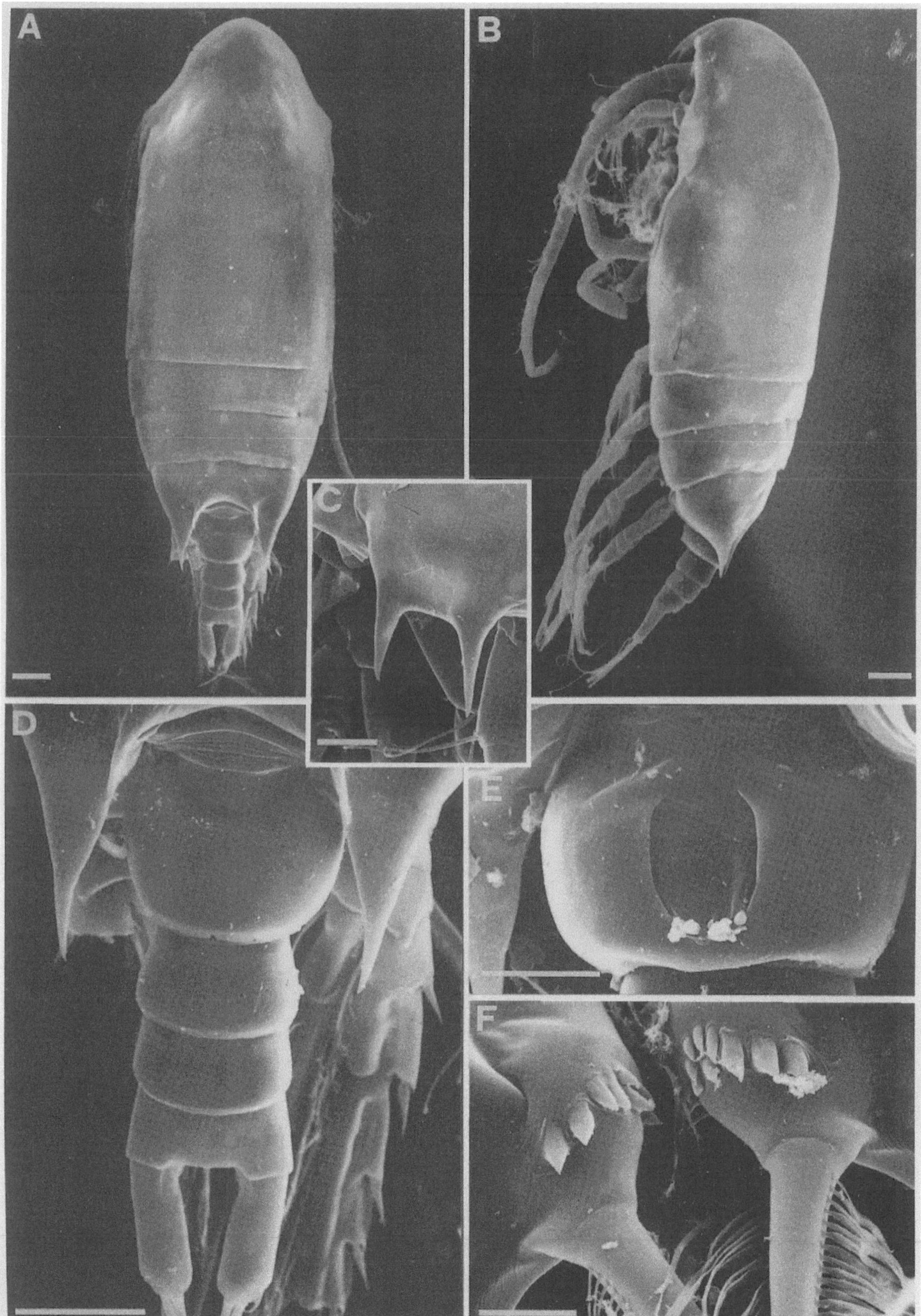

Fig. 3.22.2A–F. *Aetideus australis*. Female: **A** whole animal dorsal view; **B** whole animal lateral view; **C** rostrum; **D** abdomen dorsal view; **E** genital somite ventral view; **F** row of strong spinules at base of internal setae of P4 B1. *Bars*: **A**, **B**, **D** 100 µm; **C**, **E** 50 µm; **F** 10 µm

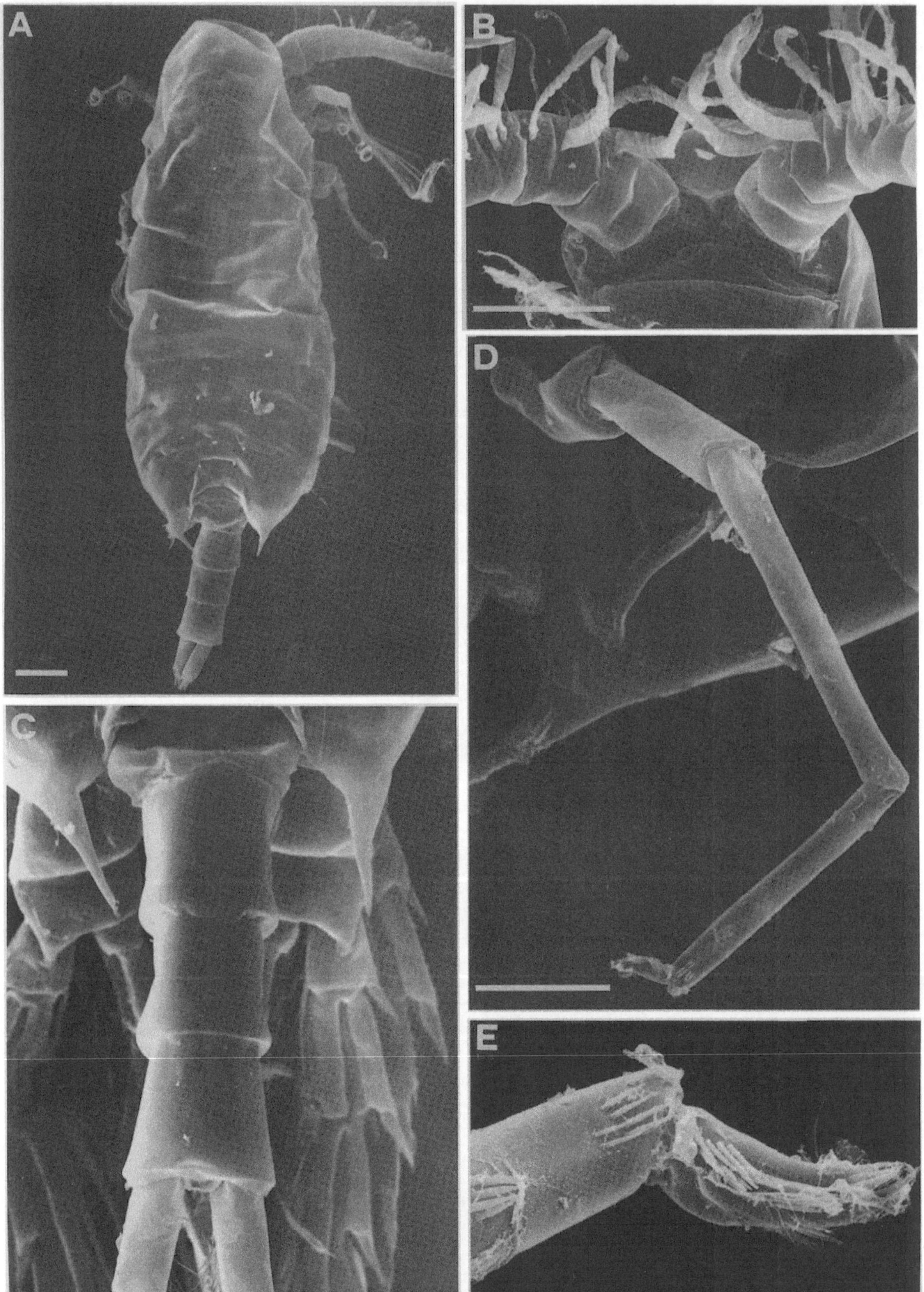

Fig. 3.22.3A–E. *Aetideus australis*. Male: **A** whole animal dorsal view; **B** forehead ventral view; **B** abdomen dorsal view; **D** P5; **E** last joint of P5. *Bars*: **A, B** 100 µm; **C, D** 50 µm; **E** 10 µm

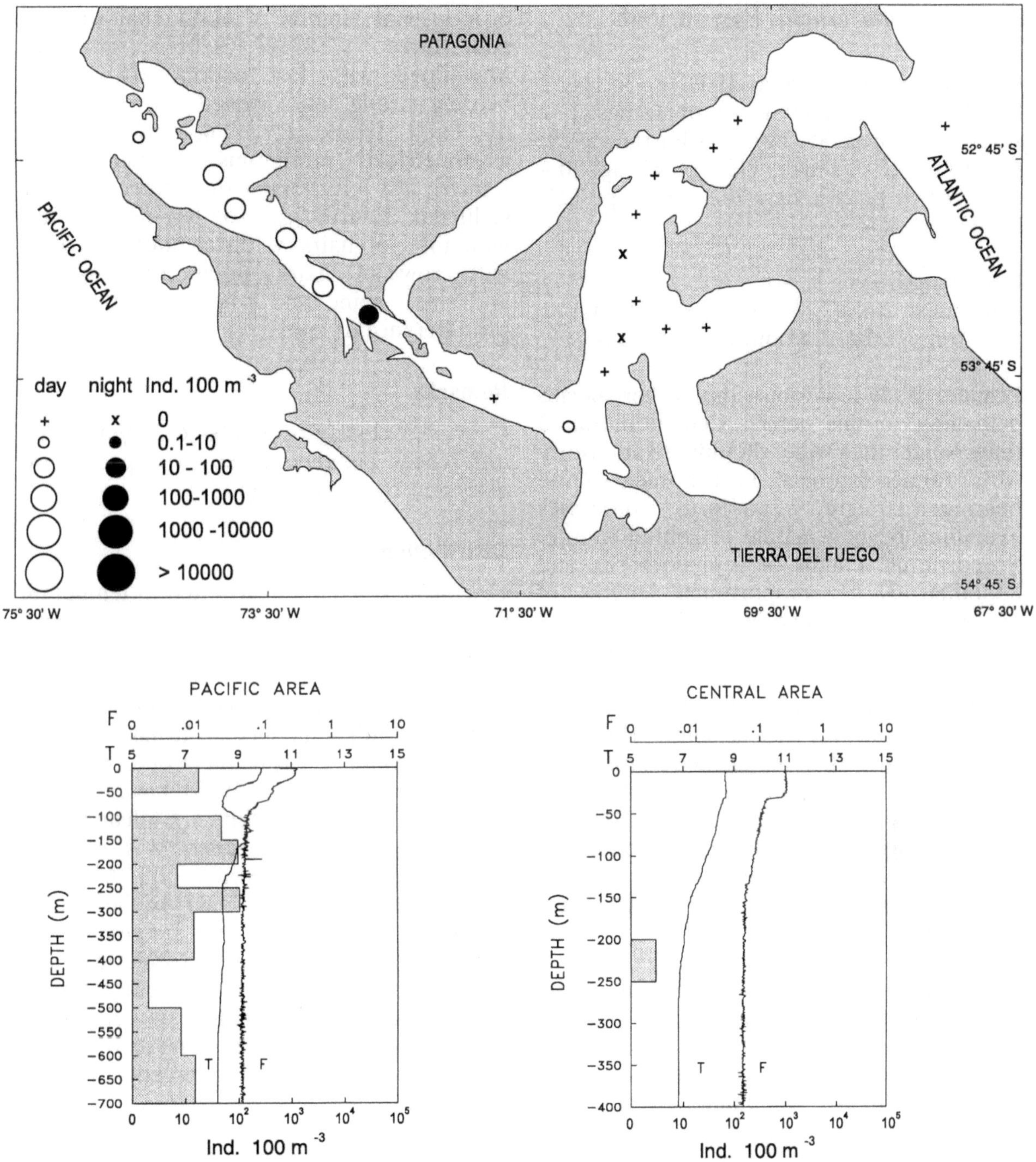

PACIFIC AREA	CENTRAL AREA
Depth range of occurrence = **0 - 700** m	Depth range of occurrence = **200 - 250** m
Population core = **250 - 300** m	Population core = **200 - 250** m
Temperature (T) = **8.20 - 9.80** °C	Temperature (T) = **6.70 - 9.01** °C
Oxygen = **348.4 - 437.2** μmol dm^{-3}	Oxygen = **346.1 - 407.3** μmol dm^{-3}
Chl fluorescence (F) = **0.04 - 0.39**	Chl fluorescence (F) = **0.05 - 0.27**

Fig. 3.22.4. *Aetideus australis*. Distribution in the Straits of Magellan in February–March 1991

3.23 *Chiridius gracilis* Farran 1908

Chiridius gracilis Farran, 1908: p. 30, pl. 2 figs. 1–3 – Scott, 1909: p. 42, pl. 11 figs. 1–9 – With, 1915: p. 85, fig. 21 – Farran, 1929: p. 229, fig. 6 – Tanaka, 1957: p. 48, fig. 30 – Park, 1975: p. 283, figs. 7, 8; 1978: p. 122, fig. 13.

Body length –
females: 2.65–2.88 mm (13 specimens)
males: 2.29–2.43 mm (5 specimens)

Female. Body less robust than other species belonging to this genus. Cephalothorax 3 times longer than wide, shouldered anteriorly with rounded forehead. Last thoracic somite bearing straight, spiniform processes extending beyond middle of genital somite. Proportional lengths of cephalothorax and abdomen 77:23; proportional lengths of abdominal somites and furca 37:22:19:8:14 = 100. Genital somite as wide as long and swollen at mid-length. P2 B1 produced into bulge on inner distal end, bearing row of long hairs and one strong spine. Re2 with row of small teeth along distal outer margin. Re3 with 3 strong outer spines with wing-like processes on either basal side. Between these spines, margins of joint moderately bulged in the middle as compared to *C. subgracilis*. Terminal spine of exopod bearing about 24 teeth on external edge.

Male. Same body shape as female, with shorter spiniform processes on postero-lateral corners of last thoracic somite. Proportional lengths of cephalothorax and abdomen 76:24. Abdomen 4-segmented; proportional lengths of abdominal somites and furca 17:27:22:19:2:13 = 100. Maxilliped with last endopodal segment bearing strong seta ornated with two rows of small teeth. P5 long and slender, asymmetrical, uniramous. Right leg 4-jointed; last joint longer than preceding and styliform. Fourth joint of left leg ornated with tufts of hairs at distal end; last joint with row of hairs along inner border and small spine and small spatula-shaped protuberance at apex.

Remarks

C. gracilis closely resembles *C. poppei*. The differences between these two species are discussed by Vervoort (1957).

Distribution

C. gracilis is widely distributed in temperate waters of the Southern hemisphere. Farran (1929) reported its occurrence off New Zealand and Park (1978) recorded it north of the Antarctic convergence in sub-Antarctic waters of the Pacific Ocean. It has also been reported south of Tasmania (Vervoort 1957). Marin and Antezana (1985) have recently recorded it for the first time in Chilean fjords. Only one female specimen was sampled previously in the Straits of Magellan (Mazzocchi and Ianora 1991). In the present study, *C. gracilis* and *C. subgracilis* are grouped together in the distribution maps. They were only found at Pacific stations where they were sampled at all depths except the surface layer. The two species were also sporadically sampled off Cape Froward.

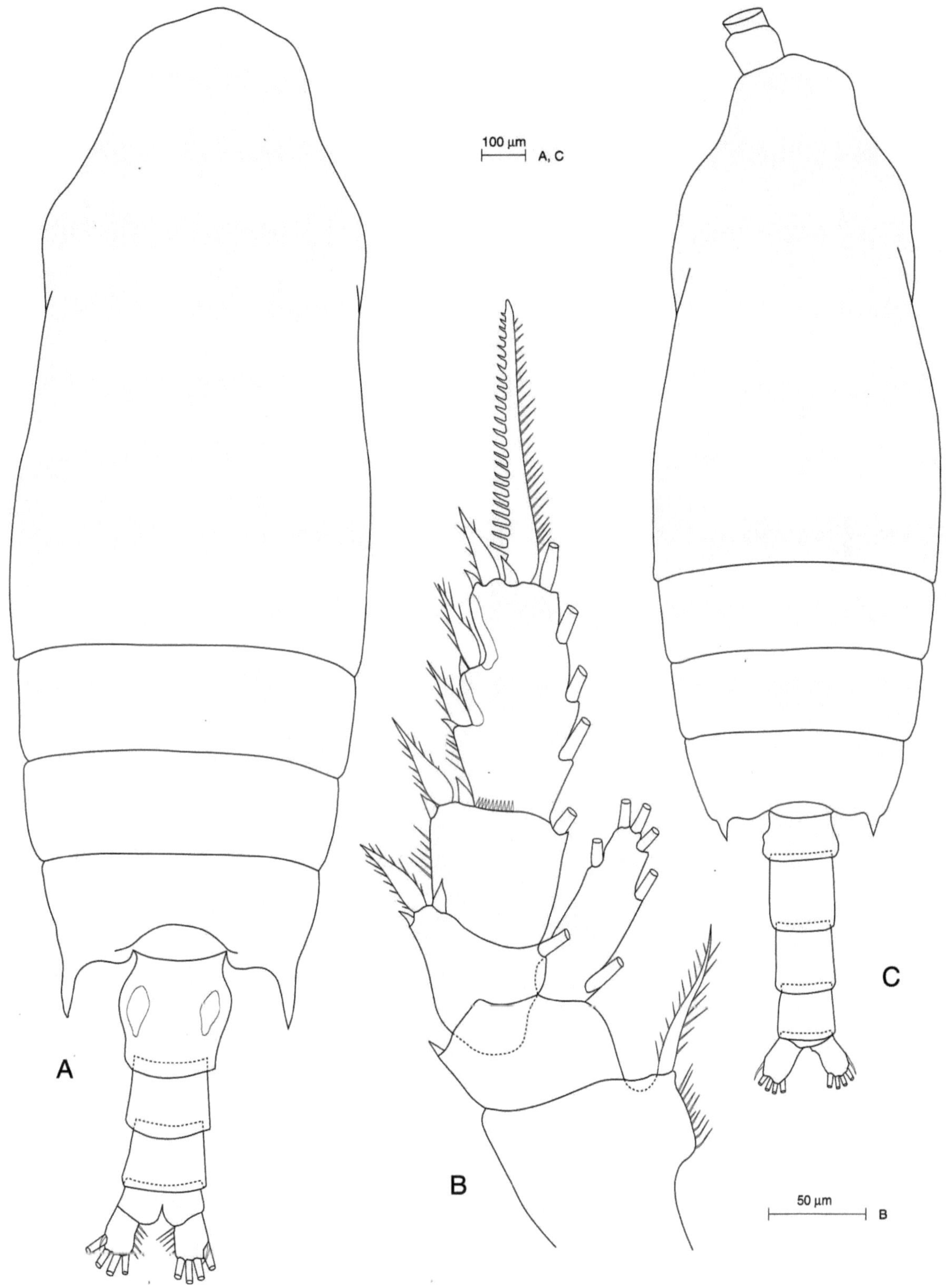

Fig. 3.23.1A–C. *Chiridius gracilis*. Female: **A** whole animal dorsal view; **B** P2. Male: **C** whole animal dorsal view

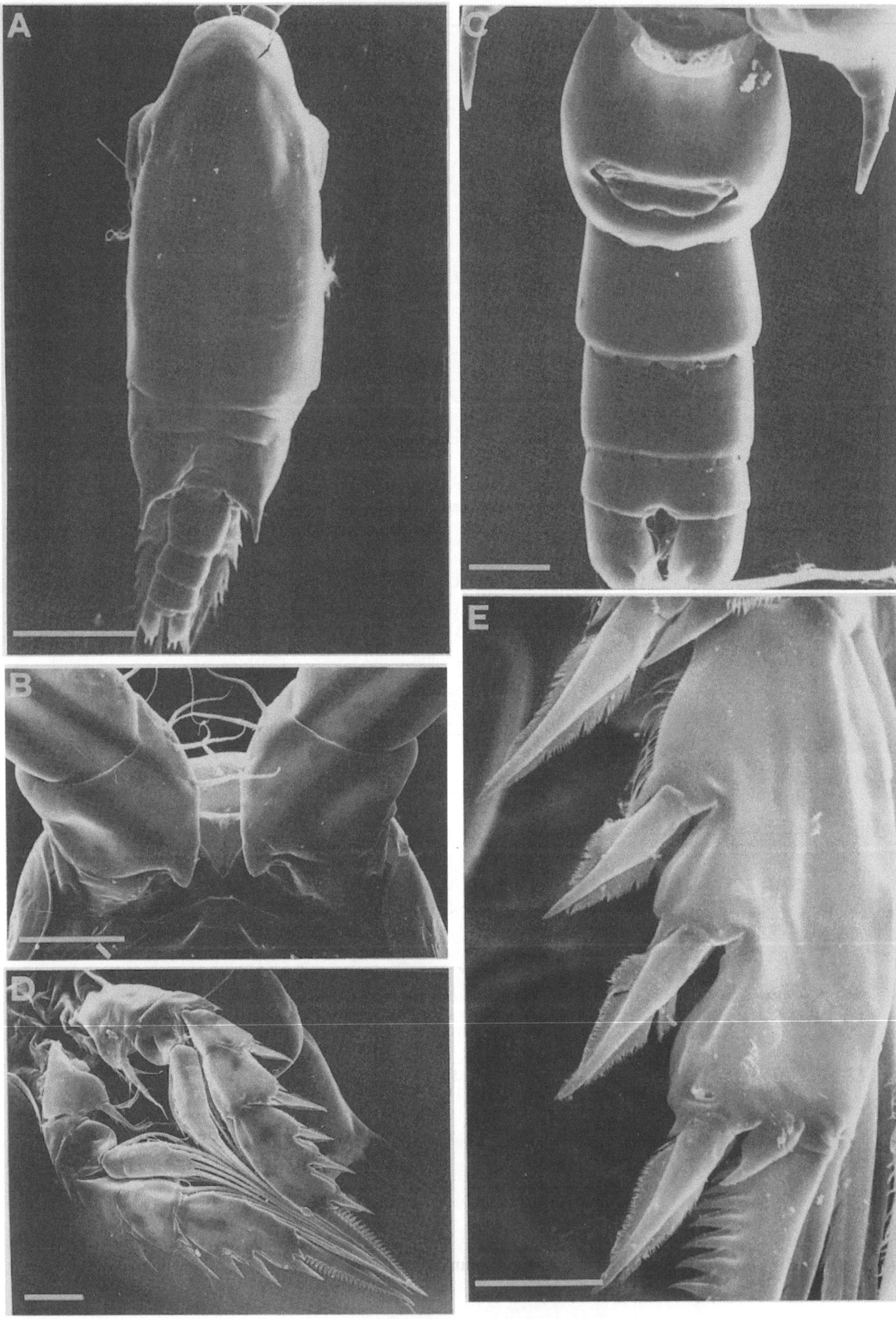

Fig. 3.23.2A–E. *Chiridius gracilis*. Female: **A** whole animal dorsal view; **B** forehead ventral view; **C** abdomen ventral view; **D** P2; **E** last joint of P2. *Bars*: **A** 500 μm; **B**, **C**, **D** 100 μm; **E** 50 μm

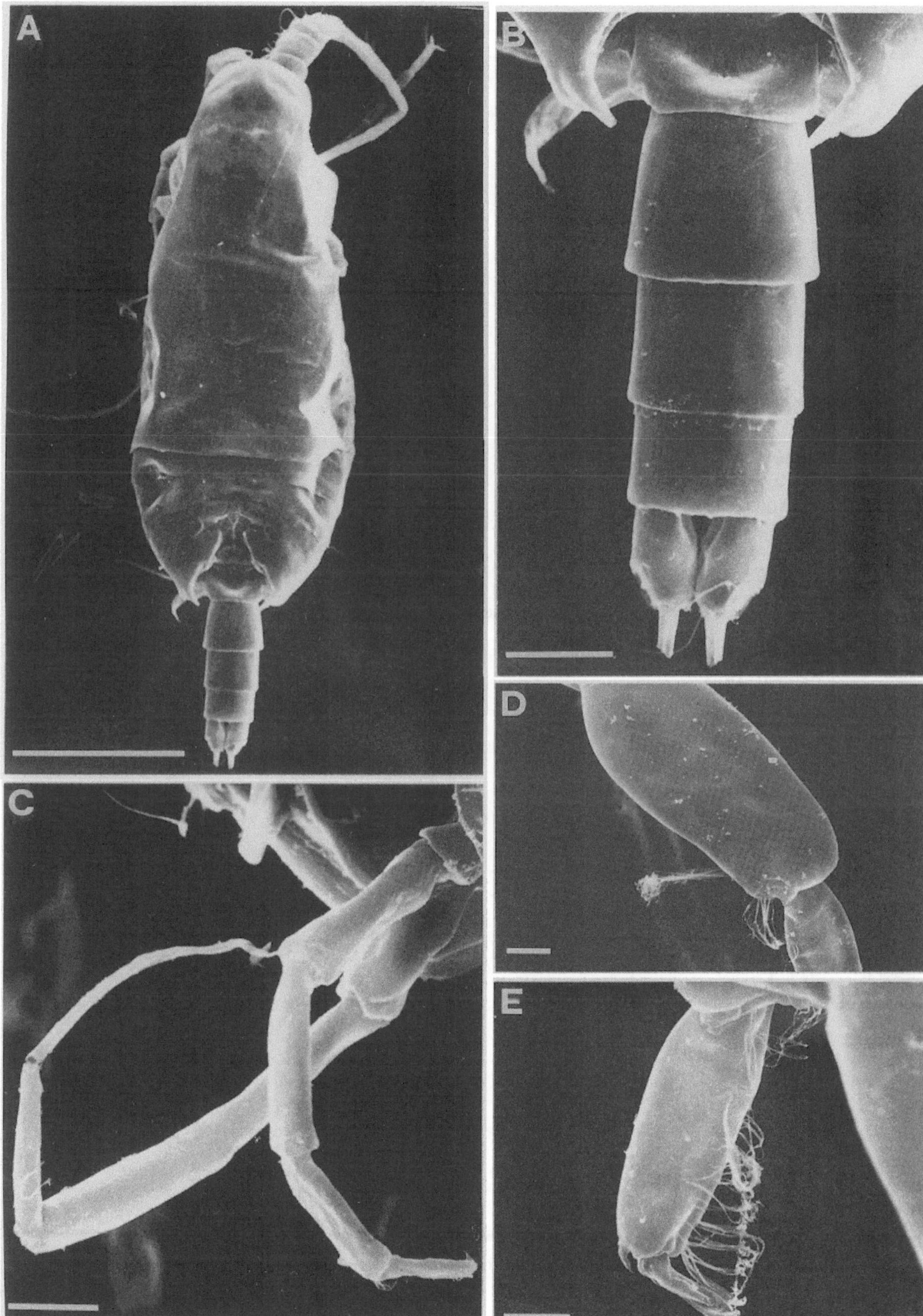

Fig. 3.23.3A–E. *Chiridius gracilis*. Male: **A** whole animal dorsal view; **B** abdomen dorsal view; **C** P5; **D** penultimate segment left P5; **E** last segment left P5. *Bars*: **A** 500 μm; **B** 100 μm; **C** 50 μm; **D**, **E** 10 μm

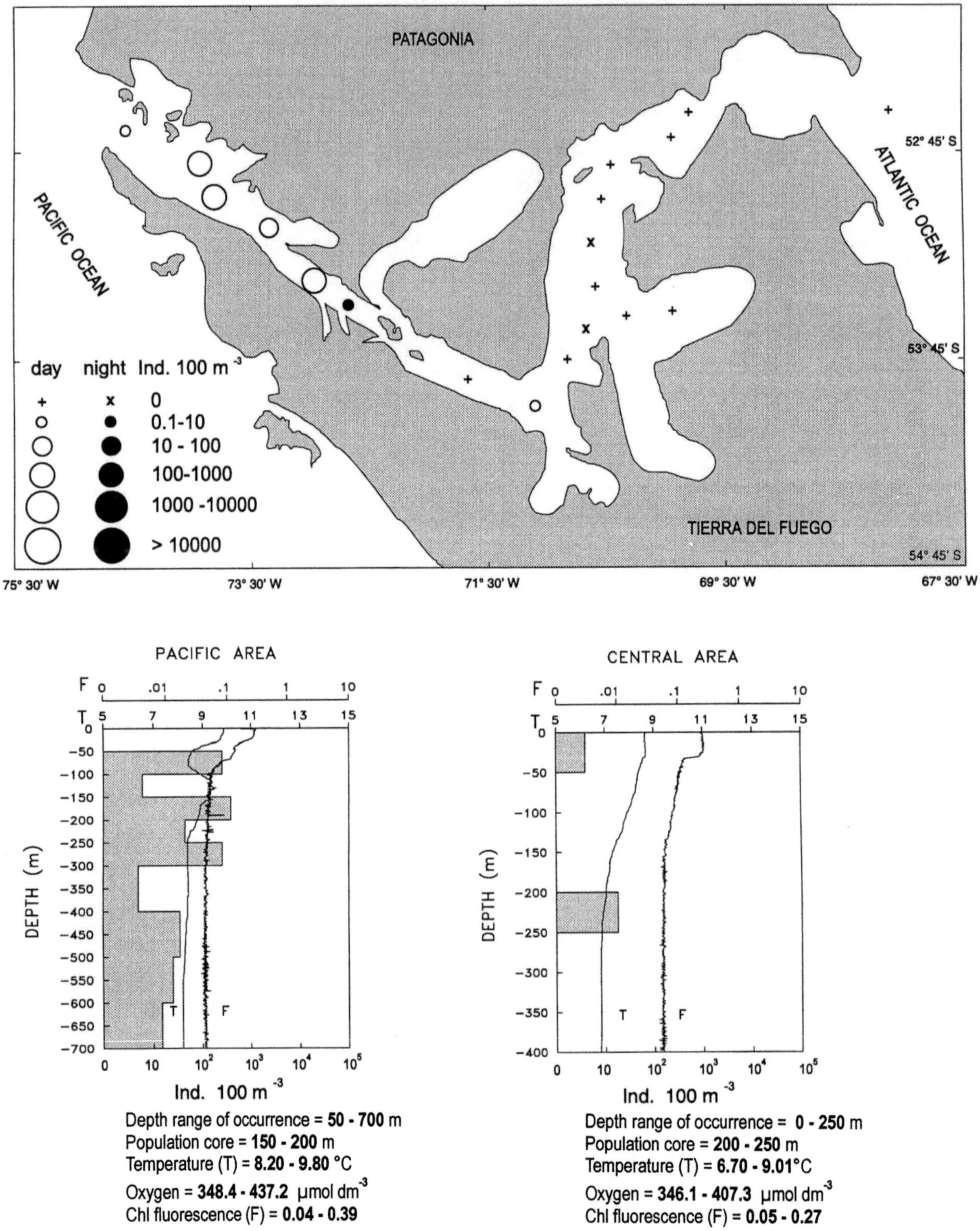

Fig. 3.23.4. *Chiridius gracilis* + *C. subgracilis*. Distribution in the Straits of Magellan in February–March 1991

3.24 *Chiridius subgracilis* Park 1975

Chirdius subgracilis Park, 1975: p. 285, figs. 9, 10 – Bradford and Jillett, 1980: p. 23, fig. 12.

Body length –
females: 2.35–2.72 mm (12 specimens)
males: 2.00–2.16 mm (5 specimens)

Female. Same body shape as *C. gracilis* with the same proportional lengths between the cephalothorax and abdomen (77:23). Proportional lengths of abdominal somites and furca 39:23:17:8:13 = 100. In P2, outer margin of Re3 with 3 strong spines bordered by wing-like processes as in *C. gracilis*, but in this case the margins between these spines are conspicuously bulged in the middle.

Male. General shape somewhat more slender than *C. gracilis*. Proportional lengths of cephalothorax and abdomen 70:30. In the maxilliped, none of the setae on the last endopodal segment bears hairs or spinules. Proportional lengths of abdominal somites and furca 18:26:22:20:2:12 = 100. P5 as in *C. gracilis*.

Remarks

According to Park (1975), this species is clearly distinct from *C. gracilis* in the female which is smaller in size and bears a posteriorly located genital swelling in dorsal view. At first glance, our specimens do not show any evident differences in morphology or size from *C. gracilis*. They differ only for the presence of conspicuous lobes on the external margin of the P2 Re3; this feature is easier to appreciate after dissection. The males of these two species are easier to distinguish from one another. The male of *C. subgracilis* has a longer abdomen compared to the cephalothorax. After dissection, it is also possible to see the differences in the last endopodal segment of the maxilliped (absence of spinules on the distal setae in *C. subgracilis*).

Distribution

C. subgracilis was reported by Park (1975) as the most common species belonging to this genus in his plankton collections from the Gulf of Mexico. Bradford and Jillett (1980) report it as a deep-water species, living generally below 250 m, in the New Zealand region of the SW Pacific. This represents a first record for the Straits of Magellan. See *C. gracilis* for the distribution range in the present study.

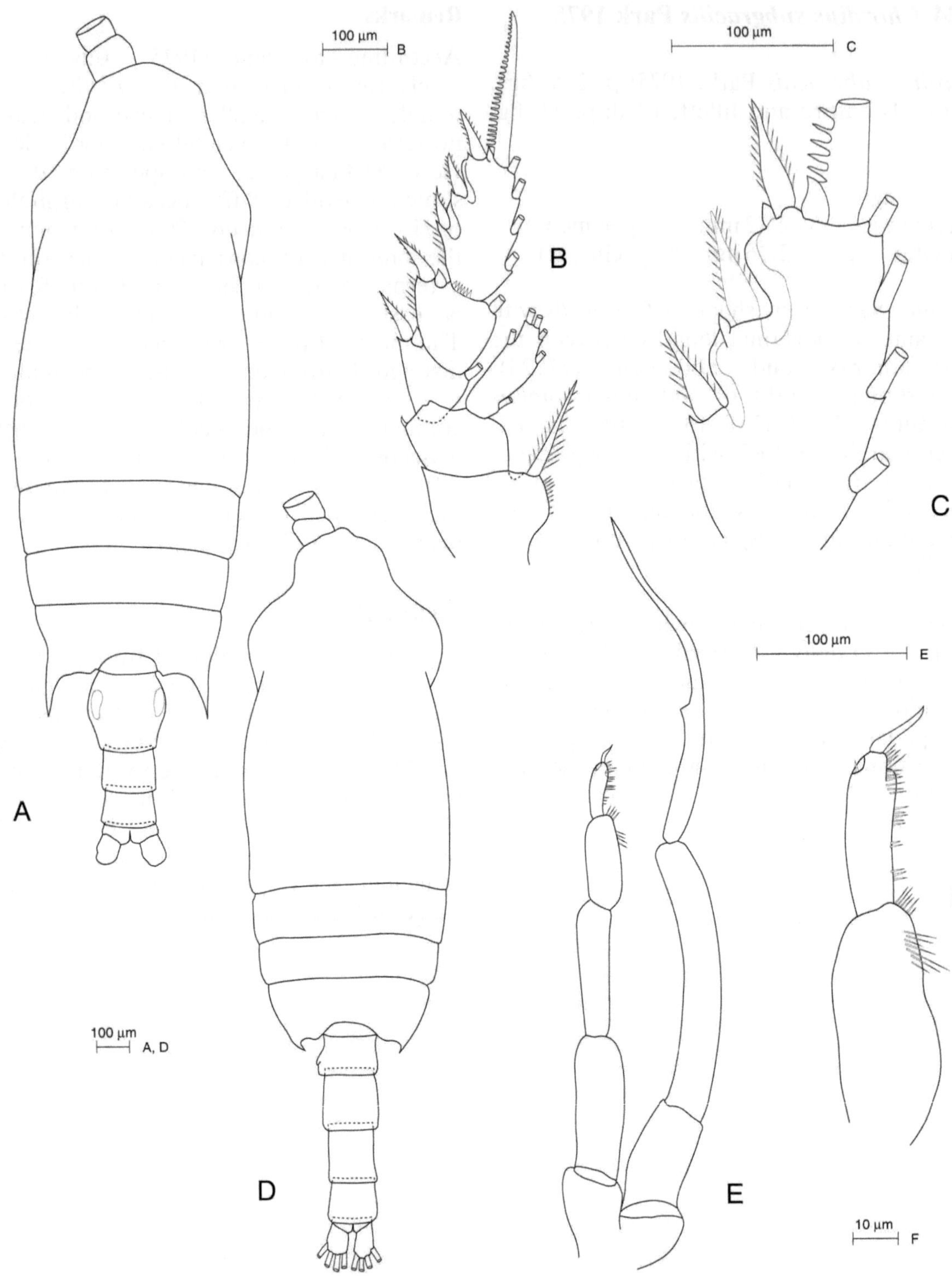

Fig. 3.24.1A–F. *Chiridius subgracilis*. Female: *A* whole animal dorsal view; **B** P2; **C** Re3 of P2. Male: **D** whole animal dorsal view; **E** P5; **F** last two joints of left P5

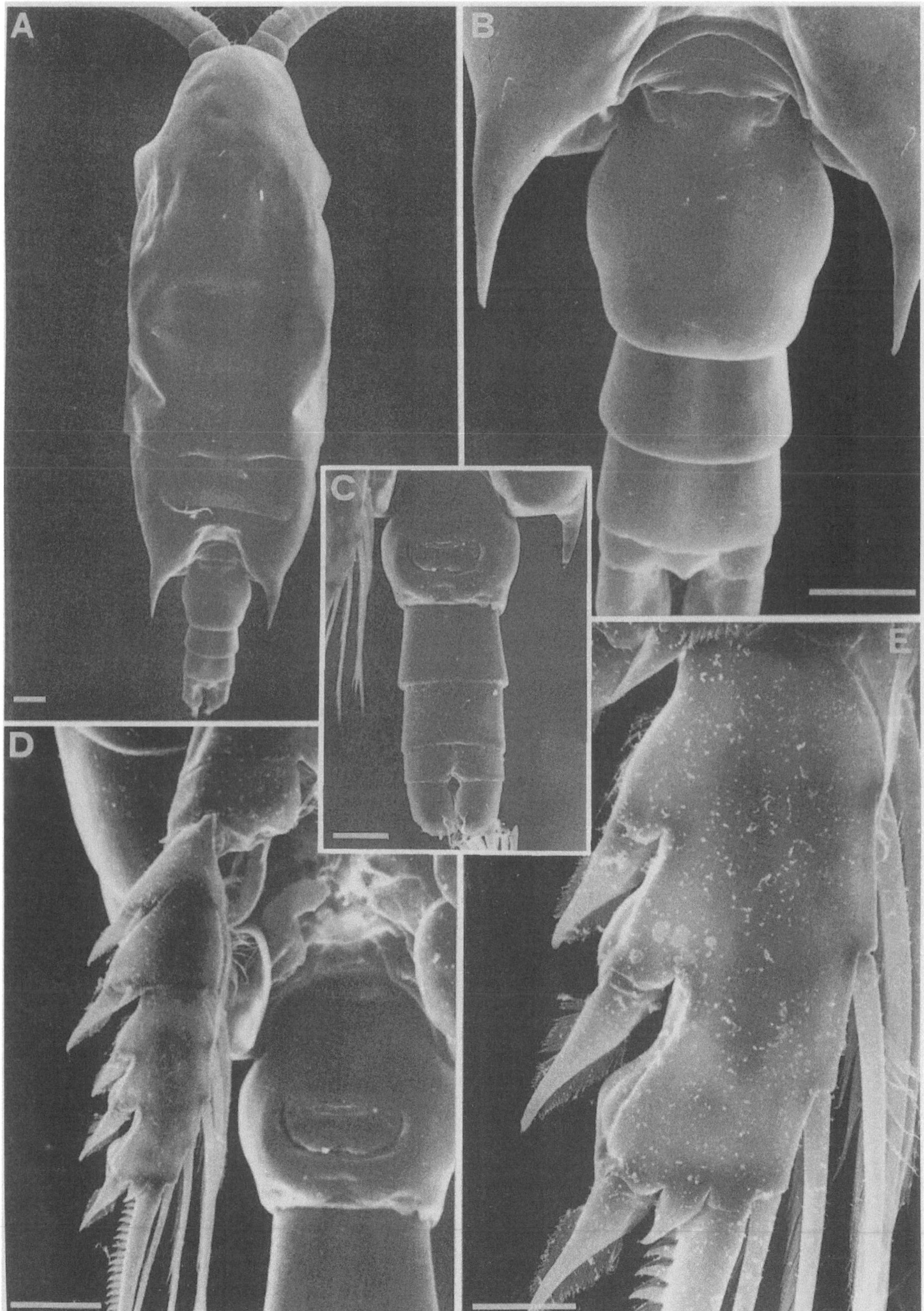

Fig. 3.24.2A–E. *Chiridius subgracilis*. Female: **A** whole animal dorsal view; **B** abdomen dorsal view; **C** abdomen ventral view; **D** P2; **E** Re3 of P2. *Bars*: **A**, **B**, **C**, **D** 100 µm; **E** 50 µm

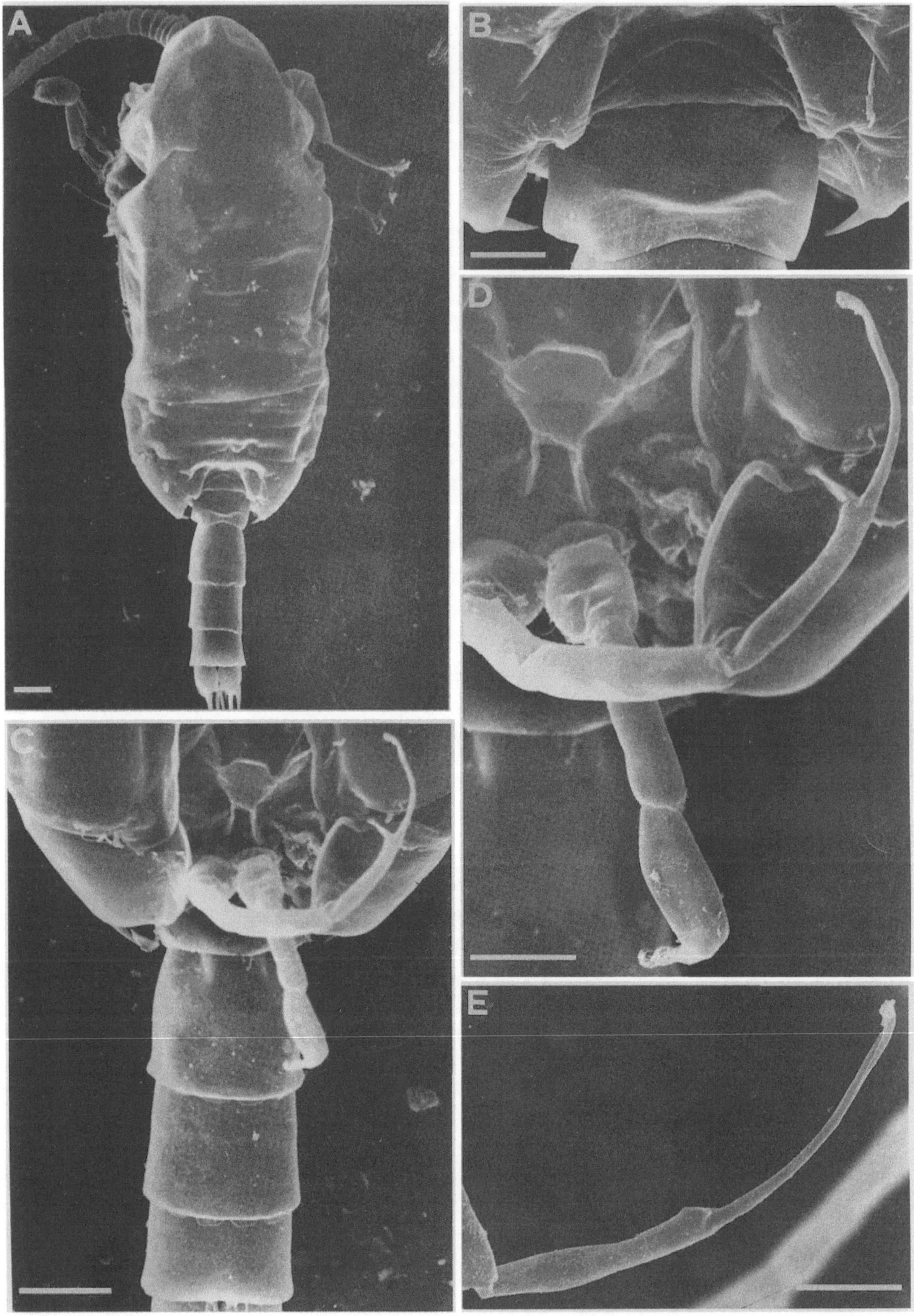

Fig. 3.24.3A–E. *Chiridius subgracilis*. Male: **A** whole animal dorsal view; **B** detail of posterior corners of last thoracic somite in dorsal view; **C** abdomen ventral view; **D** P5; **E** terminal joint of right P5. *Bars*: **A**, **C** 100 μm; **B**, **D**, **E** 50 μm

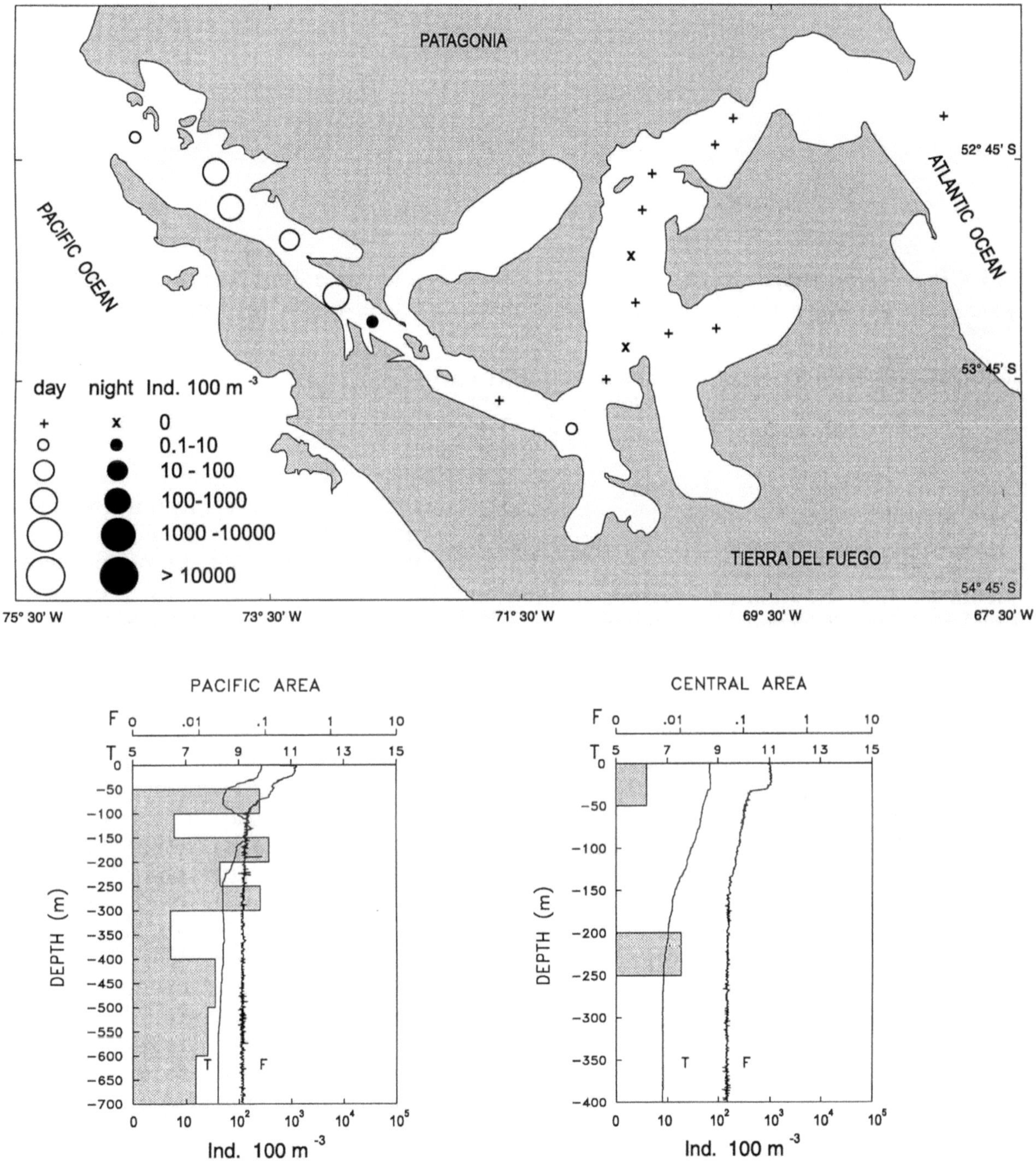

Depth range of occurrence = **50 - 700** m
Population core = **150 - 200** m
Temperature (T) = **8.20 - 9.80** °C
Oxygen = **348.4 - 437.2** μmol dm^{-3}
Chl fluorescence (F) = **0.04 - 0.39**

Depth range of occurrence = **0 - 250** m
Population core = **200 - 250** m
Temperature (T) = **6.70 - 9.01**°C
Oxygen = **346.1 - 407.3** μmol dm^{-3}
Chl fluorescence (F) = **0.05 - 0.27**

Fig. 3.24.4. *Chiridius subgracilis* + *C. gracilis*. Distribution in the Straits of Magellan in February–March 1991

3.25 *Clausocalanus brevipes* Frost and Fleminger 1968

Clausocalanus brevipes Frost and Fleminger, 1968: p. 70, pls. 56–59 – Ramirez, 1969: p. 52 figs. 47, 48 – Heron and Bowman, 1971: p. 159, figs. 116–129.
Clausocalanus arcuicornis Ramirez, 1966: p. 10, pl. 2 figs. 1–8.

Body length –
females: 1.47–1.80 mm (13 specimens)
males: 1.27–1.59 mm (8 specimens)

Female. Cephalothorax 2.5 times longer than wide. Rostrum slender and curved in lateral view. A1 reaching end of genital somite. Proportional lengths of cephalothorax and abdomen 74:26; proportional lengths of abdominal somites and furca 35:23:18:10:14 = 100. Seminal receptacle conspicuous in ventral view. In lateral view, dorsal lobe of seminal receptacle large and bulb-like, constricted in region of attachment to ventral lobe. Third joint of P5 more than 2 times as long as first joint; terminal points of P5 spinulated along both margins.

Male. Body more robust than female. Rostrum knob-like in lateral view. Proportional lengths of cephalothorax and abdomen 69:31. Proportional lengths of abdominal somites and furca 7:41:21:18:3:10 = 100. Left P5 much shorter than urosome.

Remarks

Females of *C. brevipes* are easily distinguished from *C. ingens* for the characteristic shape of the seminal receptacle. Males of *C. brevipes* are the only ones in this genus having a P5 with a left ramus shorter than the abdomen. Our specimens differ from those described by Frost and Fleminger only by the somewhat larger size (on average, both females and males are 0.2 mm larger).

Distribution

According to Frost and Fleminger (1968) *C. brevipes* is characteristic of sub-Antarctic waters. It was reported for the first time for Chilean fjords by Marin and Antezana (1985) but had previously been recorded by Ramirez (1969, 1970, 1971) for the Argentinian continental shelf, both in surface and deep waters. In the present study, it was frequently sampled throughout the study area with the exception of the Primera Angostura and more Atlantic stations. It was mainly found in the 0–300 m layer of the Pacific and Central areas of the Straits of Magellan and in the entire water column in the Atlantic area.

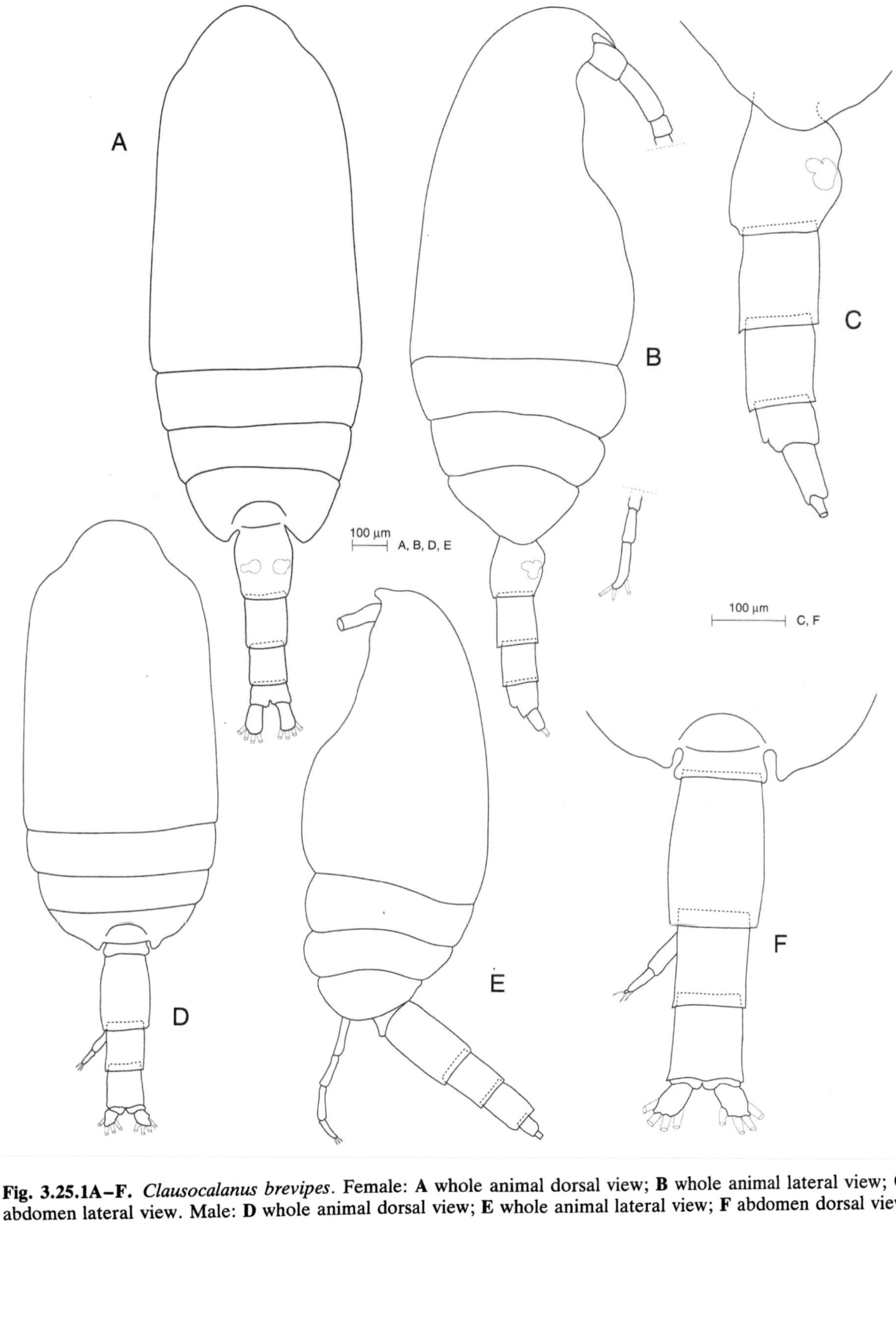

Fig. 3.25.1A–F. *Clausocalanus brevipes*. Female: **A** whole animal dorsal view; **B** whole animal lateral view; **C** abdomen lateral view. Male: **D** whole animal dorsal view; **E** whole animal lateral view; **F** abdomen dorsal view

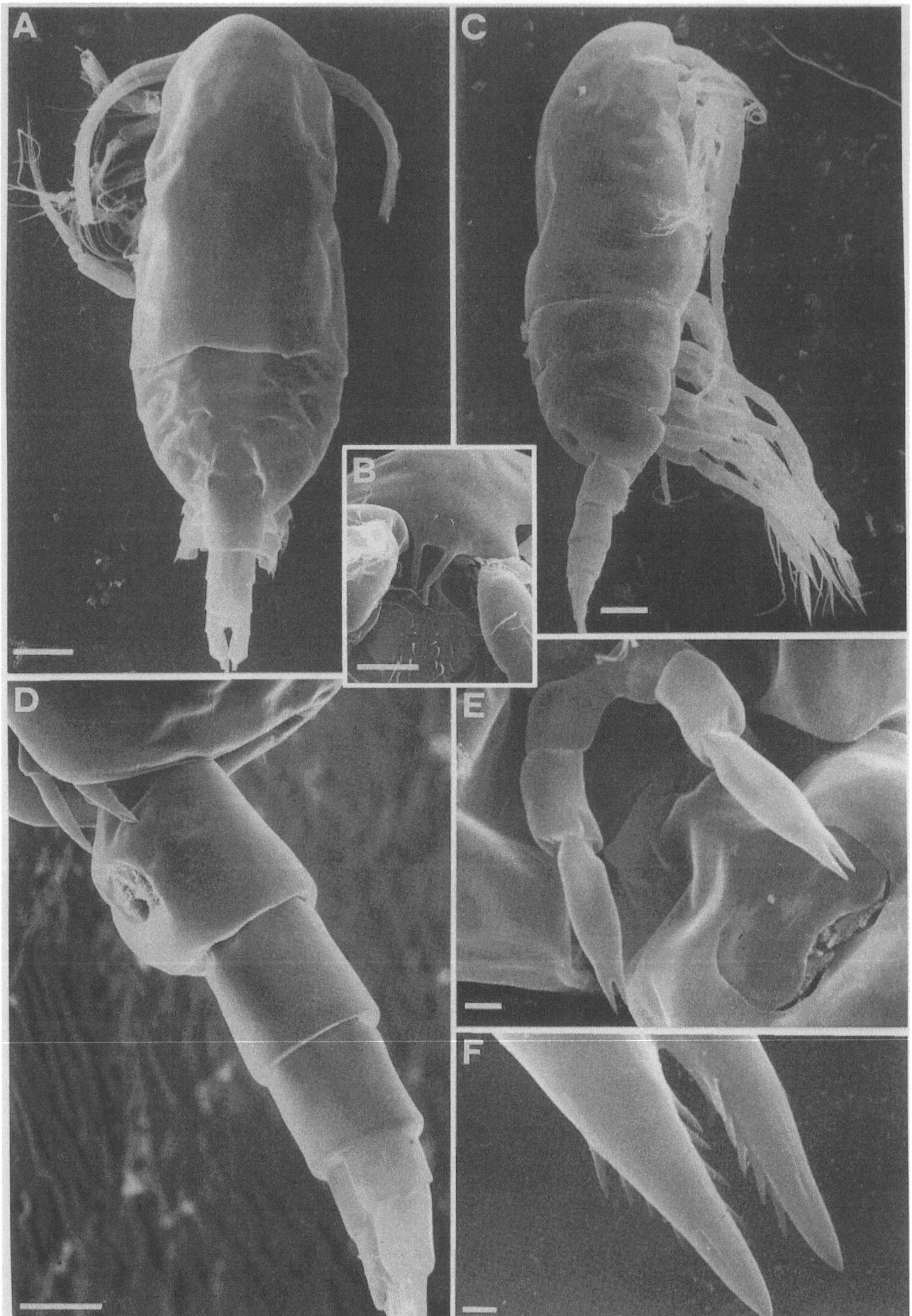

Fig. 3.25.2A–F. *Clausocalanus brevipes*. Female: **A** whole animal dorsal view; **B** rostrum; **C** whole animal lateral view; **D** abdomen lateral view; **E** P5; **F** detail of terminal points of P5. *Bars*: **A**, **C** 100 µm; **B**, **D** 50 µm; **E** 10 µm; **F** 1 µm

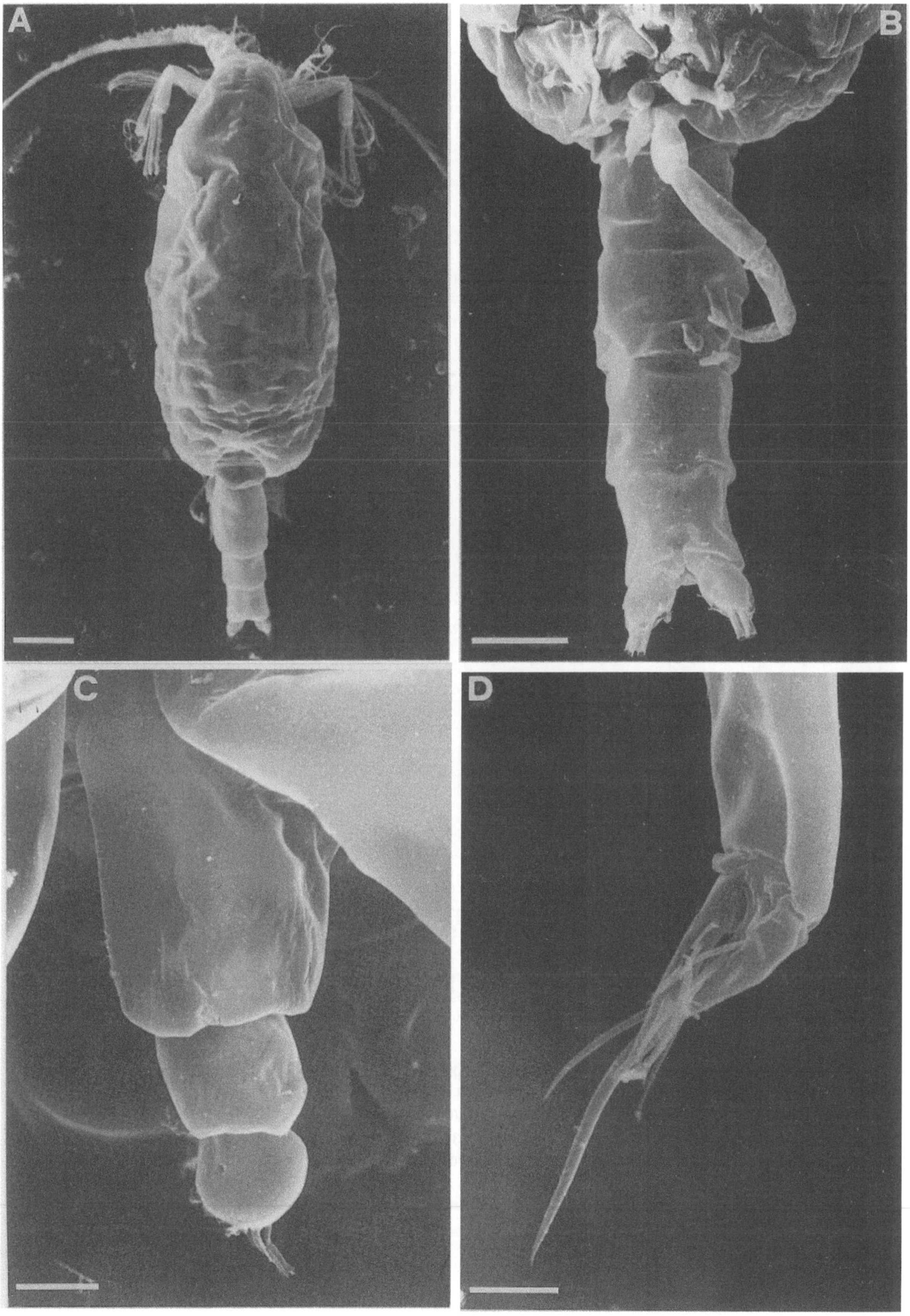

Fig. 3.25.3A–D. *Clausocalanus brevipes*. Male: **A** whole animal dorsal view; **B** abdomen ventral view and P5; **C** right P5; **D** last joint of left P5. *Bars*: **A** 100 µm; **B** 50 µm; **C, D** 5 µm

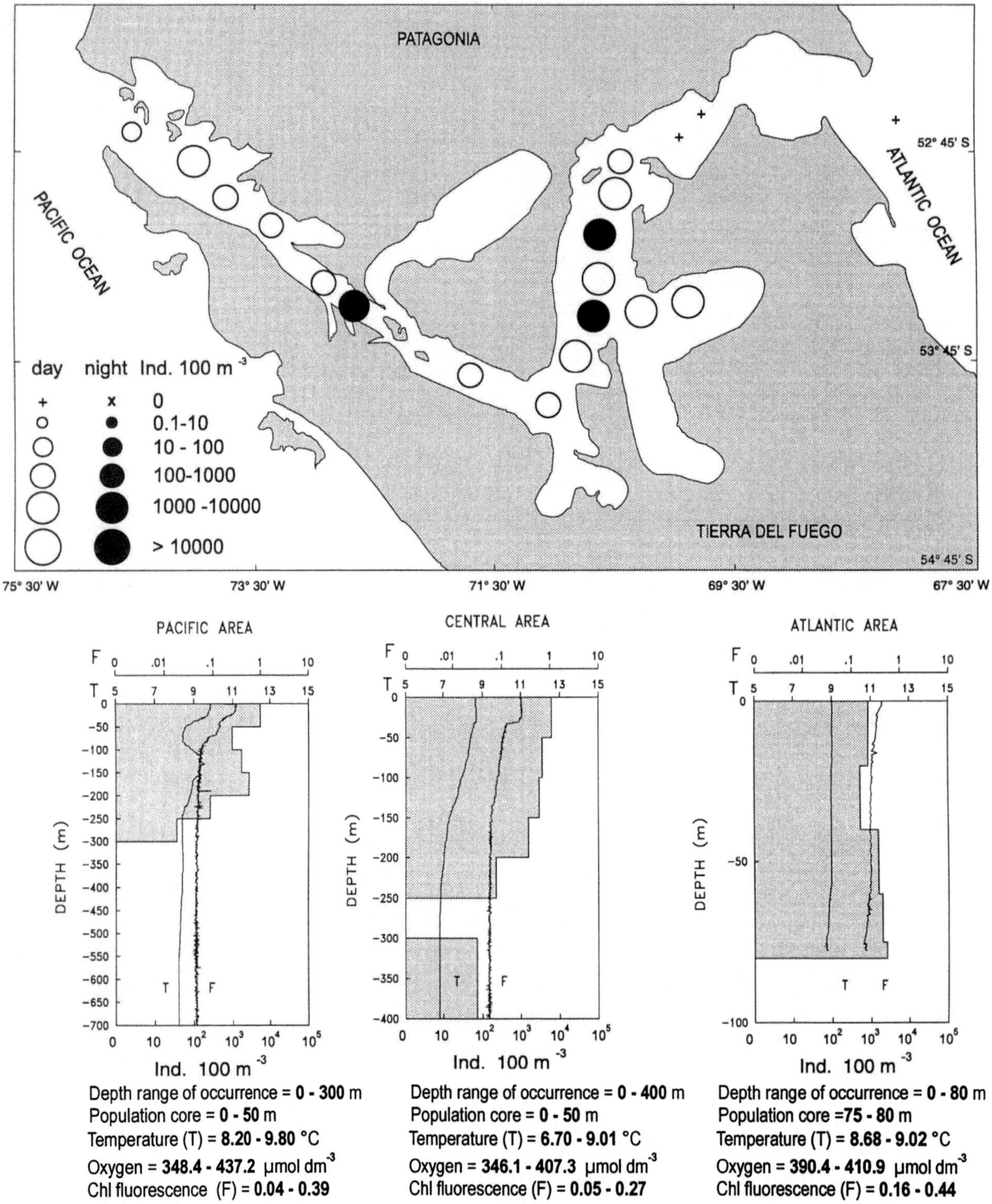

Fig. 3.25.4. *Clausocalanus brevipes*. Distribution in the Straits of Magellan in February–March 1991

3.26 *Clausocalanus ingens* Frost and Fleminger 1968

Clausocalanus ingens Frost and Fleminger, 1968: p. 39, pls. 19–23 – Ramirez, 1969: p. 52, pl. 8 figs. 49, 50, 51.
Clausocalanus laticeps Tanaka, 1960a: p. 32, pl. 11 figs. 1–14.

Body length –
females: 1.67–1.84 mm (2 specimens)

Female. Body shape similar to *C. brevipes*. Cephalothorax 2.6 times longer than wide. Rostrum slender and curved in lateral view. Al almost reaching end of cephalothorax. Proportional lengths of cephalothorax and abdomen 75:25; proportional lengths of abdominal somites and furca 36:20:17:12:15 = 100. In lateral view, ventral lobe of seminal receptacle small; dorsal lobe broad and long, directed dorsally. P5 small, 3-jointed; terminal points of last joint smooth and lacking spinules.

Remarks

Our specimens concord perfectly with those described by Frost and Fleminger (1968). We did not find adult males.

Distribution

C. ingens is generally present in warm and temperate waters of the Southern hemisphere. It was reported for the first time in Chilean fjords by Marin and Antezana (1985). In the present study, the species was found from 150–200 m at a single station in the Pacific area and from 0–50 m at two Central stations in the Straits of Magellan.

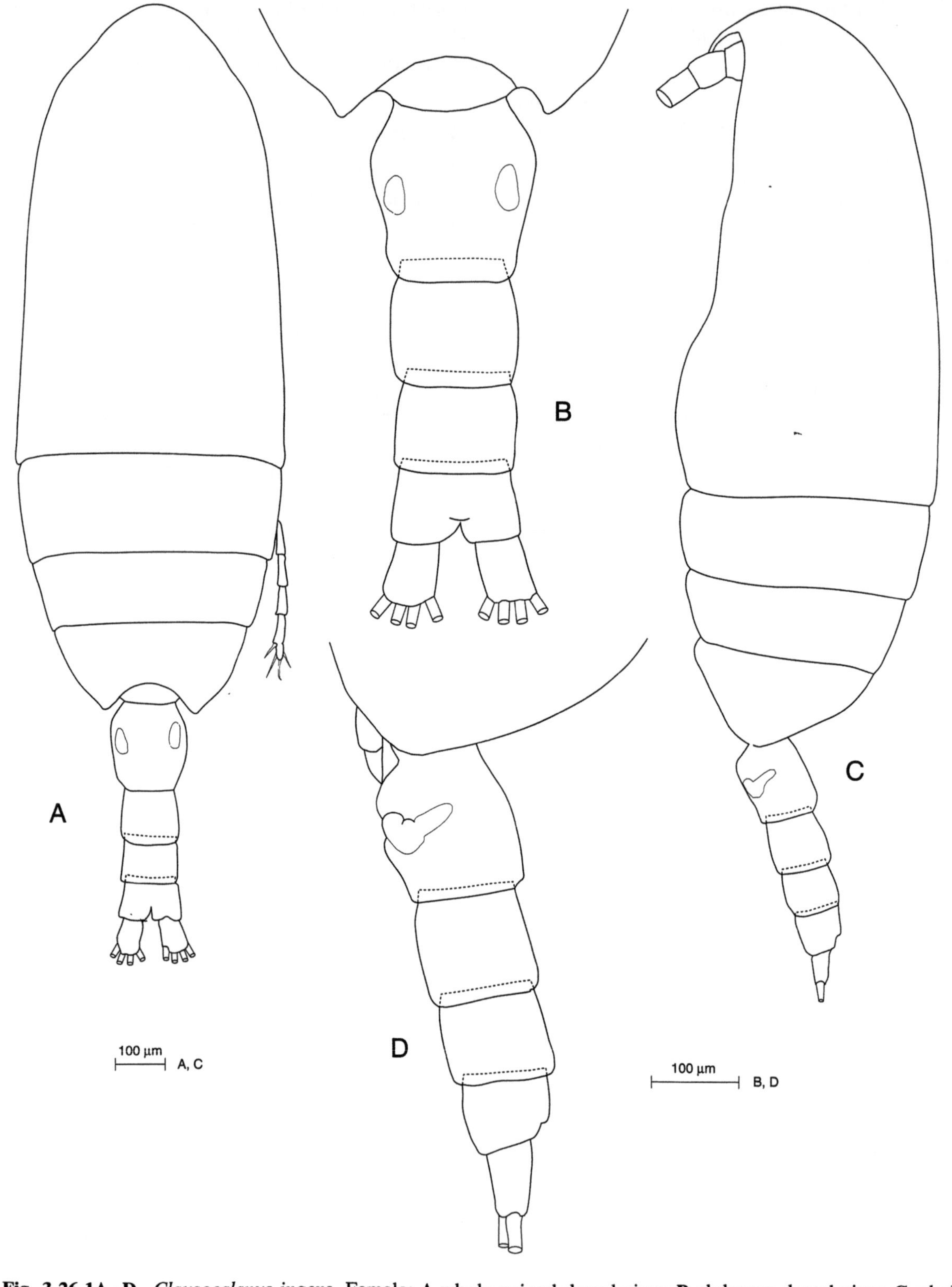

Fig. 3.26.1A–D. *Clausocalanus ingens*. Female: **A** whole animal dorsal view; **B** abdomen dorsal view; **C** whole animal lateral view; **D** abdomen lateral view

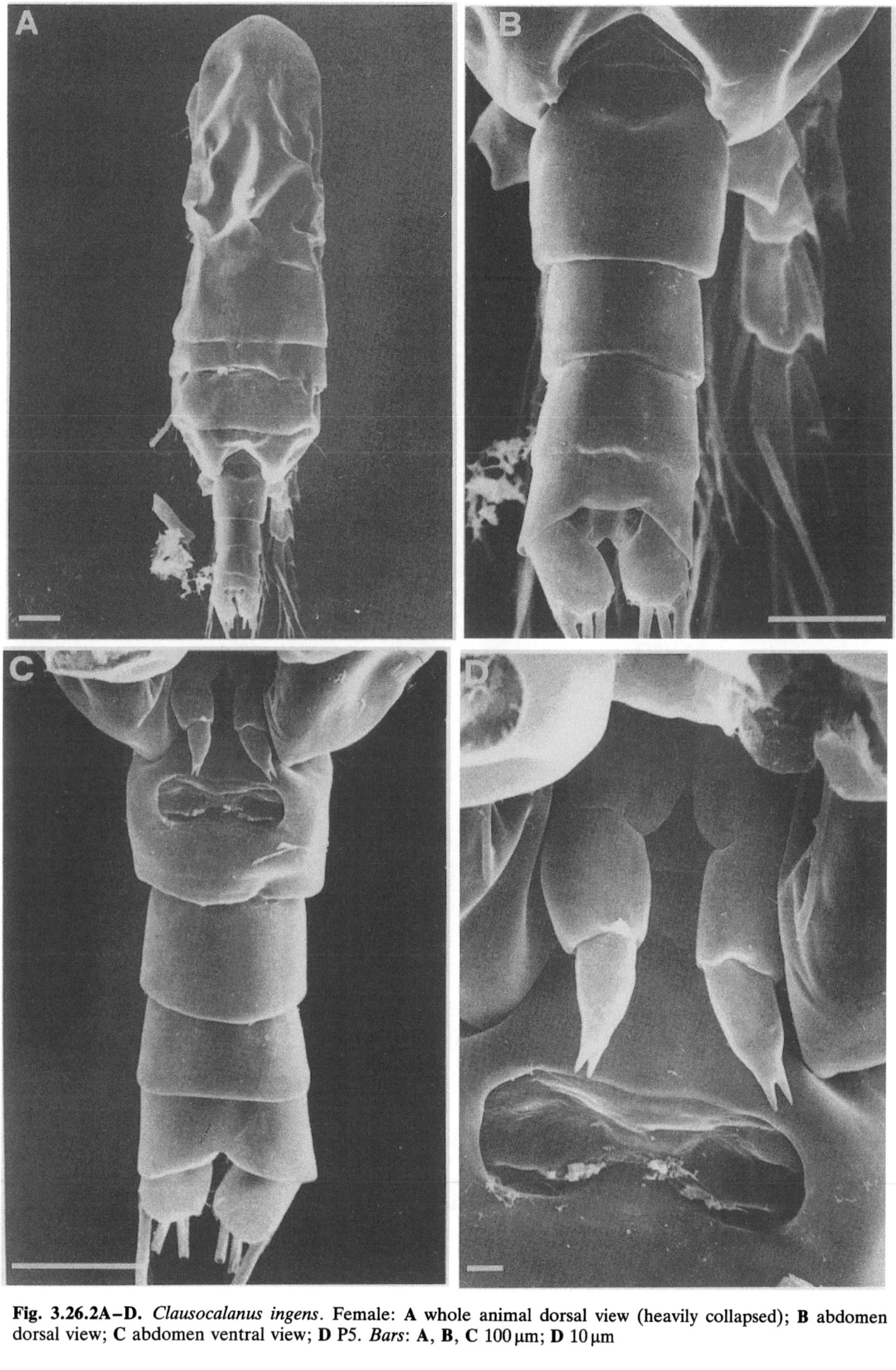

Fig. 3.26.2A–D. *Clausocalanus ingens*. Female: **A** whole animal dorsal view (heavily collapsed); **B** abdomen dorsal view; **C** abdomen ventral view; **D** P5. *Bars*: **A**, **B**, **C** 100 μm; **D** 10 μm

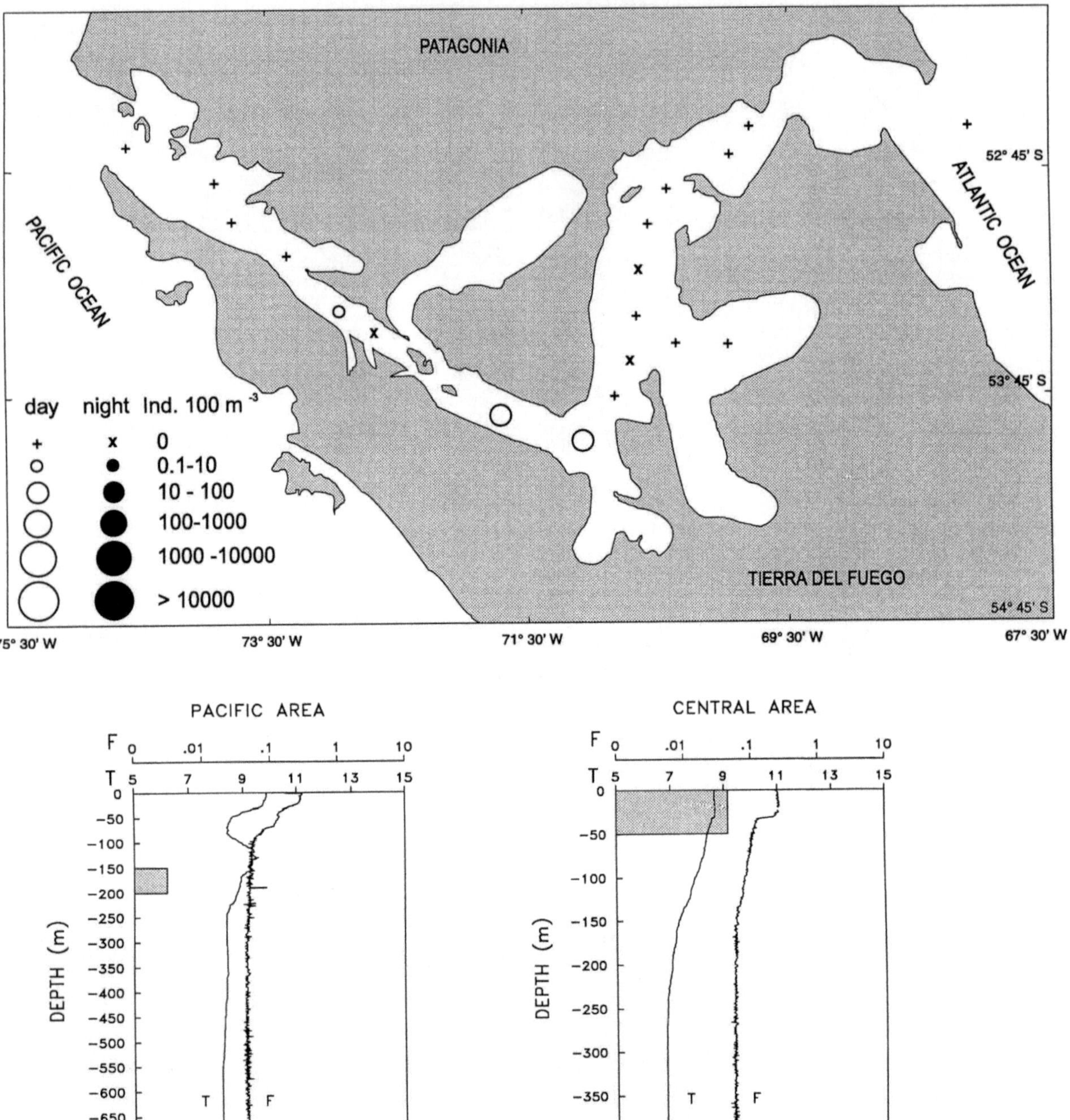

PACIFIC AREA	CENTRAL AREA
Depth range of occurrence = **150 - 200** m	Depth range of occurrence = **0 - 50** m
Population core = **150 - 200** m	Population core = **0 - 50** m
Temperature (T) = **8.20 - 9.80** °C	Temperature (T) = **6.70 - 9.01** °C
Oxygen = **348.4 - 437.2** μmol dm^{-3}	Oxygen = **346.1 - 407.3** μmol dm^{-3}
Chl fluorescence (F) = **0.04 - 0.39**	Chl fluorescence (F) = **0.05 - 0.27**

Fig. 3.26.3. *Clausocalanus ingens*. Distribution in the Straits of Magellan in February–March 1991

3.27 *Clausocalanus laticeps* Farran 1929

Clausocalanus laticeps Farran, 1929: p. 224, fig. 4–Vervoort, 1965a: p. 384, pl. 1 – Frost and Fleminger, 1968: p. 42, pls. 24–28 – Ramirez, 1969: p. 53, figs. 53, 55 – Heron and Bowman, 1971: p. 157, figs. 100–115– Ramirez and Dinofrio, 1976: p. 6, pl. 3 fig. 4, pl. 5 fig. 14.

Body length –
females: 1.27–1.49 mm (4 specimens)

Female. Body stout. Cephalothorax 2.3 times longer than wide. Dorsal region of head strongly vaulted in lateral view. Rostrum short, usually straight, directed ventroposteriad in lateral view. A1 not reaching end of cephalothorax. Proportional lengths of cephalothorax and abdomen 72:28; proportional lengths of abdominal somites and furca 34:19:16:15:16 = 100. In lateral view, ventral lobe of seminal receptacle large, dorsal lobe slender and digitiform. Points of last joint of P5 smooth and lacking spinules.

Remarks

The vaulted forehead is usually a distinctive character of *C. laticeps* females (Frost and Fleminger 1968). We did not record males.

Distribution

C. laticeps is considered by Frost and Fleminger (1968) a sub-Antarctic species. It has been reported for Chilean fjords (Marin and Antezana 1985) and offshore Argentinian waters (Ramirez 1970, 1971). In the present study, only a few individuals were sampled from 100–150 m off Cape Froward.

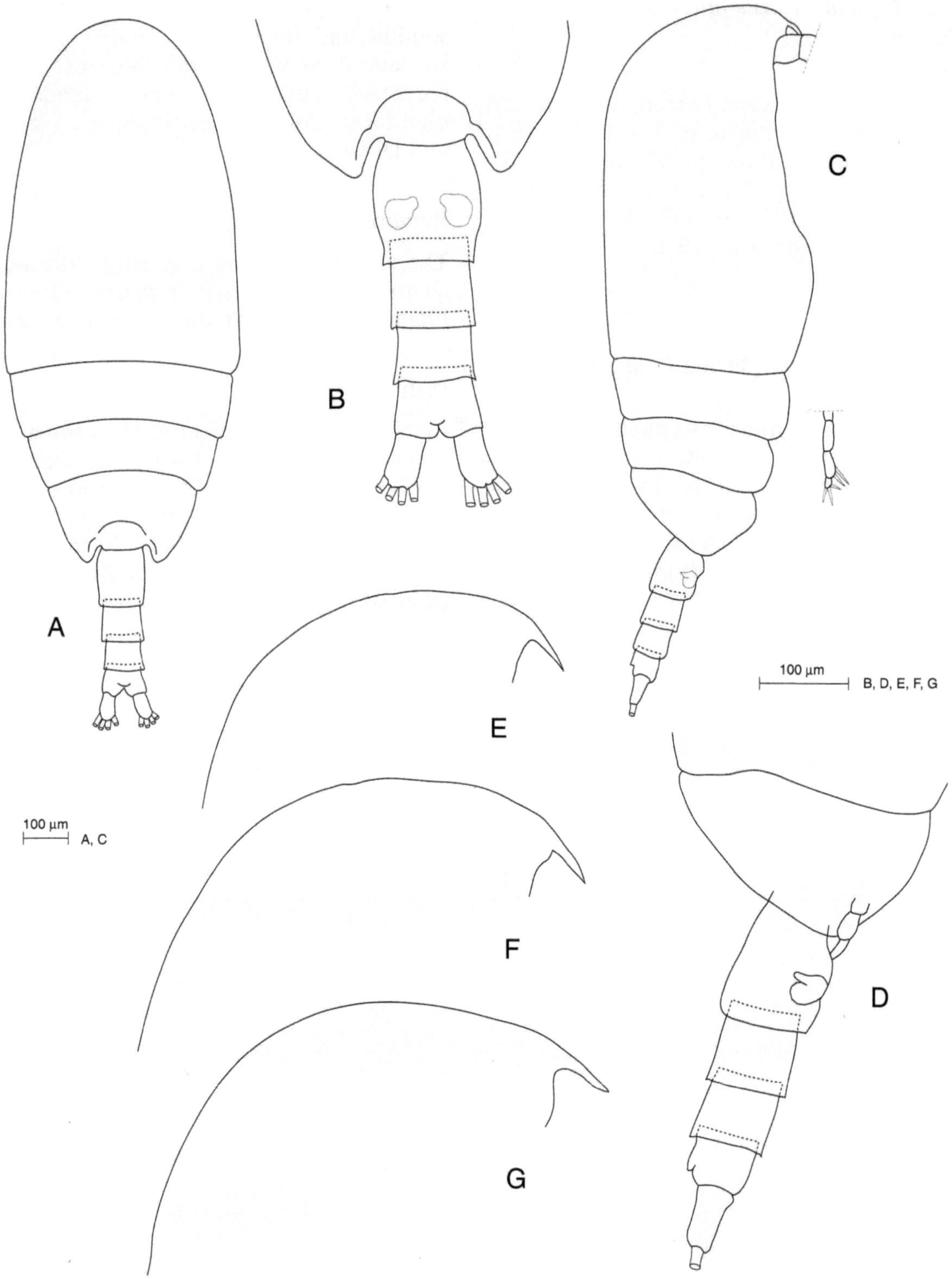

Fig. 3.27.1A–G. *Clausocalanus laticeps*. Female: **A** whole animal dorsal view; **B** abdomen dorsal view; **C** whole animal lateral view; **D** abdomen showing characteristic shape of genital pore; **E, F, G** differences in the curvature of the rostrum in lateral view

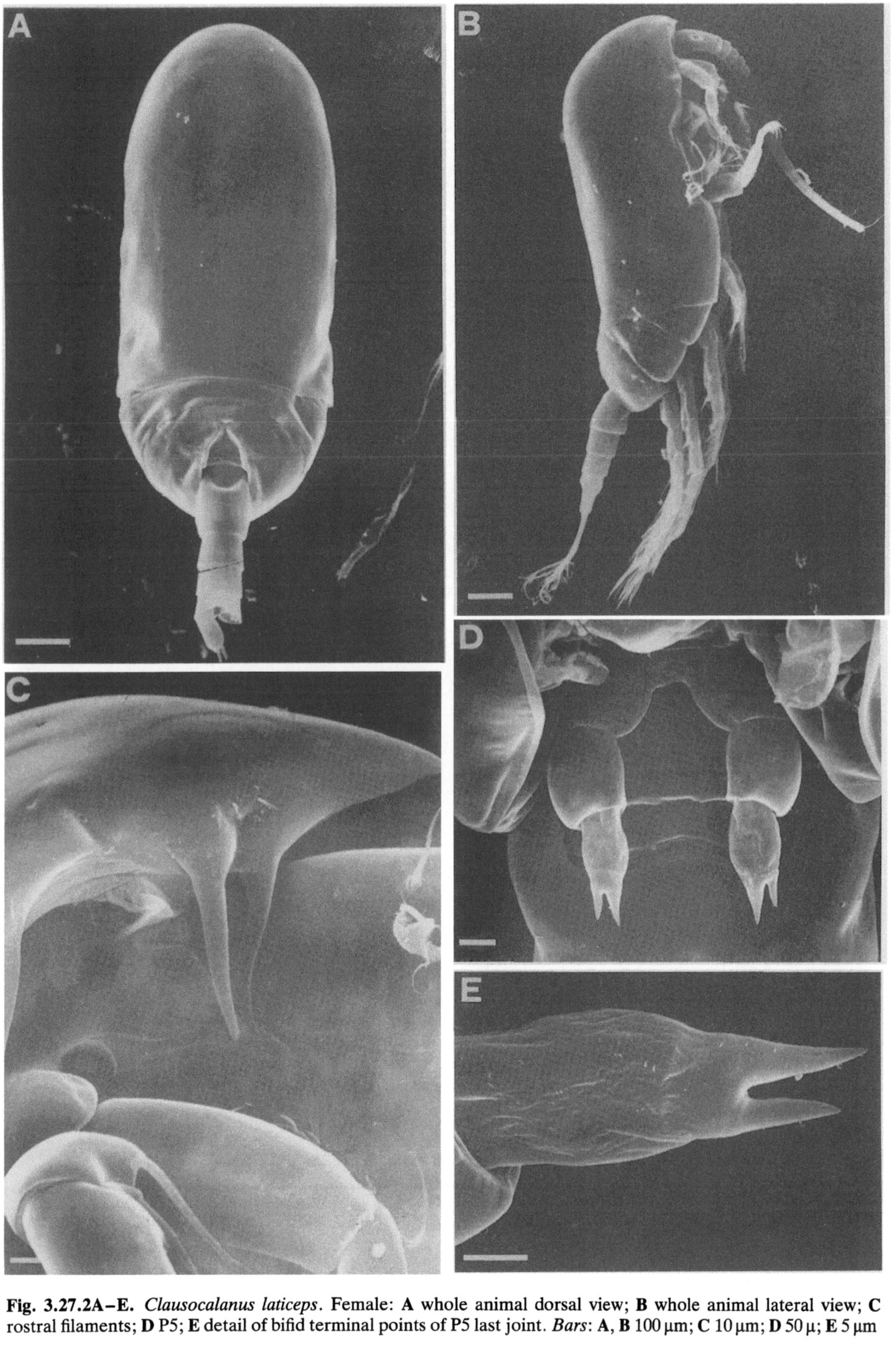

Fig. 3.27.2A–E. *Clausocalanus laticeps*. Female: **A** whole animal dorsal view; **B** whole animal lateral view; **C** rostral filaments; **D** P5; **E** detail of bifid terminal points of P5 last joint. *Bars*: **A**, **B** 100 μm; **C** 10 μm; **D** 50 μ; **E** 5 μm

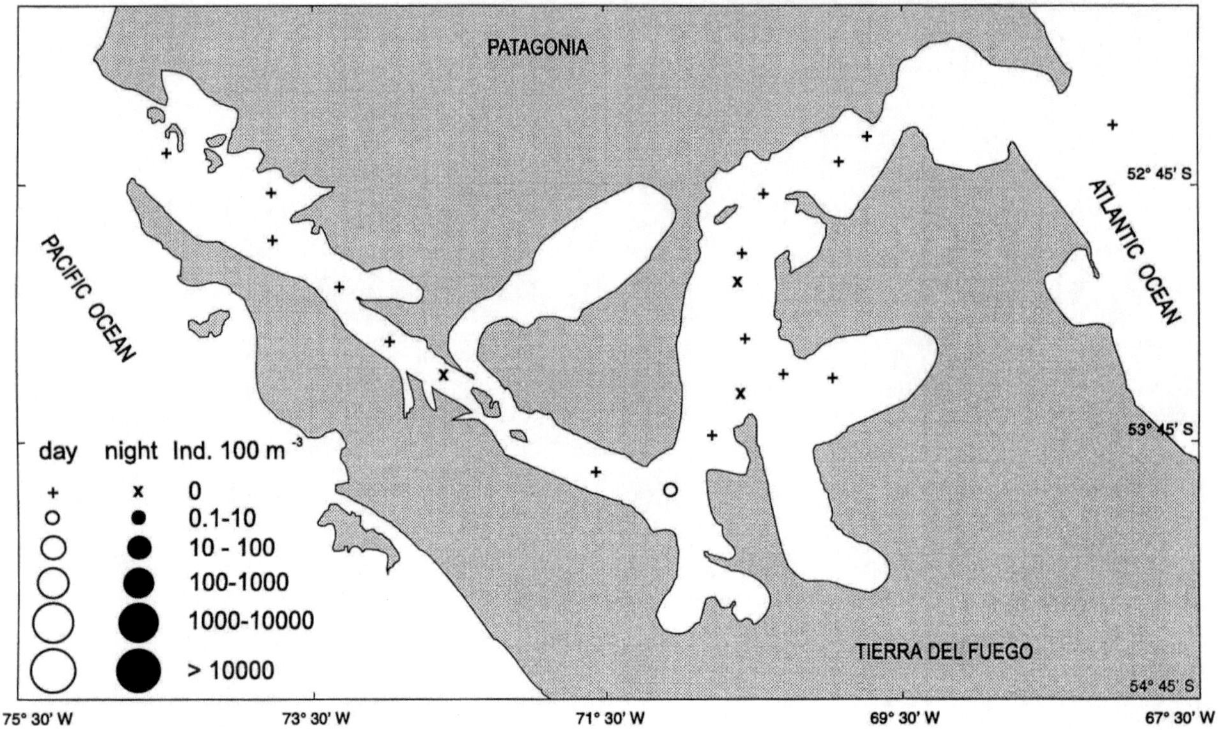

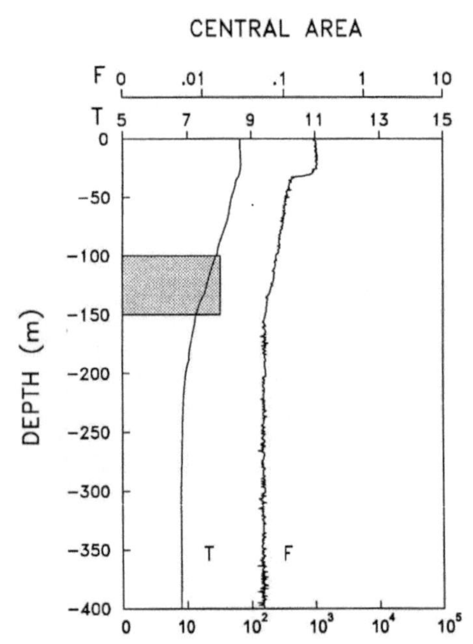

Depth range of occurrence = **100 - 150** m
Population core = **100 - 150** m
Temperature (T) = **6.70 - 9.01** °C
Oxygen = **376.1 - 407.3** μmol dm^{-3}
Chl fluorescence (F) = **0.05- 0.27**

Fig. 3.27.3. *Clausocalanus laticeps*. Distribution in the Straits of Magellan in February–March 1991

3.28 *Ctenocalanus citer* Heron and Bowman 1971

Ctenocalanus citer Heron and Bowman, 1971: p. 163, figs. 1, 16–18, 31–36, 54–58, 71–77, 94–99, 130–150.

Body length –
females: 1.06–1.45 mm (60 specimens)
males: 1.10–1.45 mm (30 specimens)

Female. Body slender with cephalothorax 2.7 times longer than wide. Proportional lengths of cephalothorax and abdomen 75:25; proportional lengths of abdominal somites and furca 34:18:14:16:18 = 100. In lateral view, ventral surface of genital somite with small protuberance or lapett-like structure overhanging seminal receptacles. P4 with 3-jointed endopod and exopod. Re1, Re2 with one outer spine; Re3 with 1 long terminal spine, 4 inner setae and 3 outer spines, first 2 of these spines with comb-like teeth on medial margin. P5 vestigial.

Male. Cephalothorax 2.3 times longer than wide. Proportional lengths of cephalothorax and abdomen 65:35; proportional lengths of abdominal somites and furca 12:32:18:14:10:14 = 100. P5 asymmetrical and uniramous. Left P5 4-jointed with group of long bristles on inner margin of last joint; distal margin of penultimate joint with tuft of long bristles. Right P5 1-jointed with 2 small terminal spines.

Remarks

C. citer is very similar to *C. vanus*, differing mainly for the shorter A1 which reaches the posterior margin of the caudal rami (A1 of *C. vanus* exceeds the caudal rami by several segments) and for the male with a 1-jointed right P5 (male of *C. vanus* without right P5) (Heron and Bowman 1971). The left P5 of our male specimens is consistently similar in morphology whereas the right P5 is highly variable in size, either rounded or pointed with one or two small apical spines. Our specimens concord with the description of Heron and Bowman; the only difference is in the number of teeth on the external spine of P4 Re3. Also, our female and male specimens are somewhat larger than those reported by these authors.

Distribution

C. citer is very similar to the tropical, subtropical congeneric species *C. vanus* which has a circumglobal distribution. The species was first described in samples collected in the Pacific sector of the Antarctic and sub-Antarctic regions. It has since been cited by Boysen-Ennen and Piatkowski (1988) as a common species in the Weddell Sea, Antarctica. It is probable that its presence in the Southern Oceans has often been confused with that of *C. vanus*. Mazzocchi and Ianora (1991) reported it as quite frequent in previous collections from the Straits of Magellan. In the present study, *C. citer* was one of the most frequently sampled species throughout the Straits of Magellan. It was found at all depths but was most abundant in the upper 150 m.

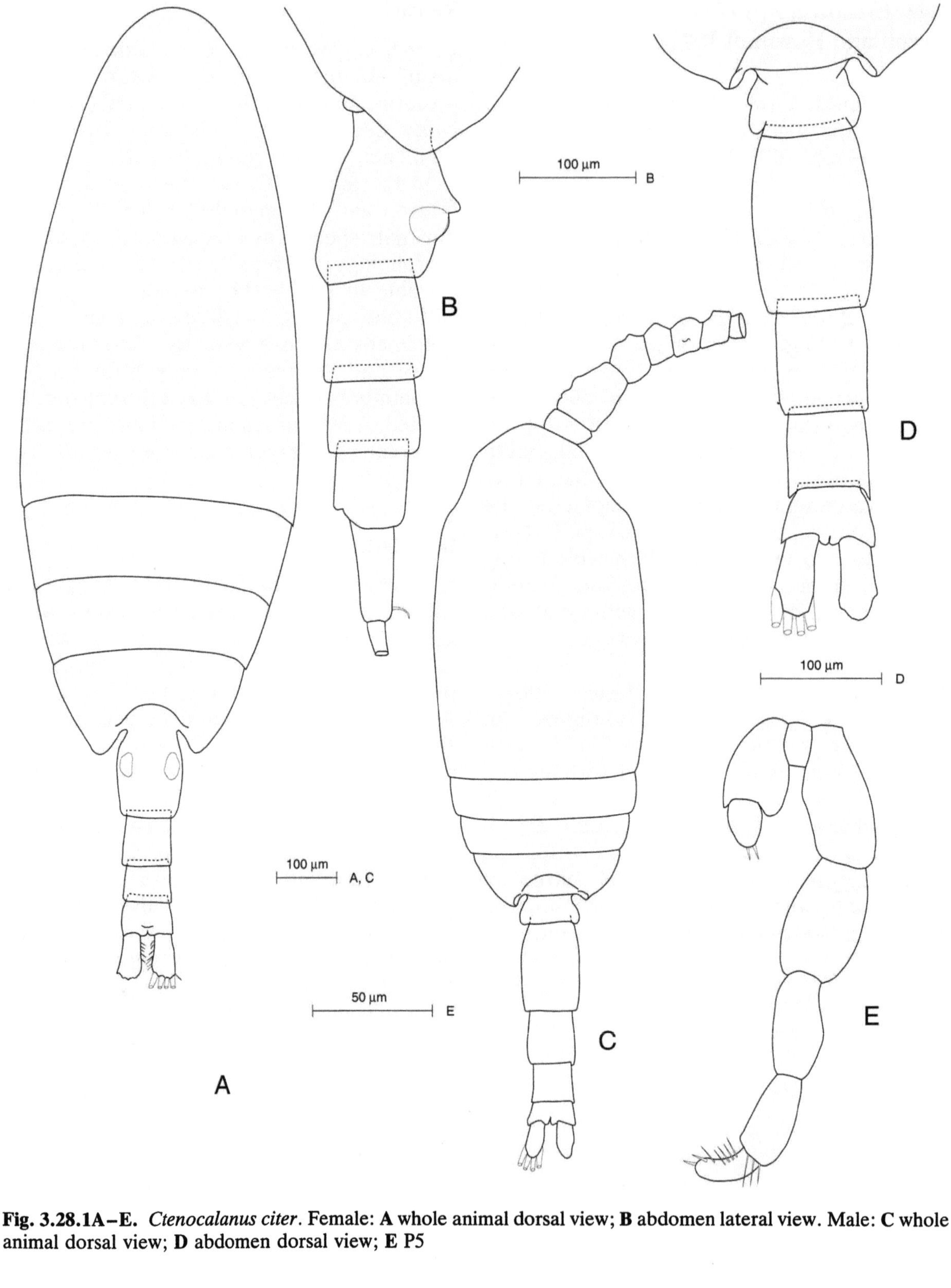

Fig. 3.28.1A–E. *Ctenocalanus citer*. Female: **A** whole animal dorsal view; **B** abdomen lateral view. Male: **C** whole animal dorsal view; **D** abdomen dorsal view; **E** P5

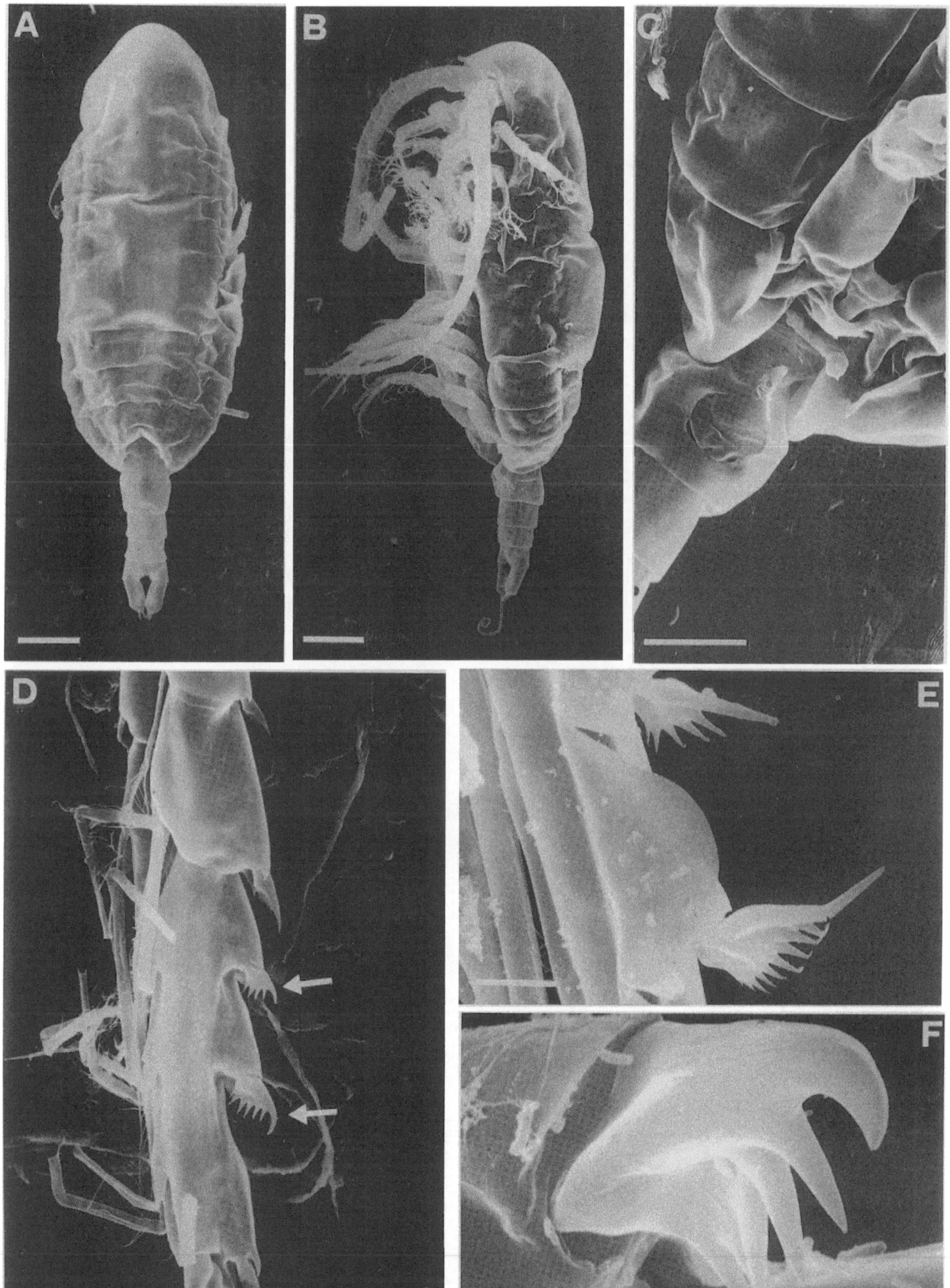

Fig. 3.28.2A–F. *Ctenocalanus citer*. Female: **A** whole animal dorsal view; **B** whole animal lateral view; **C** vestigial P5 and genital segment ventral view; **D** P4; **E**, **F** detail of of outer spines of P4 with comb-like ridges on medial margin (*arrows* in **D**). *Bars*: **A**, **B** 100 μm; **C** 50 μm; **D**, **E** 10 μm; **F** 5 μm

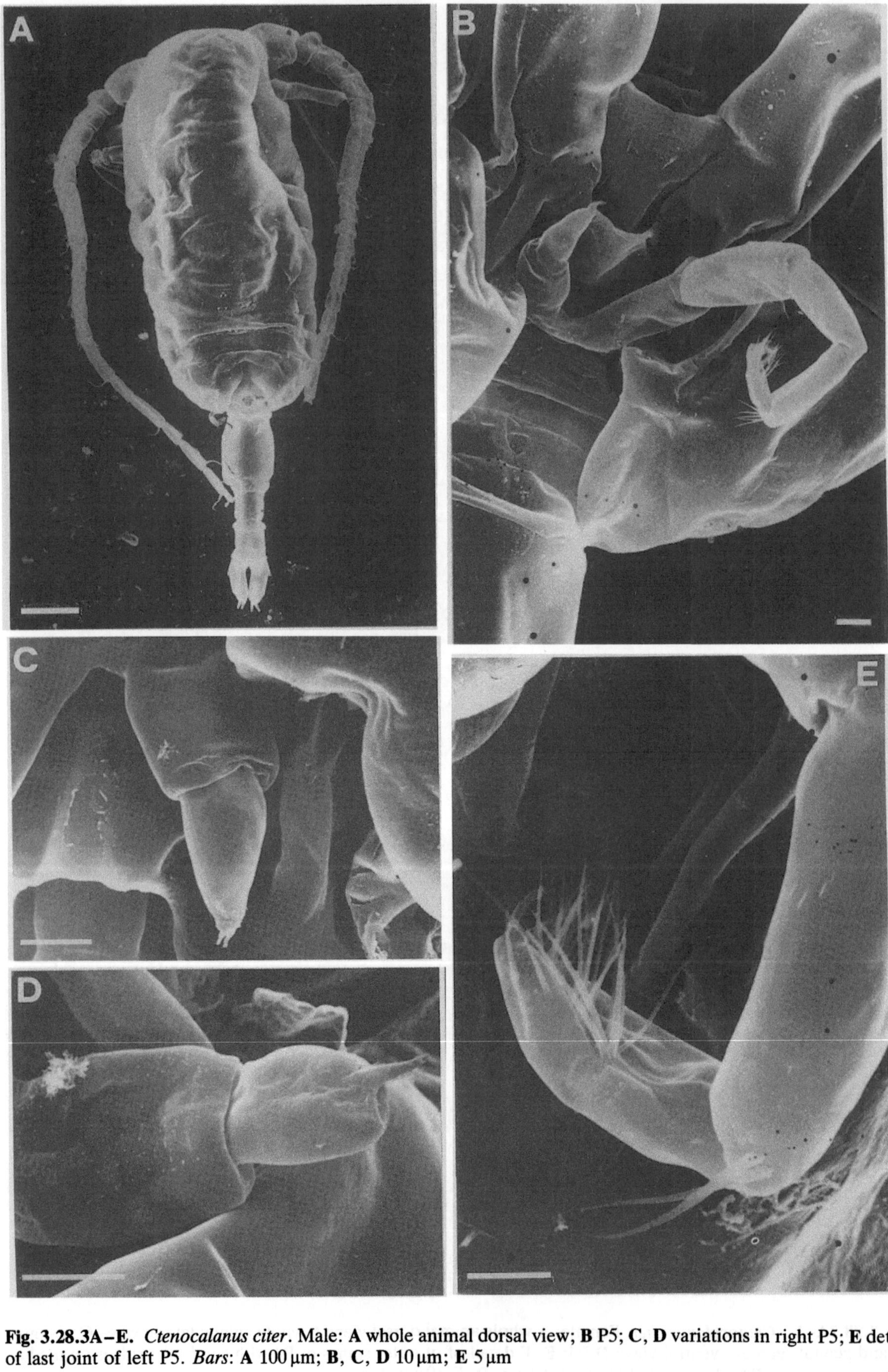

Fig. 3.28.3A–E. *Ctenocalanus citer*. Male: **A** whole animal dorsal view; **B** P5; **C**, **D** variations in right P5; **E** detail of last joint of left P5. *Bars*: **A** 100 µm; **B**, **C**, **D** 10 µm; **E** 5 µm

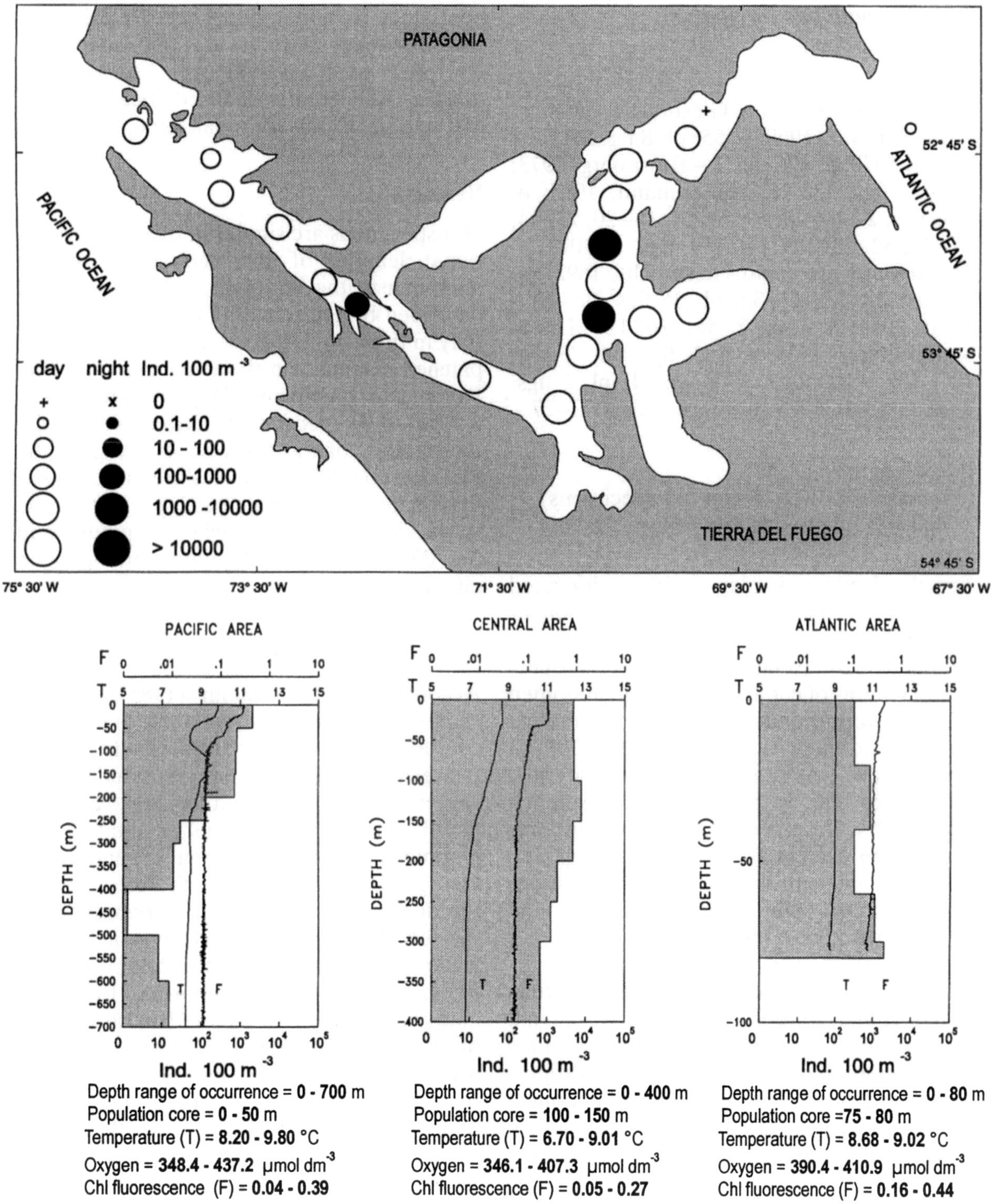

Fig. 3.28.4. *Ctenocalanus citer*. Distribution in the Straits of Magellan in February–March 1991

3.29 *Drepanopus forcipatus* Giesbrecht 1888

Drepanopus forcipatus Giesbrecht, 1888: p. 335 – Ramirez, 1969: p. 55, pl. 8 fig. 56, pl. 9 fig. 63; 1970: p. 476, fig. 2 – Björnberg, 1973: p. 317, figs. 13, 14 – Hülsemann, 1985: p. 911, figs. 2–4, 6–8, 10, 12, 14, 16–20, 23, 24, 27–29, 32, 33; 1991: p. 199, figs. 1–107.
Drepanopus pectinatus Giesbrecht, 1892: pl. 10 figs. 23–25, 27–30, pl. 36 figs. 36–38 – Scott, 1914: p. 8, pl. 2, figs. 10, 11 – Hardy and Gunther, 1935: p. 153, figs. 56, 73, 74, 100, 102 – Ramirez, 1966: p. 11, pl. 2 figs. 9–16 – Arcos, 1975: p. 13, pl. 3 figs. 22–27.

Body length –
females: 1.39–1.78 mm (35 specimens)
males: 1.12–1.35 mm (10 specimens)

Female. Body slender with elongated abdomen. Cephalothorax 2.4 times longer than wide. Last thoracic somite with short, rounded postero-lateral corners. Proportional lengths of cephalothorax and abdomen 65:35; proportional lengths of abdominal somites and furca 41:19:13:11:16 = 100. In dorsal view, genital somite swollen laterally in anterior half, with clearly visible seminal receptacles. In lateral view, swelling produced into evident bulge enveloping genital aperture on ventral side. P5 with bulbous base; terminal spine of P5 bent distally and ornated with 30 or more small teeth.

Male. Cephalothorax 2.2 times longer than wide, ending anteriorly with flat forehead. Last thoracic somite with short rounded postero-lateral corners as in female. Proportional lengths of cephalothorax and abdomen 63:37; proportional lengths of abdominal somites and furca 9:37:18:14:10:12 = 100. A1 with 3 conspicuous patches of long spinules at base. A2 with a row of curved spines at the base of its inner side. P5 asymmetrical and uniramous. Right P5 with bent, long terminal spine; left P5 3-jointed with 2 patches of spines on inner margin of Rel. Re2 with a strong spine on external margin; Re3 greatly reduced, bearing hairs and ending in thin terminal spine.

Remarks

Our specimens are similar in size and proportional lengths of the cephalothorax and abdomen to the Patagonian shelf specimens described by Hülsemann (1985). They differ only in the male left P5 which bears 2 distinct patches of spines on the inner margin of the Rel in our specimens. *D. forcipatus* is very similar to *D. pectinatus*. Differences in the morphology of the two species are given by Hülsemann (1985). Females are more similar than males. In the male of *D. pectinatus*, the A1 lacks conspicuous patches of spinules and the P5 has a small, 1-jointed endopod on either side.

Distribution

According to Hülsemann (1985), *D. forcipatus* is confined to Atlantic and Pacific coastal shelf areas of southern South America. South of the Antarctic Convergence, the same author reports other populations of this species as limited geographically to a small area around South Georgia. The Convergence therefore divides the distribution range of *D. forcipatus* from that of *D. pectinatus* which lives in inshore waters south of the Antarctic Polar Front. *D. forcipatus* has been reported by many authors as widely distributed in shelf areas off Chile and Argentina. In the Straits of Magellan, Mazzocchi and Ianora (1991) recorded its occurrence at all stations but one, with highest densities at the three most Atlantic sites. In the present study, the species was widely distributed throughout the Straits. It was sampled up to 500 m but reached maximum abundances in the upper 100 m at Pacific and Central stations.

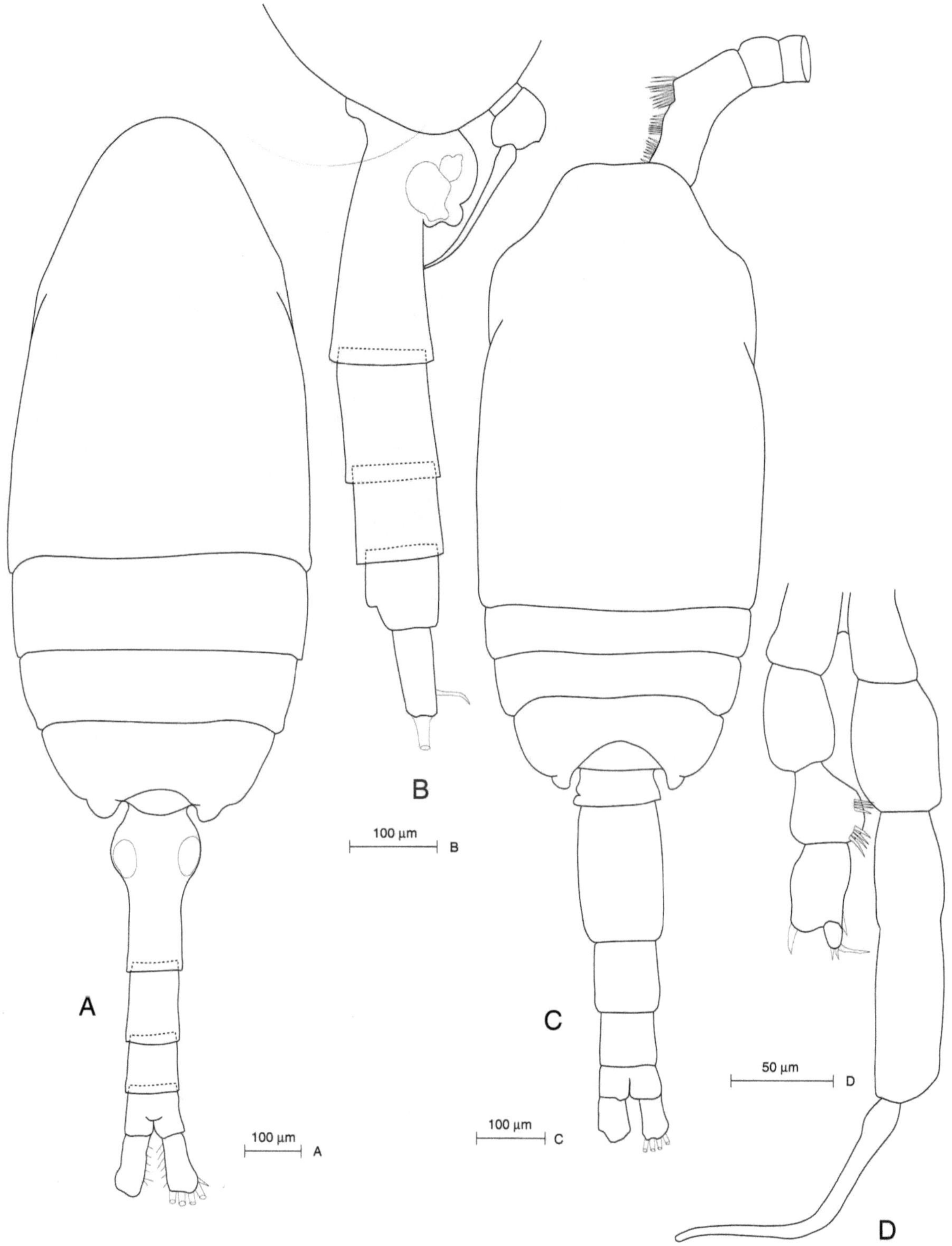

Fig. 3.29.1A–D. *Drepanopus forcipatus*. Female: **A** whole animal dorsal view; **B** abdomen lateral view. Male: **C** whole animal dorsal view; **D** P5

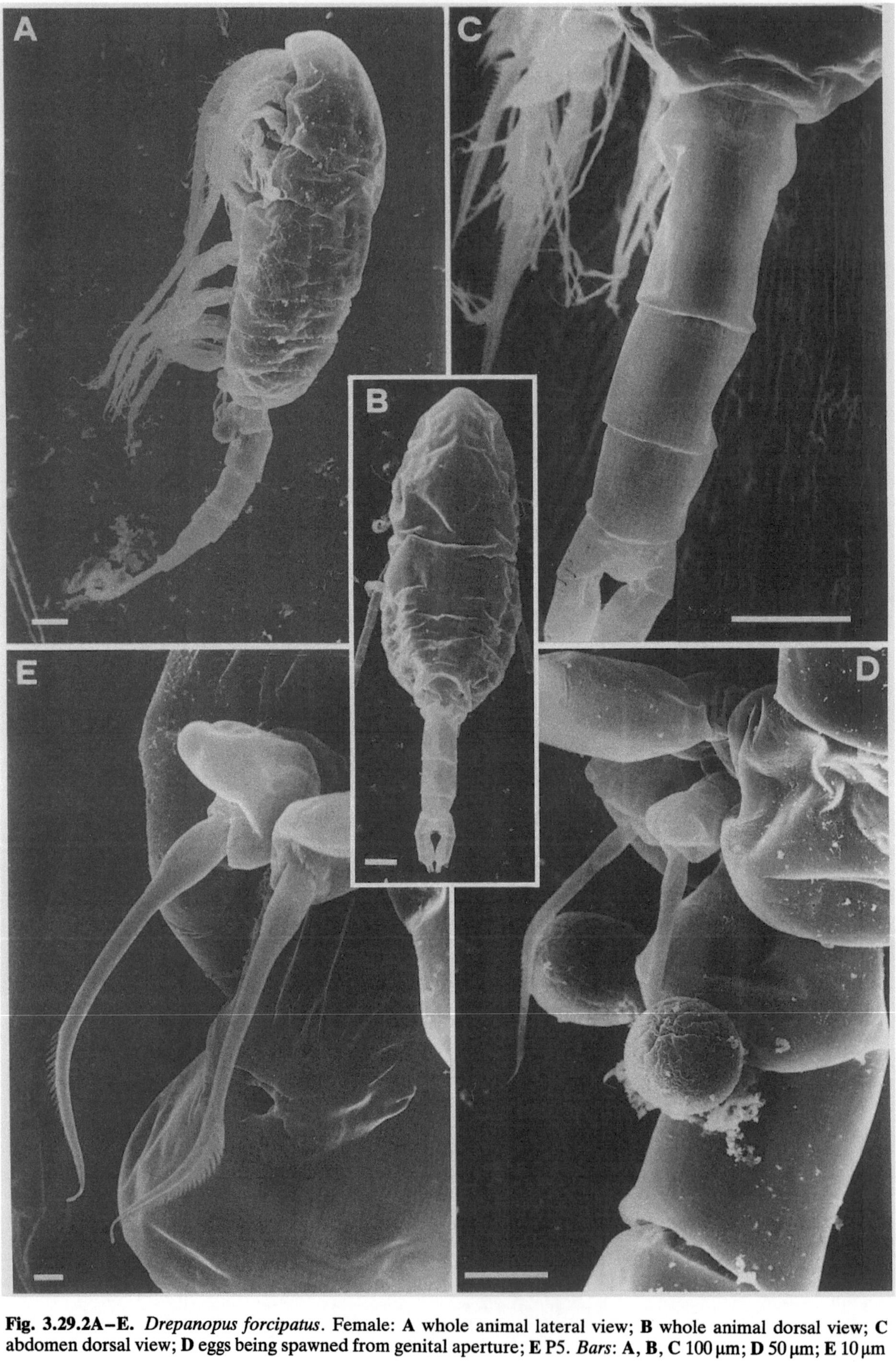

Fig. 3.29.2A–E. *Drepanopus forcipatus*. Female: **A** whole animal lateral view; **B** whole animal dorsal view; **C** abdomen dorsal view; **D** eggs being spawned from genital aperture; **E** P5. *Bars*: **A**, **B**, **C** 100 μm; **D** 50 μm; **E** 10 μm

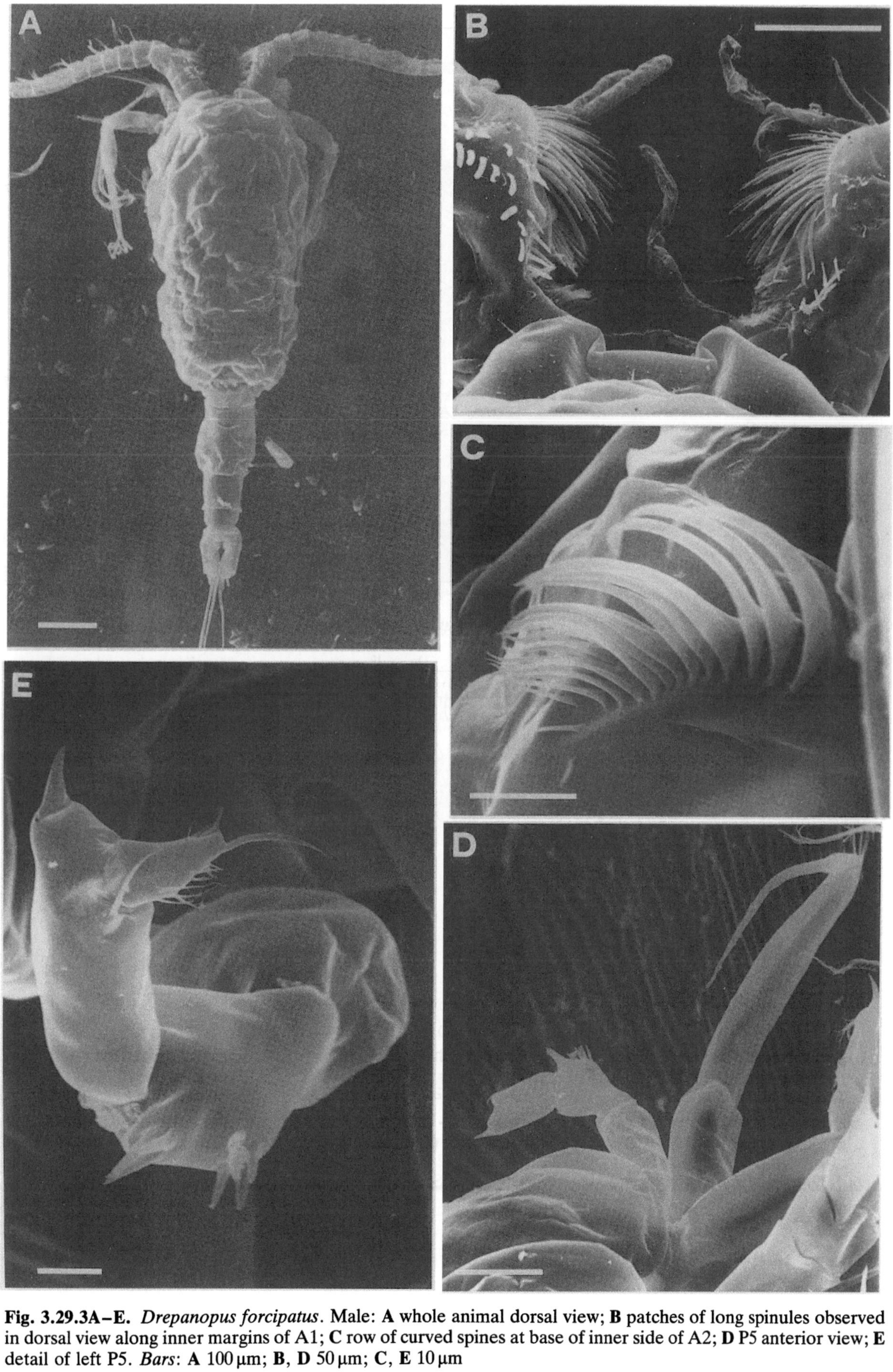

Fig. 3.29.3A–E. *Drepanopus forcipatus*. Male: **A** whole animal dorsal view; **B** patches of long spinules observed in dorsal view along inner margins of A1; **C** row of curved spines at base of inner side of A2; **D** P5 anterior view; **E** detail of left P5. *Bars*: **A** 100 µm; **B**, **D** 50 µm; **C**, **E** 10 µm

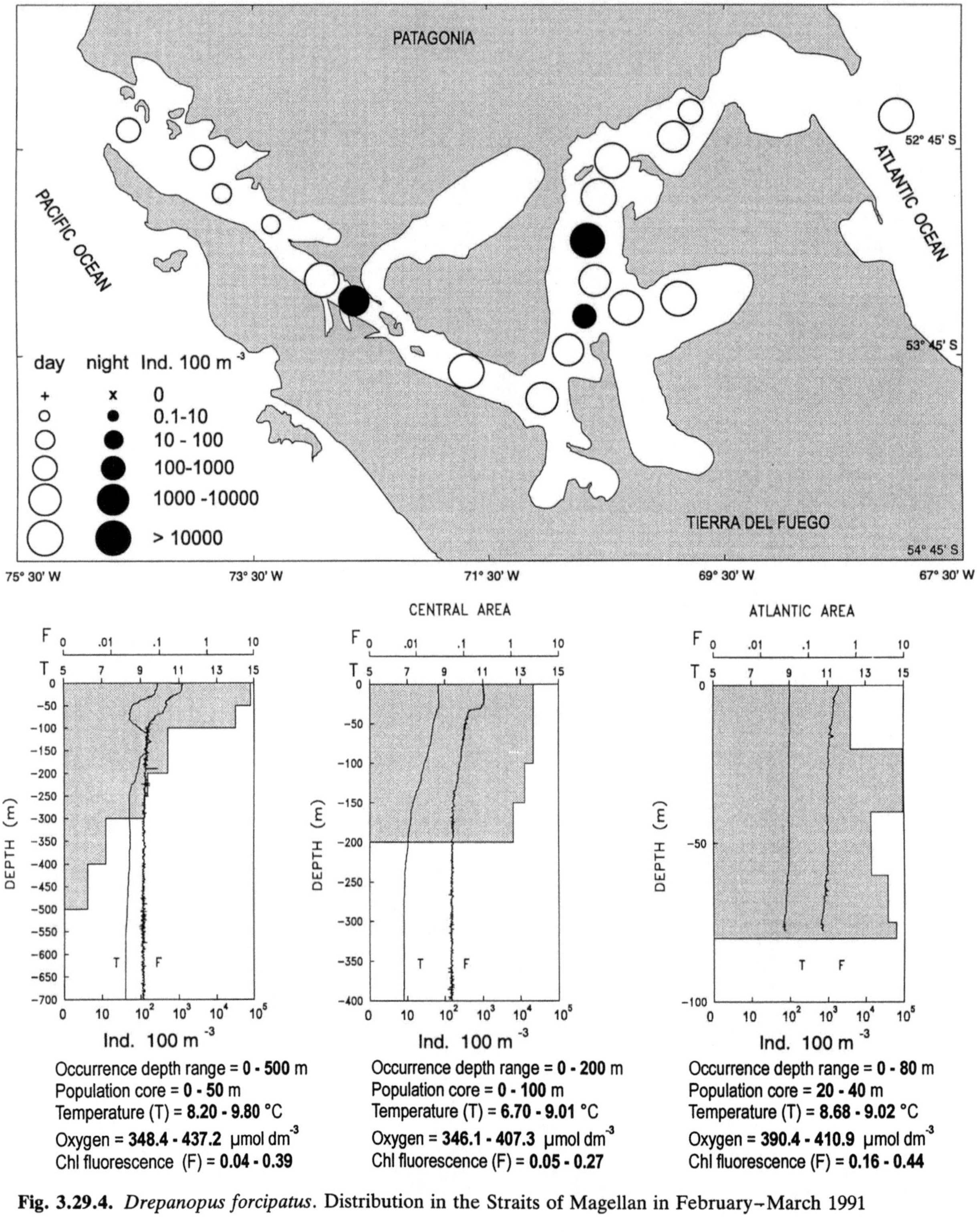

Fig. 3.29.4. *Drepanopus forcipatus*. Distribution in the Straits of Magellan in February–March 1991

3.30 *Microcalanus pygmaeus* Sars 1900

Pseudocalanus pygmaeus Sars, 1900: p. 73, pl. 21 figs. 1–14 – Giesbrecht, 1902: p. 20, pl. 2 figs. 1–5.
Microcalanus pygmaeus Rose, 1933: p. 80, fig. 35 – Brodskii, 1950: p. 116, fig. 37 – Farran and Vervoort, 1951c: p. 3, figs. 3, 4 – Tanaka, 1956b: p. 385, fig. 12; 1960a: p. 35, pl. 13 figs. 1–9 – Bradford, 1971b: p. 16, figs. 10–11 – Brodskii et al., 1983: p. 226, fig. 106.
Spinocalanus longicornis (male) Sars, 1900: p. 77, pl. 22 figs. 13, 14.

Body length –
females: 0.69–0.86 (20 specimens)
males: 0.64–0.84 (17 specimens)

Female. Body small in size. Cephalothorax 2.3 times longer than wide, widest in anterior third and tapering towards head which is somewhat triangular. Postero-lateral corners of last thoracic somite with short rounded processes. A1 reaching half genital somite. Proportional lengths of cephalothorax and abdomen 75:25; proportional lengths of abdominal somites and furca 33:19:16:17:15 = 100. Genital somite conspicuously dilated in middle with distinct pear-shaped seminal receptacles in dorsal view. Exopods of P2–P4 with serrated apical spine. P5 absent.

Male. Body shape slightly different from female, with more pronounced triangular head and more slender urosome. Cephalothorax 2.2 times longer than wide. Proportional lengths of cephalothorax and abdomen 70:30; proportional lengths of abdominal somites and furca 12:30:20:14:13:11 = 100. P5 asymmetrical, uniramous; left leg 5, 6-jointed, right leg 2, 3-jointed, almost half as long as left. Last joint of both right and left legs ending with two spines; left joint also bears long, sparse bristles.

Remarks

The genus was created by Sars (1903) to accommodate a small calanoid collected along the Norwegian coast, *Microcalanus pusillus*. He later added a second species, *M. pygmaeus*, which had been described as *Pseudocalanus pygmaeus*, sampled north of the New Siberian Islands. These two species resemble each other so closely that the distinction between them is very unclear (Farran 1929; Tanaka 1956b). According to Farran and Vervoort (1951c), the two species of *Microcalanus* described by Sars can best be regarded as forms of a variable *M. pygmaeus*. This is the view we adopt here since, in our specimens, although the A1 is short, as that reported for *M. pusillus*, all intermediate forms occur.

Distribution

M. pygmaeus is a cosmopolitan species, recorded from the Arctic, where it is widely distributed (Sars 1900; With 1915; Jespersen 1934), and from the Antarctic (Giesbrecht 1902; Farran 1929; Vervoort 1957; Bradford 1971b) Oceans. Vervoort (1957) summarized the Arctic and Antarctic occurrence of *M. pygmaeus*. In temperate and Antarctic regions, the species inhabits deep waters. Tanaka (1956b) recorded it in Sagami Bay (Japan) from 300–500 m. It was sampled from 1000–3000 m in the Indian Ocean (Grice and Hülsemann 1967) and from 100–2500 m in the NE Atlantic (Grice and Hülsemann 1965). Peterson et al. (1988) reported a "*Microcalanus* species" from Bahia de Concepcion (Chile). Fagetti (1962) reported *M. pusillus* for the Chilean coast. The species was previously reported as *M. pusillus* in the Straits of Magellan (Mazzocchi and Ianora 1991). In the present study, *M. pygmaeus* was sampled at all depths, mainly at Pacific and partly at Central stations.

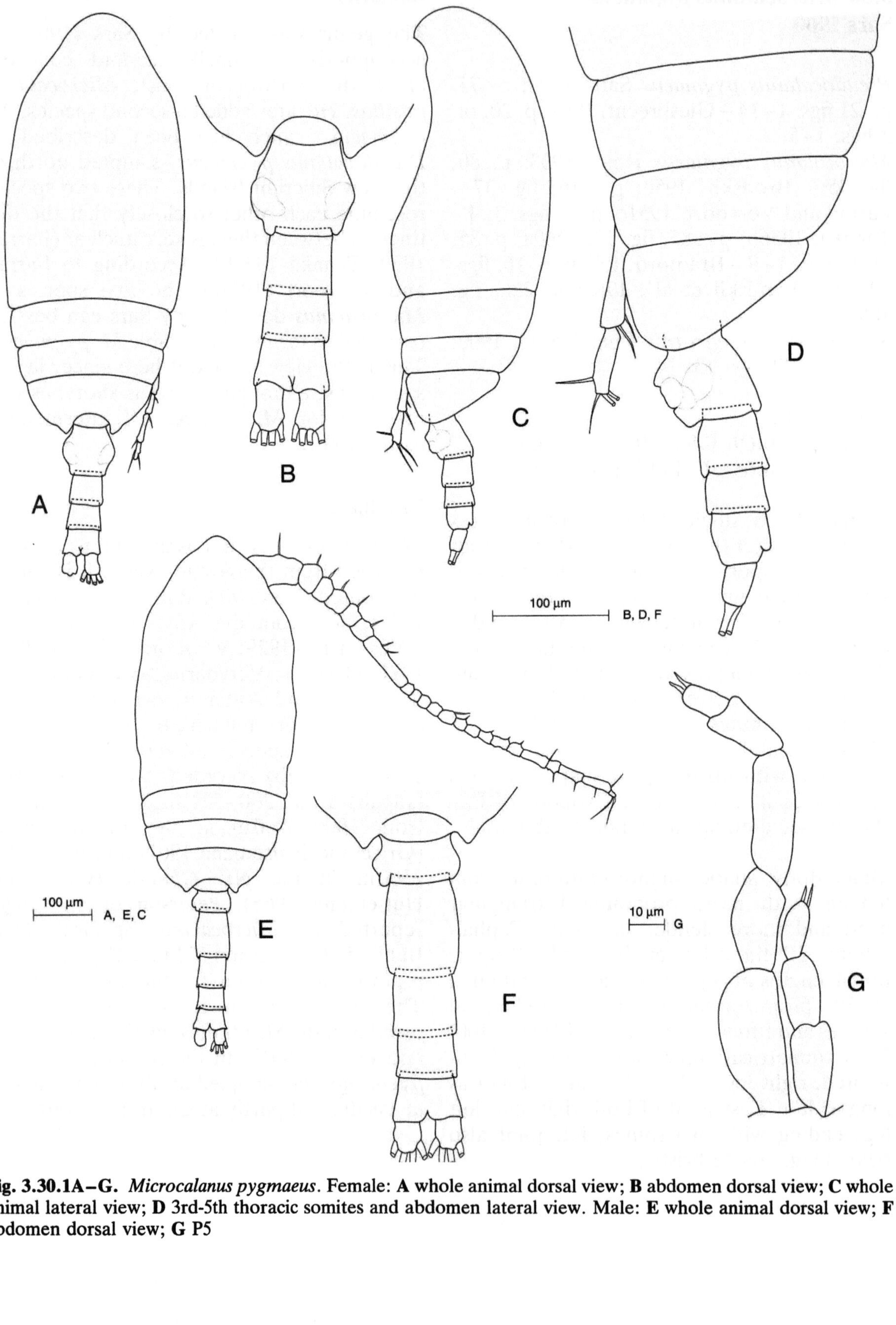

Fig. 3.30.1A–G. *Microcalanus pygmaeus*. Female: **A** whole animal dorsal view; **B** abdomen dorsal view; **C** whole animal lateral view; **D** 3rd-5th thoracic somites and abdomen lateral view. Male: **E** whole animal dorsal view; **F** abdomen dorsal view; **G** P5

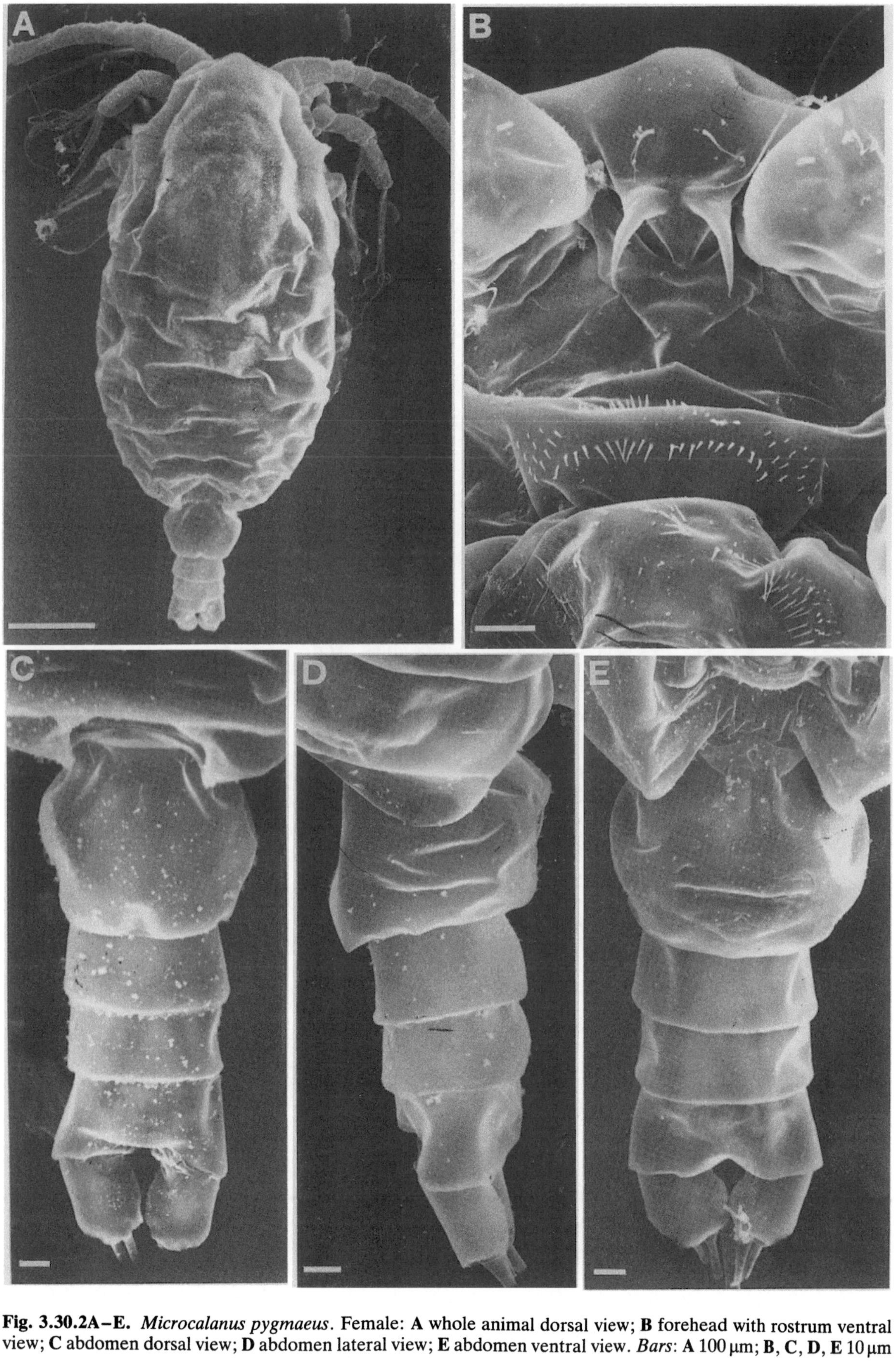

Fig. 3.30.2A–E. *Microcalanus pygmaeus*. Female: **A** whole animal dorsal view; **B** forehead with rostrum ventral view; **C** abdomen dorsal view; **D** abdomen lateral view; **E** abdomen ventral view. *Bars*: **A** 100 μm; **B**, **C**, **D**, **E** 10 μm

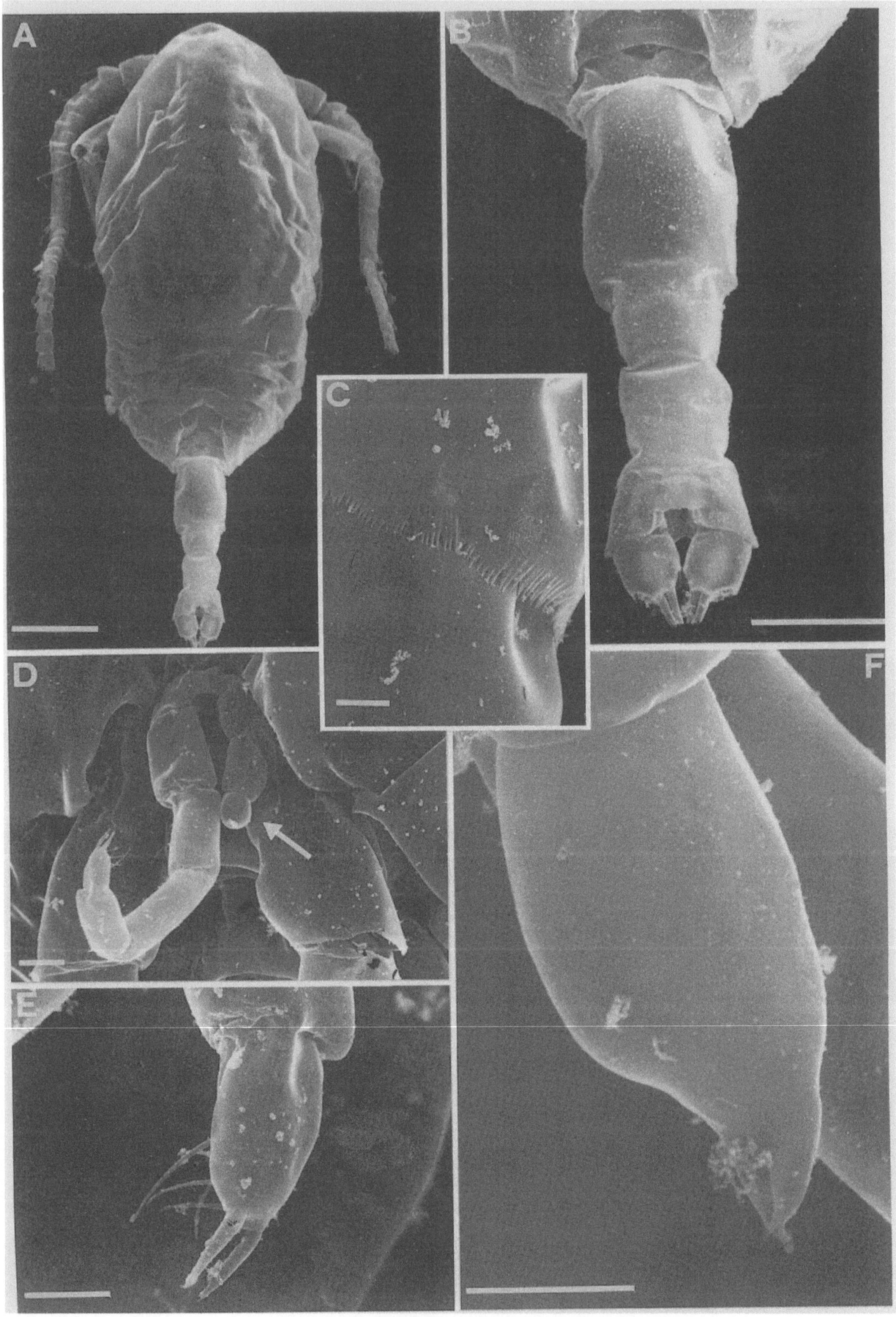

Fig. 3.30.3A–F. *Microcalanus pygmaeus*. Male: **A** whole animal dorsal view; **B** abdomen dorsal view; **C** detail of fringed margins of abdominal somites; **D** P5 posterior side; **E** last joint of left P5; **F** last joint of right P5 (*arrow* in **D**). *Bars*: **A** 100 μm; **B** 50 μm; **C**, **E**, **F** 5 μm; **D** 10 μm

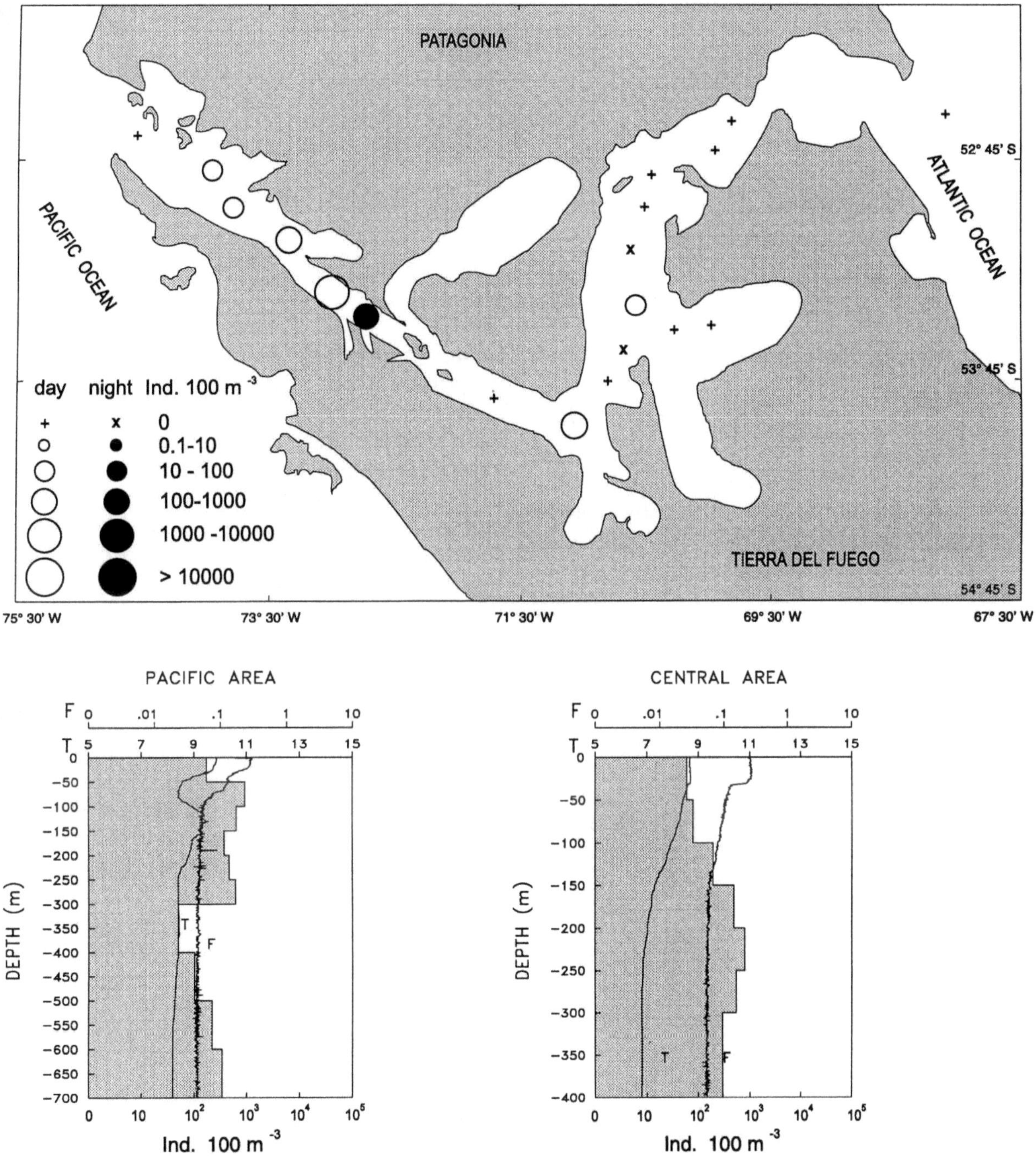

Depth range of occurrence = **0 - 700** m
Population core = **50 - 100** m
Temperature (T) = **8.20 - 9.80** °C
Oxygen = **348.4 - 437.2** µmol dm^{-3}
Chl fluorescence (F) = **0.04 - 0.39**

Depth range of occurrence = **0 - 400** m
Population core = **200 - 250** m
Temperature (T) = **6.70 - 9.01** °C
Oxygen = **346.1 - 407.3** µmol dm^{-3}
Chl fluorescence (F) = **0.05 - 0.27**

Fig. 3.30.4. *Microcalanus pygmaeus*. Distribution in the Straits of Magellan in February–March 1991

3.31 *Paraeuchaeta antarctica* Giesbrecht 1902

Euchaeta antarctica Giesbrecht, 1902: p. 21, pl. 3 figs. 1–8 – Wolfenden, 1908: p. 18, pl. 4 figs. 5, 6 – Bradford, 1971b: p. 21, figs. 64–67 – Ramirez and Dinofrio, 1976: p. 7, pl. 2a fig. 3 – Park, 1978: p. 220, figs. 77, 78 – Ferrari and Dojiri, 1987: p. 458, figs. 1–7 – Fontaine, 1988: p. 32, figs. 3–8.
Pareuchaeta antarctica Farran, 1929: p. 238.
Paraeuchaeta antarctica Bradford, 1981: p. 391, figs. 1A, 5 – Bradford et al., 1983: p. 25, fig. 10.

Body length –
females: 6.51–7.53 mm (20 specimens)
males: 5.14–5.53 mm (20 specimens)

Female. Large copepod, easily recognized by its very characteristic genital somite in lateral view. Proportional lengths of cephalothorax and abdomen 72:28 with cephalothorax 2.7 times longer than wide. Postero-lateral corners of last thoracic somite rounded and extending posteriorly, bearing tuft of long hairs. Proportional lengths of abdominal somites and furca 40:26:25:9 = 100. Genital prominence enlarged, characterized by flap-like structure (lappet) and series of median folds (see Fontaine 1988 for detailed description). 2nd and 3rd abdominal somites covered with short bristles forming characteristic checkerboard pattern on dorsal and lateral surfaces, with long hairs on ventral surface. Posterior margins of abdominal somites 2–3 bordered by row of triangular flaps.

Male. Same body shape as female with proportional lengths of cephalothorax and abdomen 71:29. Postero-lateral corners of last thoracic somite elongated and extending about 1/2 length of first abdominal somite. Proportional lengths of abdominal somite and furca 14:32:25:21:8 = 100. Abdominal somite naked with posterior margins bordered by fringe of "shark-like" teeth. Caudal rami very reduced. P5 with asymmetrical basipods, right one much shorter and stouter than left. Endopod of right leg bearing lobe-like process on external margin at 2/3 length. Re2 of right leg shorter than Rel and with enlarged flat end. Left P5 3-jointed; serrated lamella on distal part of Re2 scoop-shaped, with uniform small teeth along entire internal margin and minute teeth along distal part of external margin. Digitiform process of Re2 short, ear-shaped. Re3 longer than serrated lamella, with long spinules on distal end; conical process at base of Re3 with long hairs at base and tapering into sharp point.

Remarks

Some taxonomists consider the family Euchaetidae to be monogeneric (Vervoort 1957; Park 1978; Fontaine 1988), while others include both genera *Euchaeta* and *Paraeuchaeta* (Heptner 1971; Bradford 1974; Bradford et al. 1983; Razouls 1993). The separation of the two genera *Euchaeta* and *Paraeuchaeta* was first proposed by Scott (1909) on the basis of the structure of the apical setae of the maxillae of the females and of the left exopod in the male P5.

Distribution

P. antarctica is widely distributed in all regions of the Antarctic Ocean. It has been recorded by Fontaine (1988), often in large numbers, in deep-water samples up to 1500 m in the Southern Ocean and the Ross Sea. It was also recorded as one of the dominant deep-water copepods (1000 m) of Croker Passage, Antarctic Peninsula (Hopkins 1985). Ward and Wood (1988) found that its distribution centred around the 250–500 m depth. The species also occurs North of the Convergence in deep waters of the Atlantic, Pacific and Indian Oceans. Park (1978) reports it for the SW Atlantic and Bradford et al. (1983) for the SW Pacific off New Zealand. Marin and Antezana (1985) reported it for Chilean fjords and maintained that its presence indicates the incursion of Antarctic zooplankton into the area. The species has yet to be recorded for the Argentinian coast. Mazzocchi and Ianora (1991) recorded low densities of *P. antarctica*, mainly copepodites, in the central and western stations of the Straits of Magellan. In the present study, it was widely distributed from the surface to maximum depths throughout the Straits.

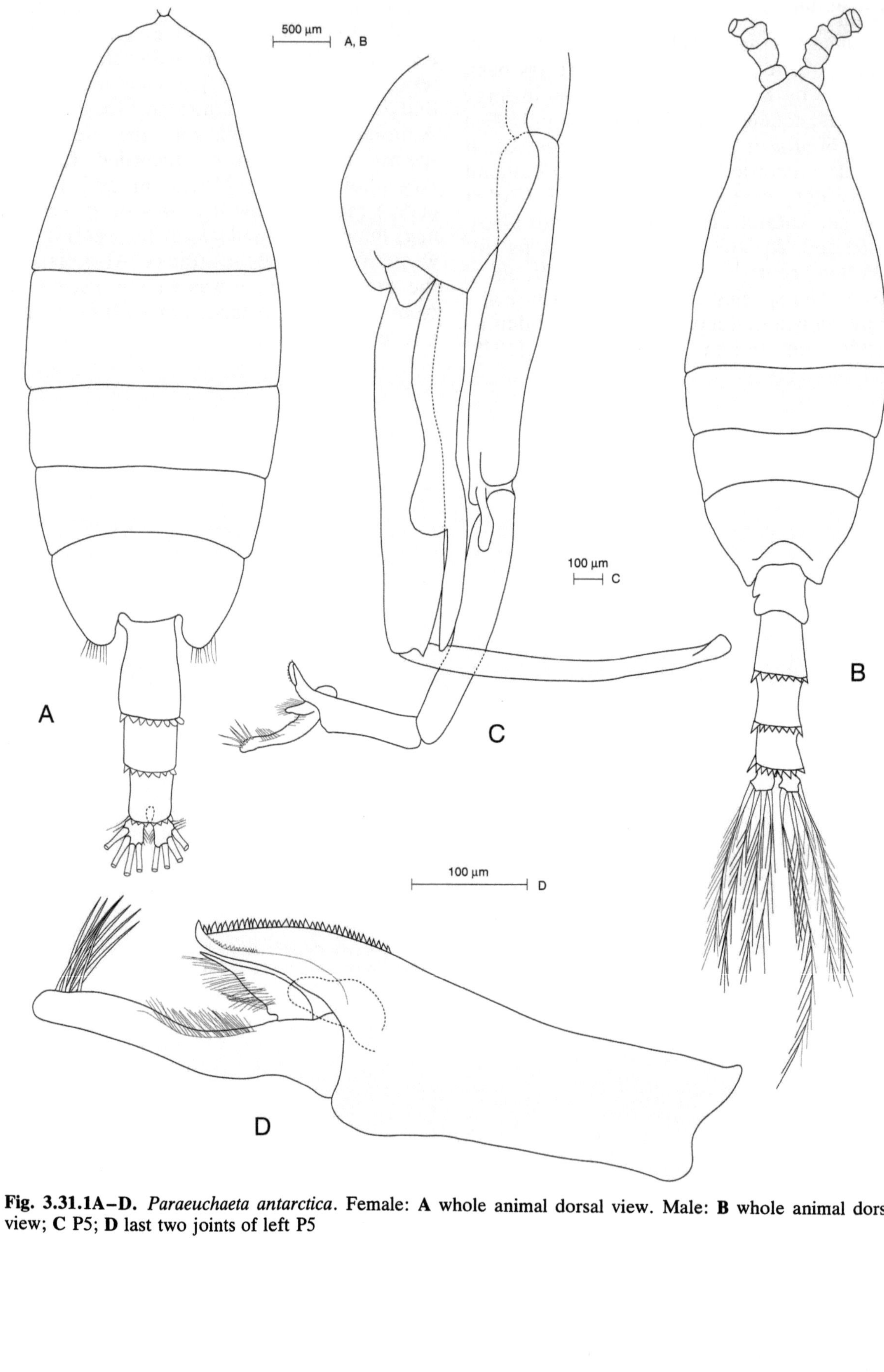

Fig. 3.31.1A–D. *Paraeuchaeta antarctica*. Female: **A** whole animal dorsal view. Male: **B** whole animal dorsal view; **C** P5; **D** last two joints of left P5

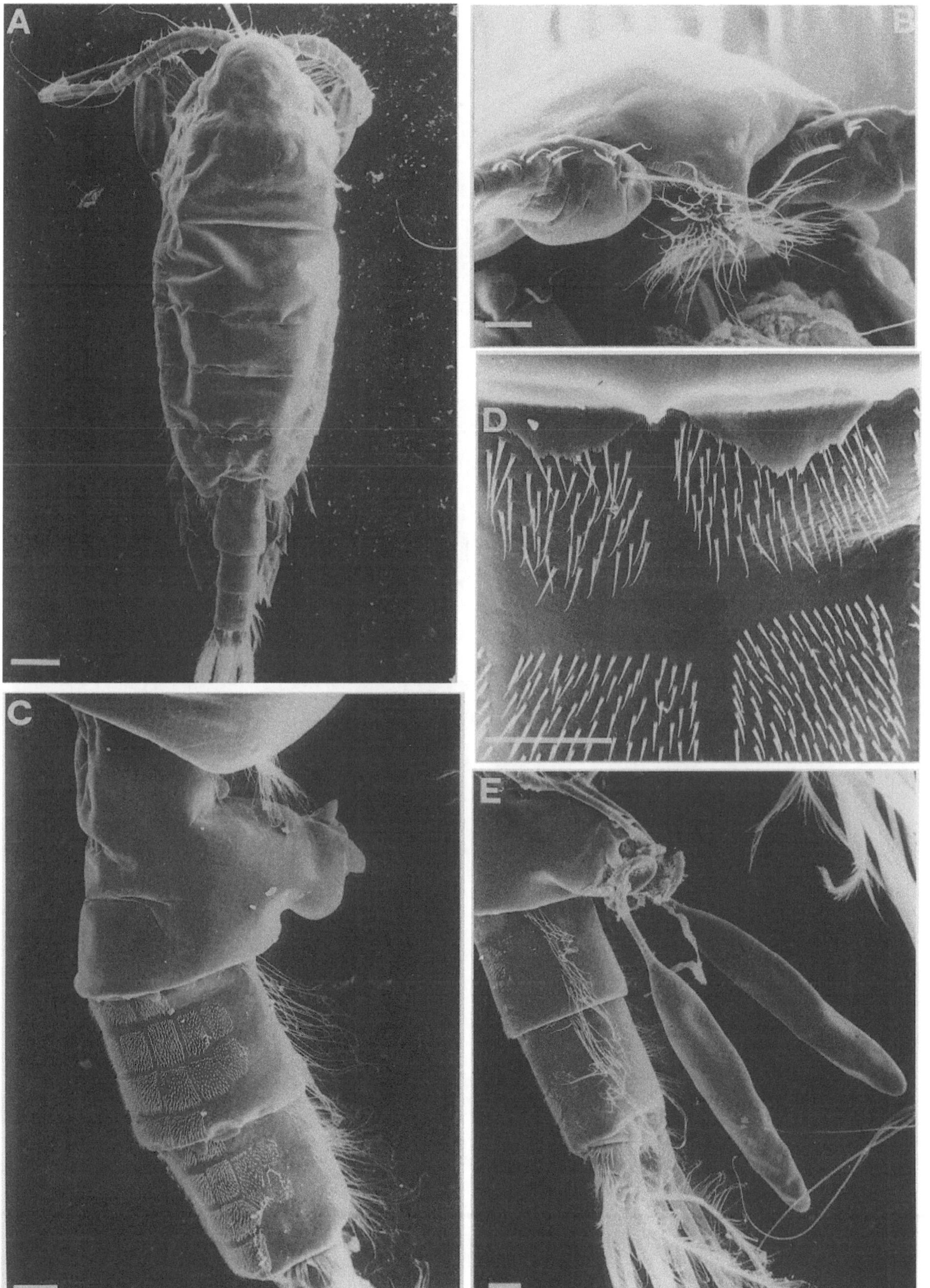

Fig. 3.31.2A–E. *Paraeuchaeta antarctica*. Female: **A** whole animal dorsal view; **B** forehead ventral view; **C** abdomen lateral view; **D** detail of spinulation on 2nd and 3rd abdominal somites; **E** abdomen with spermatophores attached to genital somite. *Bars*: **A** 500 µm; **B**, **C**, **E** 100 µm; **D** 50 µm

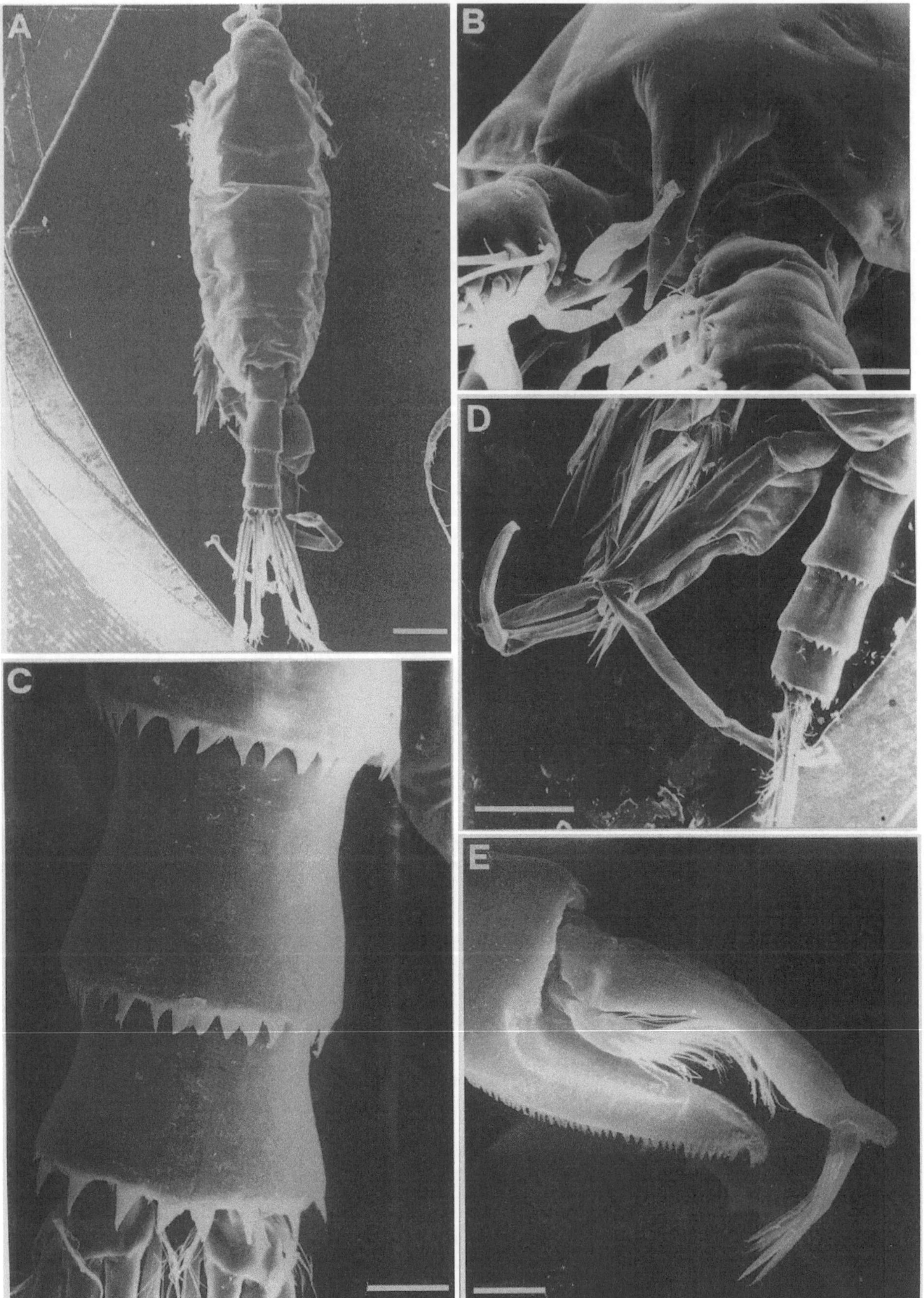

Fig. 3.31.3A–E. *Paraeuchaeta antarctica*. Male: **A** whole animal dorsal view; **B** forehead ventral view; **C** abdomen with strong teeth along posterior margins of 2nd to 4th abdominal somites; **D** P5; **E** detail of serrated lamella on distal part of Re2 and Re3 of left P5. *Bars*: **A**, **D** 500 µm; **B**, **C** 100 µm; **E** 50 µm

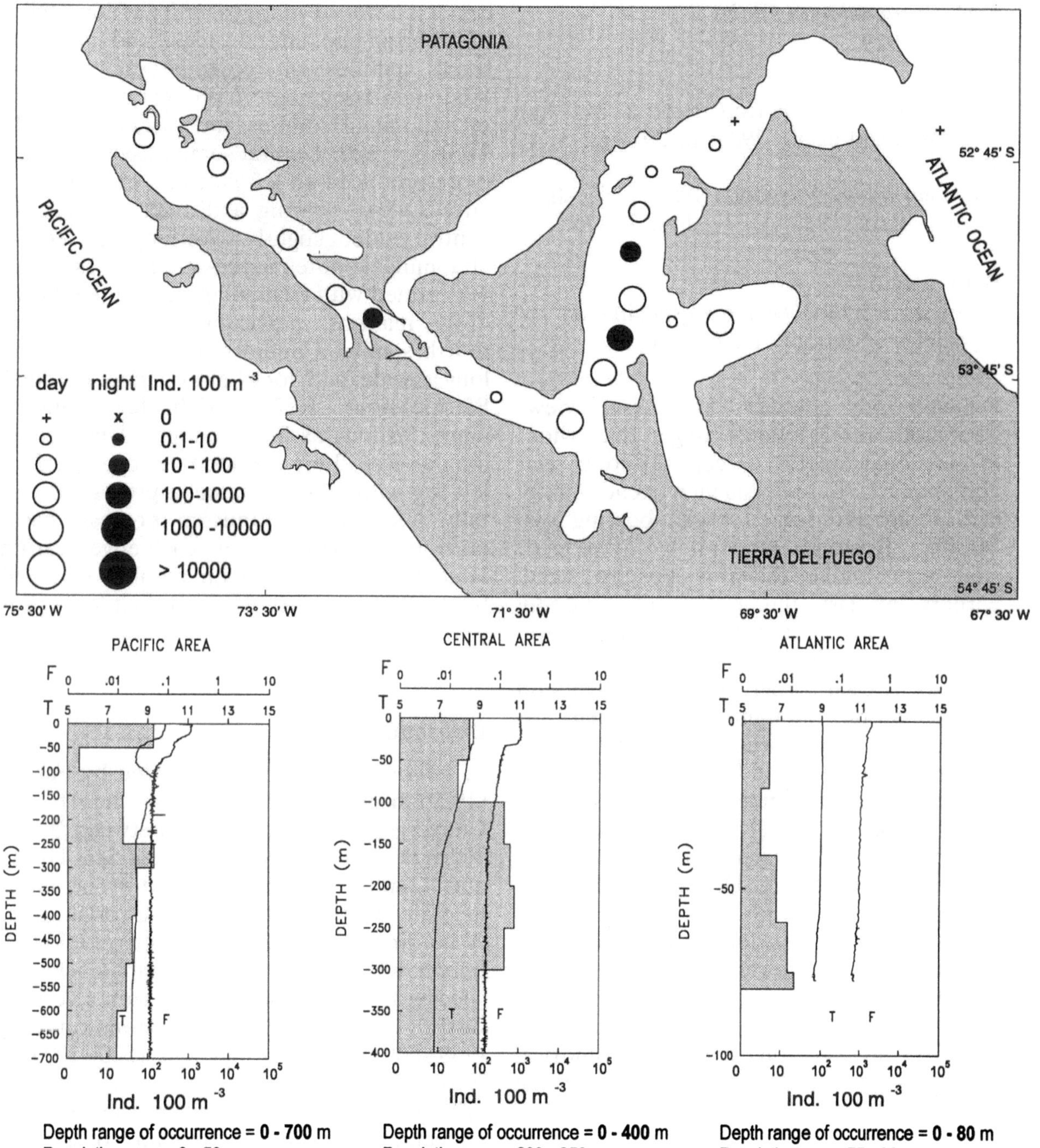

Pacific area	Central area	Atlantic area
Depth range of occurrence = **0 - 700** m	Depth range of occurrence = **0 - 400** m	Depth range of occurrence = **0 - 80** m
Population core = **0 - 50** m	Population core = **200 - 250** m	Population core =**75 - 80** m
Temperature (T) = **8.20 - 9.80** °C	Temperature (T) = **6.70 - 9.01** °C	Temperature (T) = **8.68 - 9.02** °C
Oxygen = **348.4 - 437.2** µmol dm^{-3}	Oxygen = **346.1 - 407.3** µmol dm^{-3}	Oxygen = **390.4 - 410.9** µmol dm^{-3}
Chl fluorescence (F) = **0.04 - 0.39**	Chl fluorescence (F) = **0.05 - 0.27**	Chl fluorescence (F) = **0.16 - 0.44**

Fig. 3.31.4. *Paraeuchaeta antarctica*. Distribution in the Straits of Magellan in February–March 1991

3.32 *Paraeuchaeta biloba* Farran 1929

Pareuchaeta biloba Farran, 1929: p. 242, fig. 11 – Bradford et al., 1983: p. 29, figs. 3, 4, 81.
Euchaeta biloba Vervoort, 1957: p. 79, figs. 60–68 – Park, 1978: p. 217, figs. 74–76.

Body length –
females: 5.43–5.64 mm (3 specimens)
male: 5.00 mm (1 specimen)

Female. Body slender in dorsal view. Cephalothorax 2.5 times longer than wide. Proportional lengths of cephalothorax and abdomen 62:38. In dorsal view, head slightly protruding into frontal organ bearing two sensilla. Rostrum moderately developed, slightly protruding forwards. Postero-lateral corners of last thoracic somite broadly rounded both in dorsal and lateral views. Patches of hairs of different length and width on apex of each corner; tuft of long hairs on postero-lateral margins. Abdomen 4-segmented; proportional lengths of abdominal somites and furca 38:24:24:2:12 = 100. Genital somite long and slender, symmetrical in dorsal view. Genital prominence bilobed in lateral view; hairy distal part of ventral surface. Dorsal margins of all abdominal somites with small spinules. Third somite with sparse, long hairs on dorsal surface. Anal somite very short. Caudal rami about twice as long as wide, with long hairs on both inner and outer margins. Inner apical caudal setae originating from short cylindrical processes, geniculated and much longer than terminal setae.

Male. Body more slender than female. Cephalothorax 3 times longer than wide. Proportional lengths of cephalothorax and abdomen 90:41. Head triangular in dorsal view with prominent frontal organ bearing two sensilla. Rostrum slender and slightly longer than in female. Postero-lateral borders of last thoracic somite rounded, slightly prolonged in dorsal view, with tooth-like processes on dorsal margins, tufts of long hairs on ventral and lateral sides, and patch of small spinules on dorso-lateral margins. Abdomen 5-segmented; proportional lengths of abdominal somites and furca 22:29:24:17:2:6 = 100. Genital somite asymmetrical, more produced on left side in dorsal view; in lateral view, bearing a distinct tubercle on ventral side, slightly overlapping second abdominal somite. Successive somites (2nd–4th) armed with triangular teeth along dorsal distal margins, particularly developed on fourth somite. Caudal rami more or less as long as wide, with long hairs on inner borders. P5 biramous. Re2 of right leg styliform almost as long as first; endopod 1-jointed as long as Re1. Left endopod vestigial. Re2 of left leg with serrated lamella prolonged distally into strong spiniform process; basal portion of lamella hollow, each edge bearing large tooth together with number of small teeth. Re2 also with grooved digitiform process. Re3 hollow, ending in acute point and ornated with tuft of very strong, long bristles.

Distribution

P. biloba was originally recorded by Farran (1929) in surface waters north of the Atlantic Convergence and at depths between 0 and 400 m at the entrance of the Ross Sea. Hardy and Gunther (1935) reported it at intermediate depths in the South Georgia area. According to Ward and Wood (1988), *P. biloba* was one of the dominant euchaetid species around South Georgia with a vertical distribution centered in epi-mesopelagic (250–500 m) and mesopelagic (500–1000 m) waters, both in summer and winter. Vervoort (1957) found this species in the Banzare collections between 45°10′S and 65°27′S in the Indian Ocean and also south of Tasmania. Park (1978) reported *P. biloba* as one of the most common euchaetid species in the Pacific (between 39°24′S and 68°40′S) and Atlantic (between 34°43′S and 57°04′S) Oceans. Bradford et al. (1983) recorded this species in the SW Pacific off New Zealand down to 1000 m. *P. biloba* inhabits surface and intermediate waters of Antarctic and sub-

Antarctic regions and often penetrates north beyond the sub-Antarctic, probably in the Antarctic Intermediate Current (Park 1978). In the present study, *P. biloba* was sampled only from 200–400 m at two stations in the Pacific area of the Straits of Magellan.

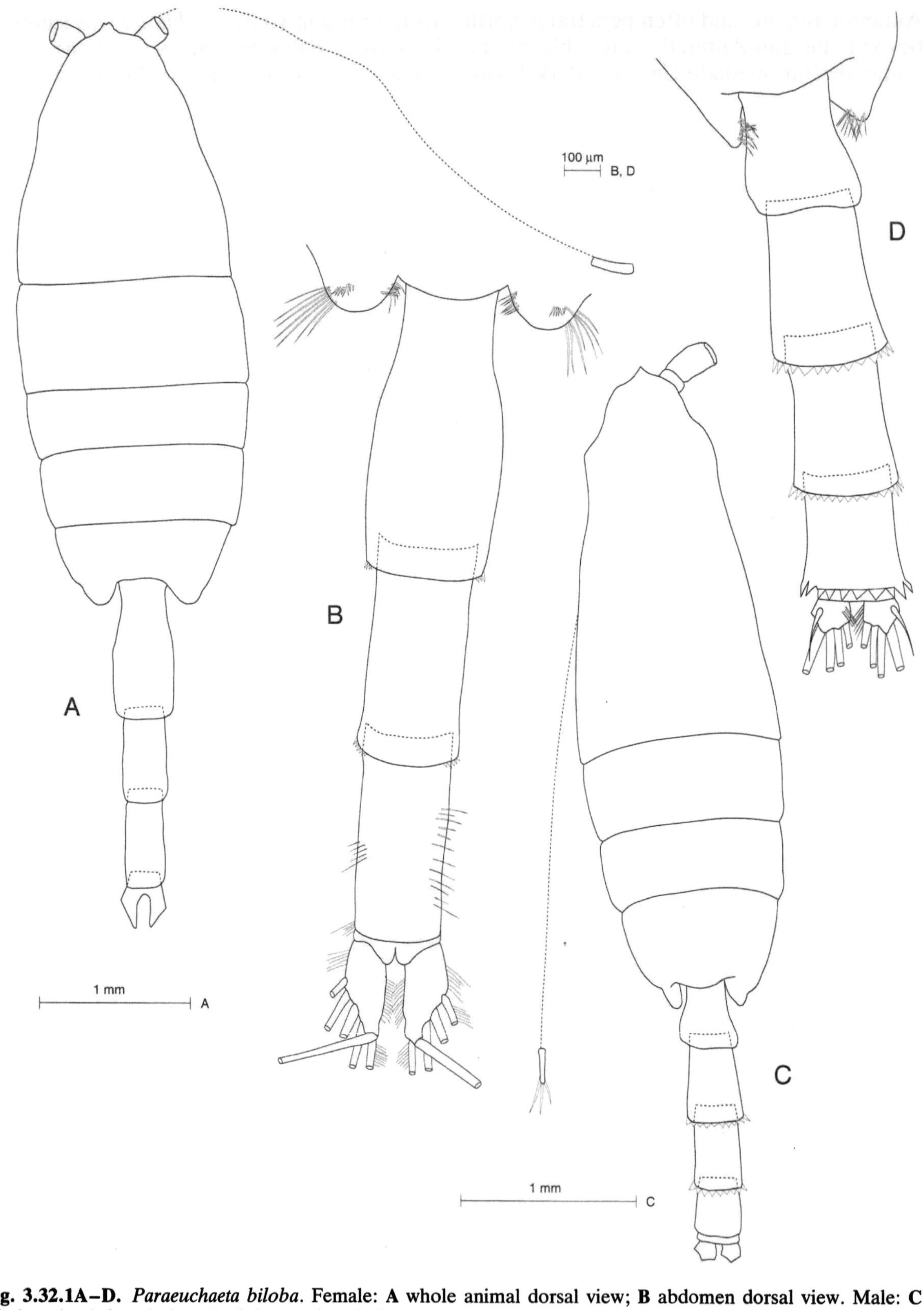

Fig. 3.32.1A–D. *Paraeuchaeta biloba*. Female: **A** whole animal dorsal view; **B** abdomen dorsal view. Male: **C** whole animal dorsal view; **D** abdomen dorsal view

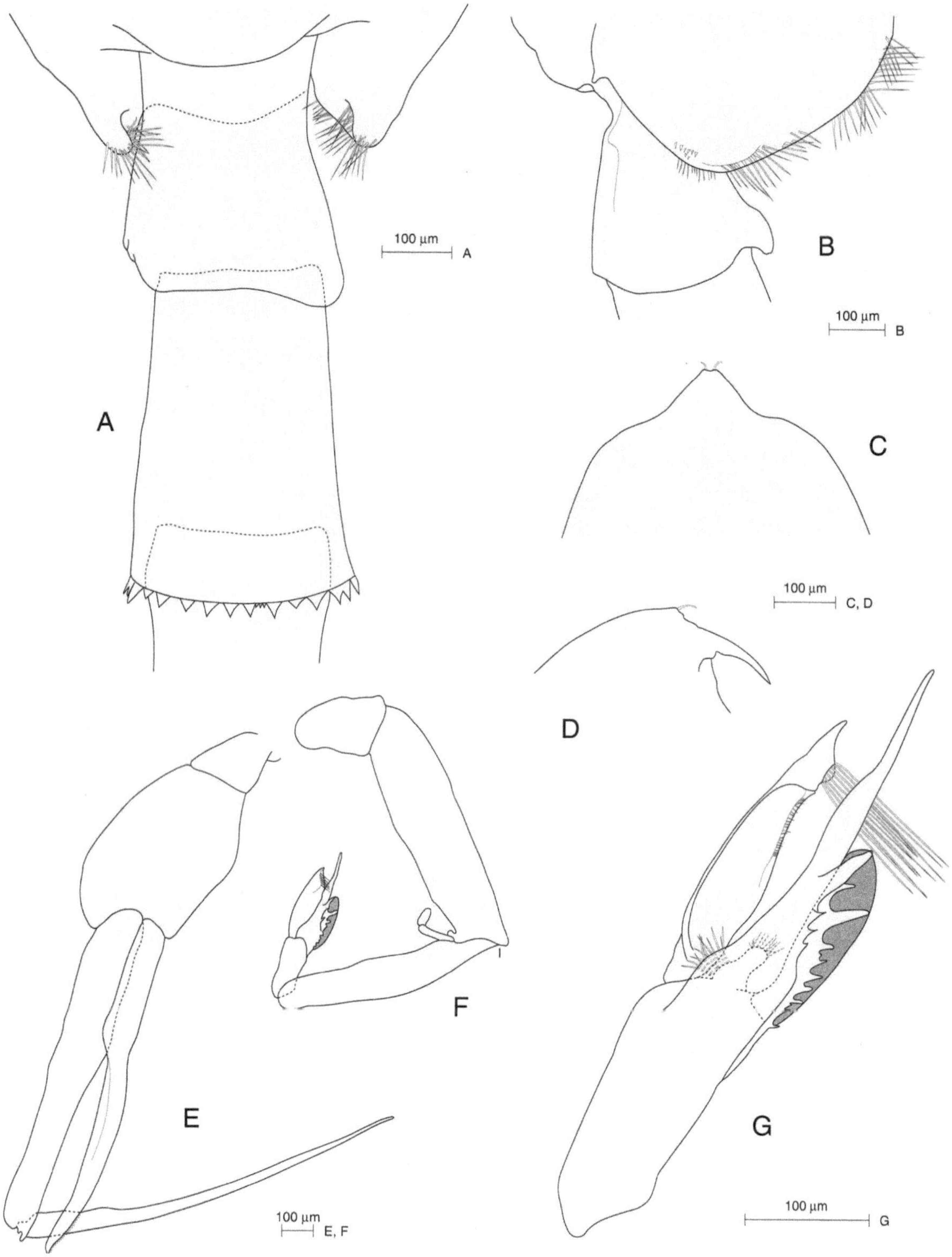

Fig. 3.32.2A–G. *Paraeuchaeta biloba*. Male: **A** 1st and 2nd abdominal somites dorsal view; **B** 1st abdominal somite lateral view; **C** forehead dorsal view; **D** forehead lateral view; **E** right P5; **F** left P5; **G** detail of last joint of left P5

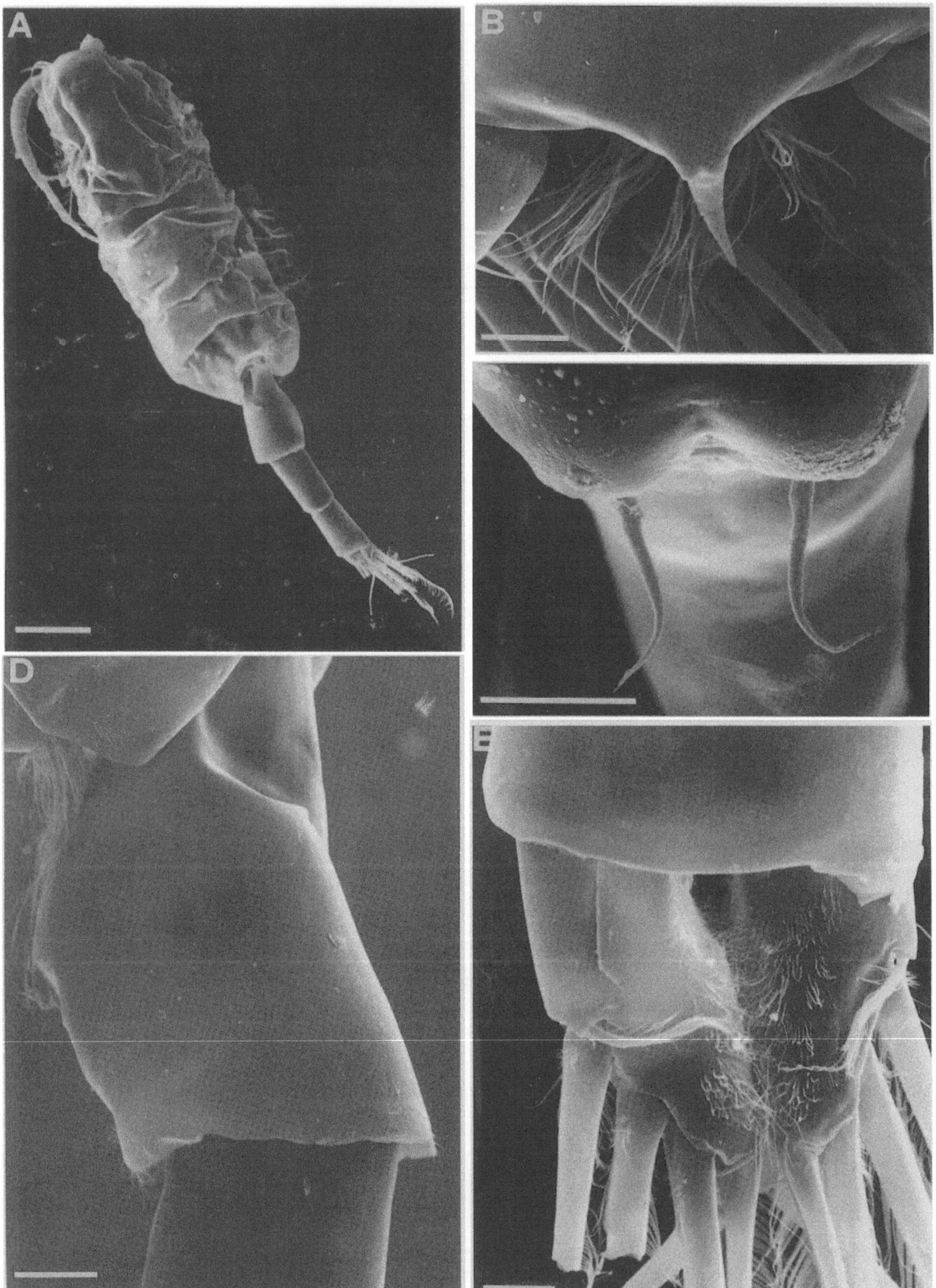

Fig. 3.32.3A–E. *Paraeuchaeta biloba*. Female: **A** whole animal dorsal view; **B** forehead ventral view; **C** sensilla on rostrum; **D** left postero-lateral corner of last thoracic somite and 1st abdominal somite; **E** furca ventral view. *Bars*: **A** 500 µm; **B**, **E** 50 µm; **C** 10 µm; **D** 100 µm

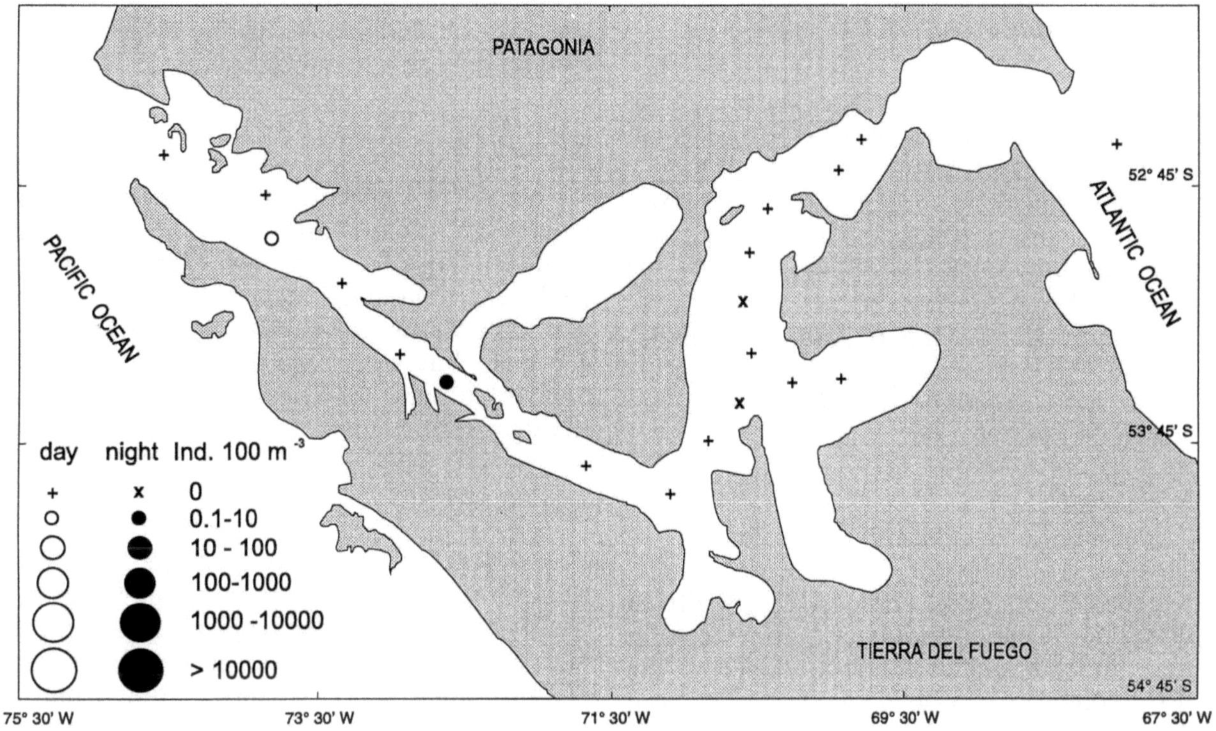

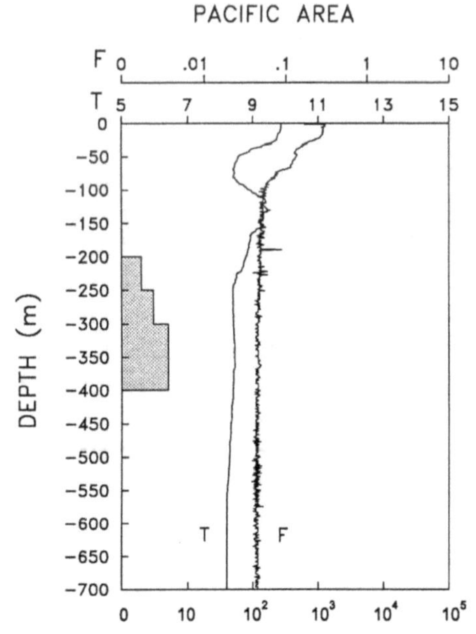

Depth range of occurrence = **200 - 400** m
Population core = **300 - 400** m
Temperature (T) = **8.20 - 9.80** °C
Oxygen = **348.4 - 437.2** µmol dm^{-3}
Chl fluorescence (F) = **0.04 - 0.39**

Fig. 3.32.4. *Paraeuchaeta biloba*. Distribution in the Straits of Magellan in February–March 1991

3.33 *Phaenna spinifera* Claus 1863

Phaenna spinifera Claus, 1863: p. 189, pl. 31 figs. 1–7 – Giesbrecht, 1892: p. 293, pl. 5 fig. 3, pl. 12 figs. 1–8, 35–37, pl. 37 figs. 17–21 – Tanaka, 1960b: p. 85, fig. 81 (immature female) – Grice, 1962: p. 203, pl. 15 fig. 1 – Owre and Foyo, 1967: p. 57, figs. 91, 292, 293 (male) – Vidal, 1968: p. 30, pl. 14 figs. 5–9 – Bradford et al., 1983: p. 67, figs 34, 35 (male) – Park, 1983: p. 318, figs. 1, 2.
For a more complete list of synonyms see Vervoort (1965b).

Body length –
females: 2.88–3.02 mm (4 specimens)

Female. Body robust and rounded. Rostrum bifurcate. Cephalothorax 1.6 times longer than wide, larger in correspondence to the first pedigerous somite. Cephalothorax about 4 times as long as abdomen. Head and first thoracic somite separated. Thoracic somites 4 and 5 incompletely separated; lateral side of somites 3 and 4 with a rather long sensillum on either side, also visible in dorsal view. Posterior corners of thorax with an angular distal projection in dorsal view and, in lateral view, with a round distal margin covering at least 1/3 the anterior genital somite. Abdomen 3-segmented. Proportional lengths of abdominal somites and furca 49 : 18 : 13 : 20 = 100. Ventral surface of abdominal somites ornated with stout setae, clearly visible in lateral view; this ornamentation shown in Giesbrecht (1892; pl. 37, fig. 19), is fragile and can easily be lost in fixed specimens. Caudal rami short, as long as wide. P5 absent.

Remarks

As noted by Vervoort (1965b), there seems to be a great variability in the length of the adult females of this species. Our specimens (only 4 adult females) are longer than generally reported in the literature (1.6–2.5 mm); but with regards to the morphological features, our specimens concord with the descriptions given in the literature.

Distribution

P. spinifera was originally described from the Mediterranean Sea but is now known to be widely distributed in the tropical, sub-tropical and temperate regions of all oceans. Vervoort (1965b) summarized its distribution from a rich, but dispersed literature, and suggested that this species is a fairly regular inhabitant of intermediate waters of the warmer parts of the Atlantic, Pacific and Indian Oceans, capable of rapid diurnal vertical migrations. In the Atlantic, this species penetrates as far south as North Island, New Zealand (Farran 1929). Björnberg (1973) and Bradford et al. (1983), respectively, found it in the eastern and western South Pacific as far south as 42°S. Park (1983) collected two females close to the southern coast of Western Australia. *P. spinifera* was reported from Chilean waters by Fagetti (1962) and Vidal (1968). In the present study, it was sampled only from 100–250 m at two stations in the Pacific area of the Straits of Magellan.

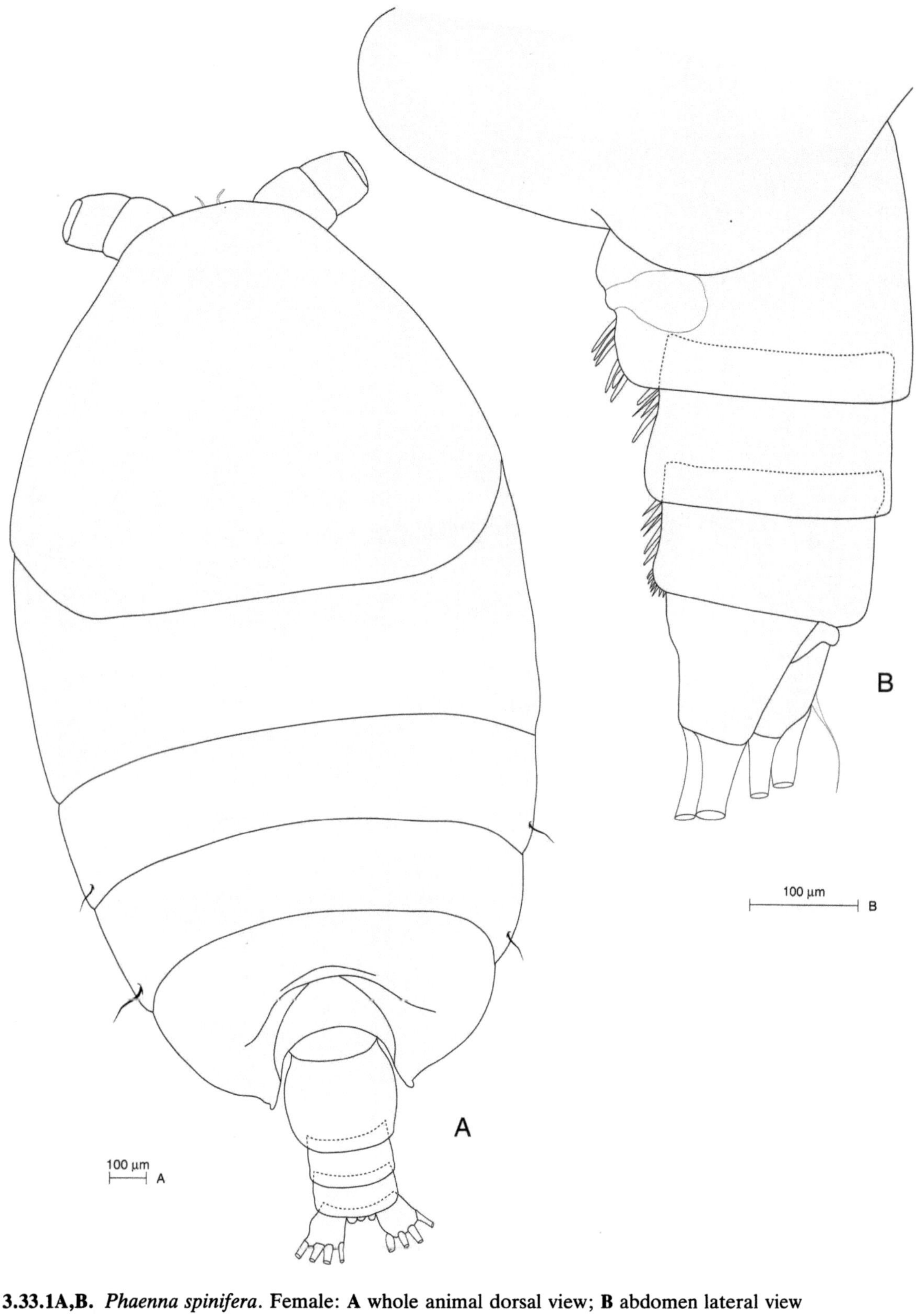

Fig. 3.33.1A,B. *Phaenna spinifera*. Female: **A** whole animal dorsal view; **B** abdomen lateral view

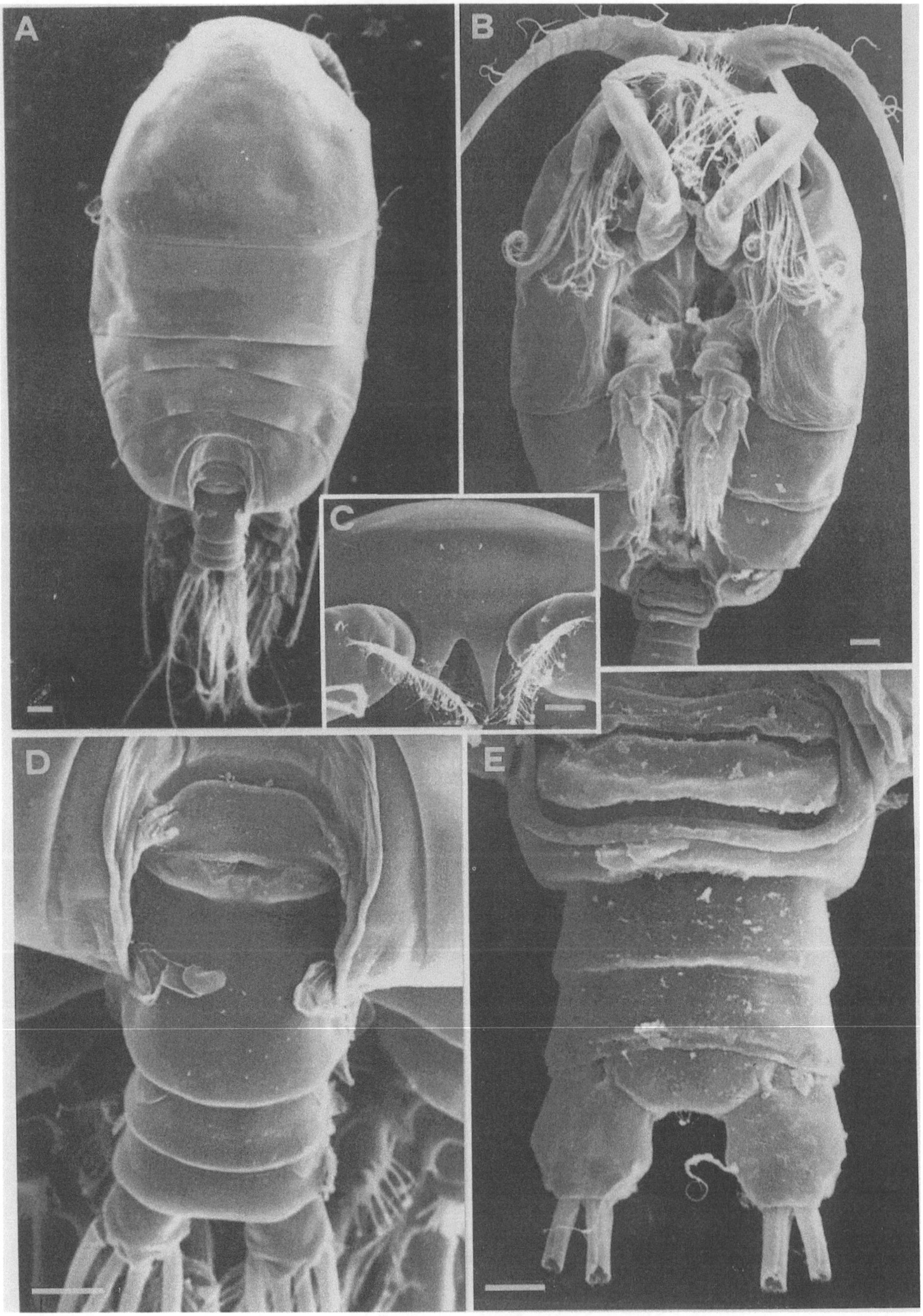

Fig. 3.33.2A–E. *Phaenna spinifera*. Female: **A** whole animal dorsal view; **B** cephalothorax ventral view; **C** bifid rostrum ventral view; **D** abdomen dorsal view; **E** abdomen ventral view. *Bars*: **A, B, D** 100 µm; **C, E** 50 µm

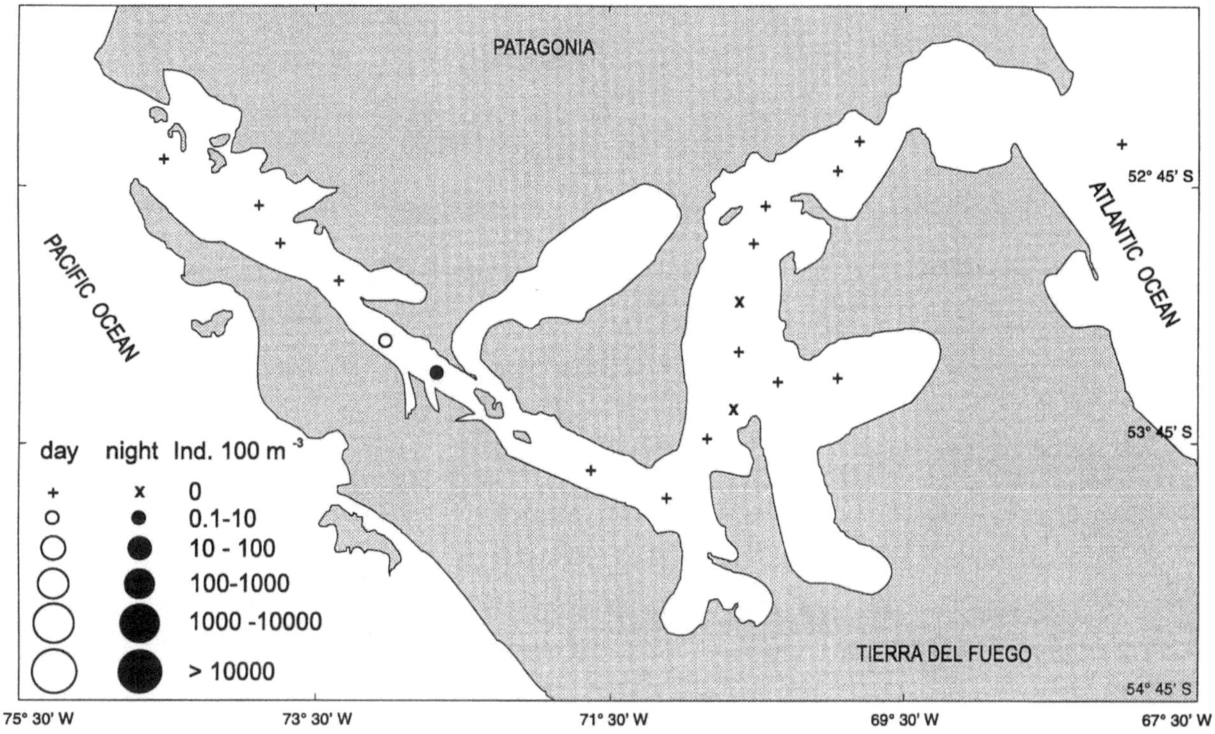

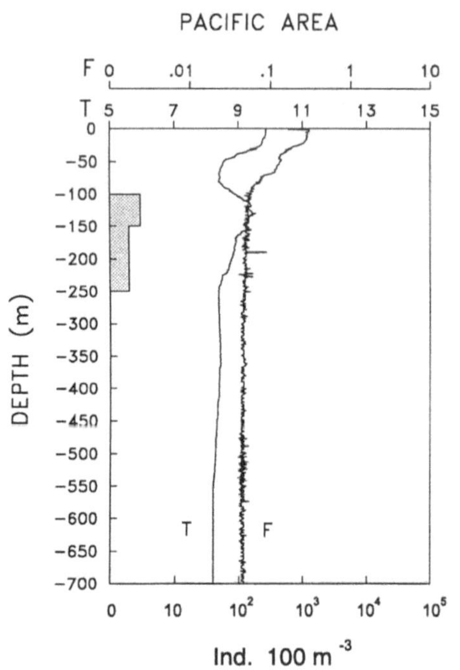

Depth range of occurrence = **100 - 250** m
Population core = **100 - 150** m
Temperature (T) = **8.20 - 9.80** °C
Oxygen = **348.4 - 437.2** µmol dm^{-3}
Chl fluorescence (F) = **0.04 - 0.39**

Fig. 3.33.3. *Phaenna spinifera*. Distribution in the Straits of Magellan in February-March 1991

3.34 *Scaphocalanus curtus* Farran 1926

Scolecithrix curta Farran, 1926: p. 259, pl. 7 figs. 1–3, 5, 6.
Scaphocalanus curtus Farran, 1929: p. 250 – Rose, 1933: p. 149, fig. 155 – Tanaka, 1961: p. 183, fig. 124 – Park, 1970: p. 497, figs. 103–112 – Hure and Scotto di Carlo, 1968: p. 2, figs. 1, 2.
Scaphocalanus subcurtus Park, 1970: p. 499, figs. 113–119 – Björnberg, 1963: p. 41, fig. 22.

Body length –
females: 1.12–1.23 mm (17 specimens)

Female. Body slender, ovoid. Cephalothorax 2.6 times longer than wide. Head fused with first thoracic somite; last two thoracic somites incompletely separated. Proportional lengths of cephalothorax and abdomen 76:24. Posterior corners of thorax blunty angular in dorsal and lateral views. A1 with transparent strip along posterior edge, starting from third segment. Proportional lengths of abdominal somites and furca 34:18:18:10:20 = 100. Caudal rami twice as long as wide. Endopod of P1 1-jointed and bearing 4 setae; Re2 with four small spines on posterior surface. P5 absent.

Remarks

This species was originally described by Farran (1926) as *Scolecithrix*. In spite of the absence of the P5, the same Author (Farran 1929) included this species in the genus *Scaphocalanus* due to its affinities with this genus. The specimens we recorded in the Straits of Magellan concord in size range and morphological details with those reported in the literature. We did not record adult males.

Distribution

S. curtus is widely distributed in all ocean areas. It has been reported from surface and intermediate waters off New Zealand (Farran 1929) and deep waters of the Bay of Biscay (Farran 1926) and Sagami Bay (Tanaka 1961). Its presence has also been recorded in the Caribbean Sea and Gulf of Mexico (Park 1970), the Gulf of Guinea (Vervoort 1965b) and in the SW Atlantic, off Brazil (Björnberg 1963). The species is very common in the Bay of Naples and in the South Adriatic (Mediterranean Sea) from 100 to 600 m (Hure and Scotto di Carlo 1968). In the present study, *S. curtus* was sampled from 100–500 m at only two stations in the Pacific area of the Straits of Magellan.

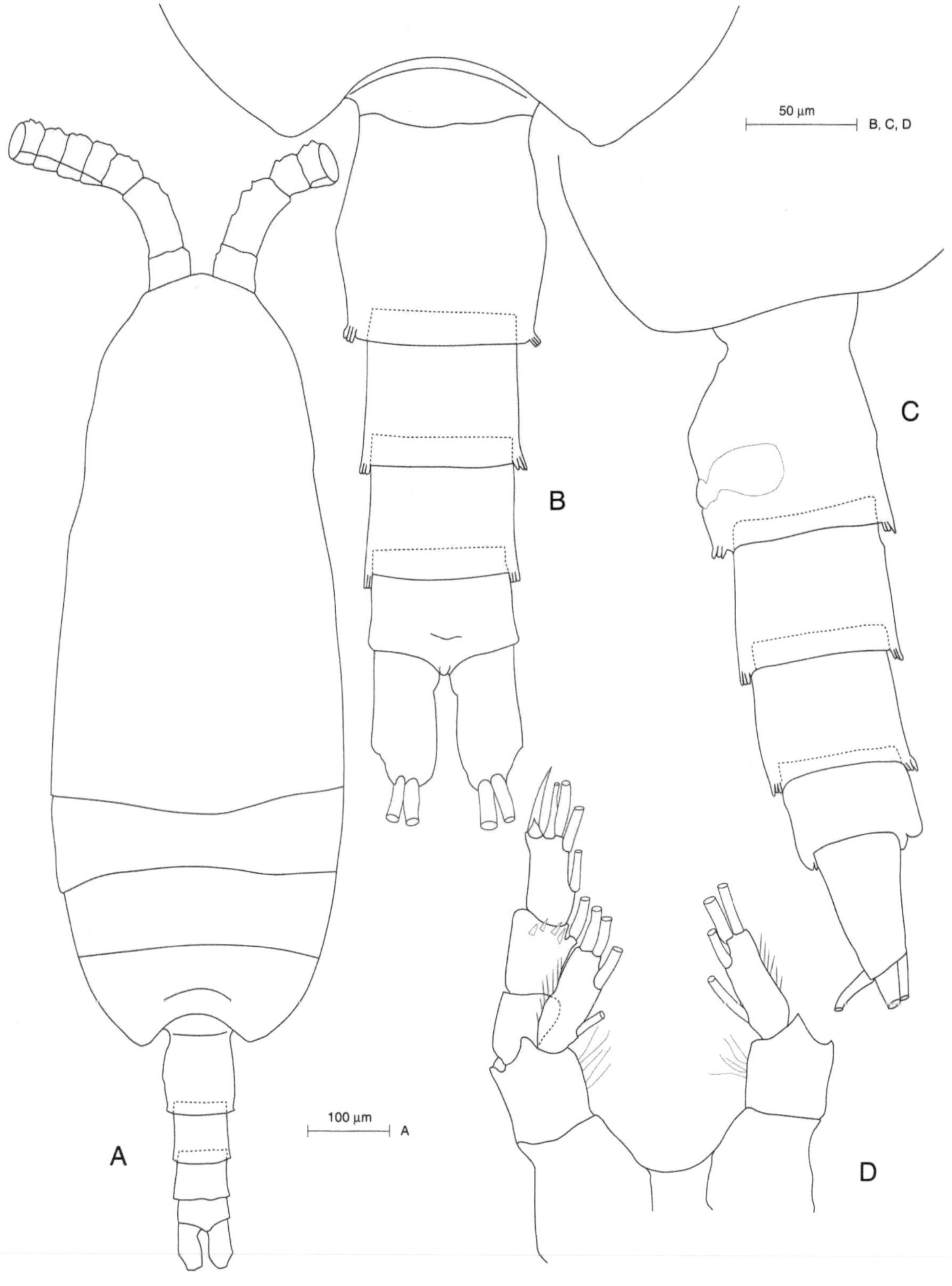

Fig. 3.34.1A–D. *Scaphocalanus curtus*. Female: **A** whole animal dorsal view; **B** abdomen dorsal view; **C** abdomen lateral view; **D** P1

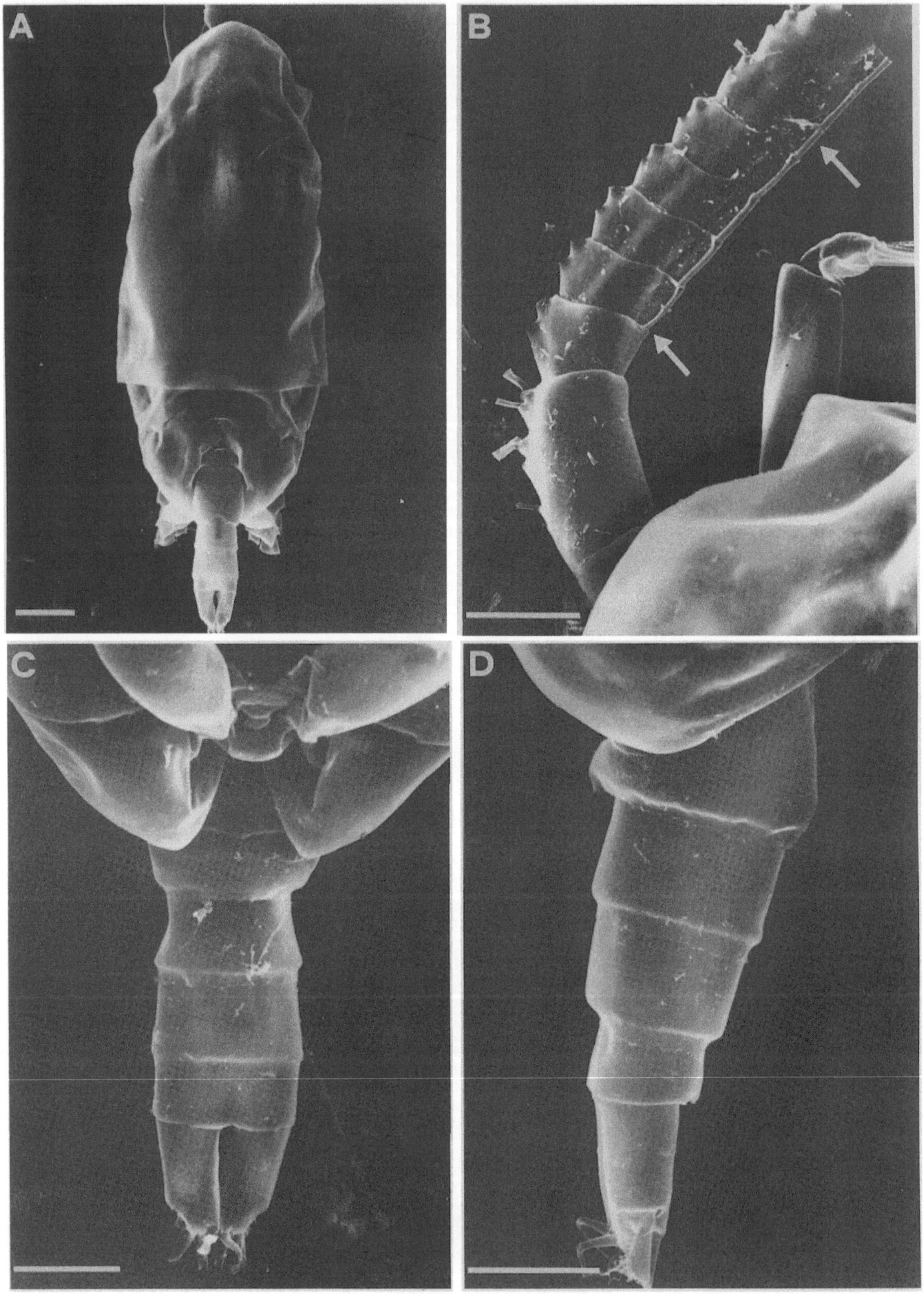

Fig. 3.34.2A–D. *Scaphocalanus curtus*. Female: **A** whole animal dorsal view; **B** A1 with transparent strip (*arrows*) along posterior edge; **C** abdomen ventral view; **D** abdomen lateral view. *Bars*: **A** 100 μm; **B**, **C**, **D** 50 μm

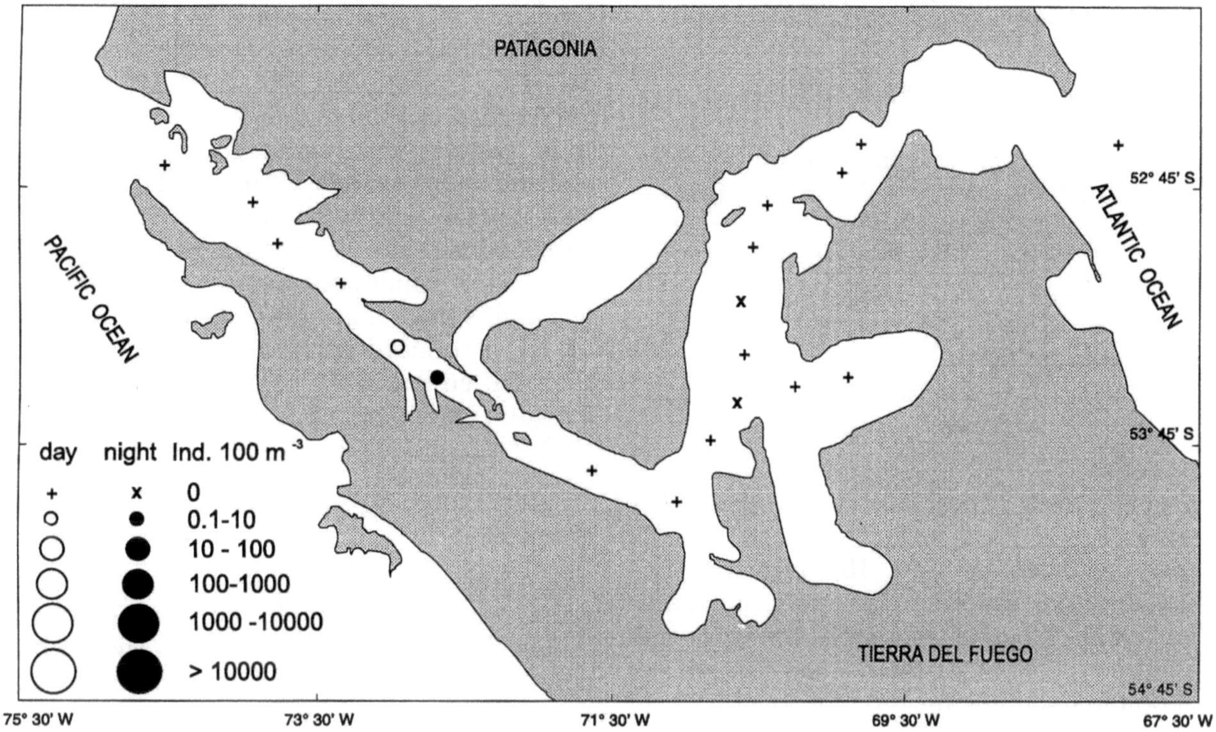

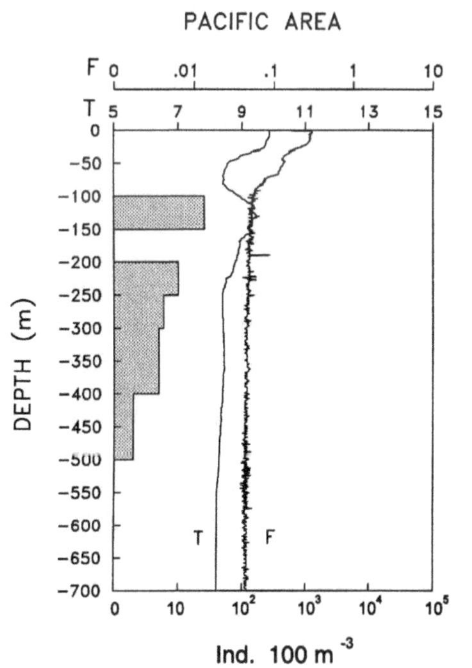

Depth range of occurrence = **100 - 500** m
Population core = **100 - 150** m
Temperature (T) = **8.20 - 9.80** °C
Oxygen = **348.4 - 437.2** μmol dm^{-3}
Chl fluorescence (F) = **0.04 - 0.39**

Fig. 3.34.3. *Scaphocalanus curtus*. Distribution in the Straits of Magellan in February–March 1991

3.35 *Scaphocalanus farrani* Park 1982

Scaphocalanus farrani Park, 1982: p. 95, figs. 11–14.
Scaphocalanus brevicornis Farran, 1929: p. 248, fig. 15 – Vervoort, 1951: p. 113, figs. 60, 61; 1957, p. 107, figs. 97–99 – Tanaka, 1960a: p. 42, pl. 18 figs. 1–6 – Bradford, 1971b: p. 23, figs. 81, 82.

Body length –
females: 2.37–2.63 mm (10 specimens)
male: 2.65 mm (1 specimen)

Female. Body robust with cephalothorax 2.8 times longer than wide. Forehead broadly rounded in dorsal view. Rostrum with two slender filaments. Abdomen quite short; proportional lengths of cephalothorax and abdomen 79:21; proportional length of abdominal somites and furca 33:20:18:12:17 = 100. Postero-lateral margins of last thoracic somite somewhat triangular in dorsal and more rounded in lateral views. Ventral surface of genital somite with slight protuberance and small seminal receptacle in lateral view. First inner lobe of maxillule with 3 posterior spines. P5 2-jointed. Distal joint with small outer spine, a medium-sized terminal spine and a longer inner spine; these latter are coarsely serrated along outer edge and bear fine teeth along inner edge. P5 can show small variations (Park 1982).

Male. Body narrowing abruptly towards the head and ending with nearly truncated forehead with slightly bulged mid-anterior edge. Cephalothorax 2.4 times longer than wide. Postero-lateral margins of last thoracic somite ending with two small rounded processes in dorsal view. Proportional lengths of cephalothorax and abdomen 67:33; proportional lengths of abdominal somites and furca 6:38:20:23:3:10 = 100. Left P5 slightly longer than exopod and ending with terminal spine. Surface of Ril ornated with small teeth at distal end. Last exopodal joint terminating with 2 spines and bordered with row of fine hairs on inner margin. Right P5 asymmetrical with short endopod ending with strong terminal spine. Inner postero-lateral margin of Rel produced into large protuberance. Exopod ending with a fine lamella with irregular inner margin, slightly shorter than Re2.

Remarks

According to Park (1982), the species previously known as *S. brevicornis* from Antarctic and sub-Antarctic Seas should be considered *S. farrani*. *S. brevicornis*, as originally described by Sars (1900) from the Arctic Ocean, is smaller in size and differs from *S. farrani* for the relative lengths and serration of the spines of P5.

Distribution

S. farrani has been previously recorded from the Antarctic and sub-Antarctic Seas as *S. brevicornis* by Farran (1929), Vervoot (1951, 1957), Tanaka (1960a) and Bradford (1971b). Park (1982) reports that *S. farrani* was one of the most common *Scaphocalanus* species in Antarctic waters. Hopkins (1985) recorded its presence from 0–1000 m in Croker Passage, Antarctic Peninsula. The species has not been previously recorded in Chilean and Argentine waters. In the present study, it was sampled from 250–500 m at three stations in the Pacific area and at one Central station in the Straits of Magellan.

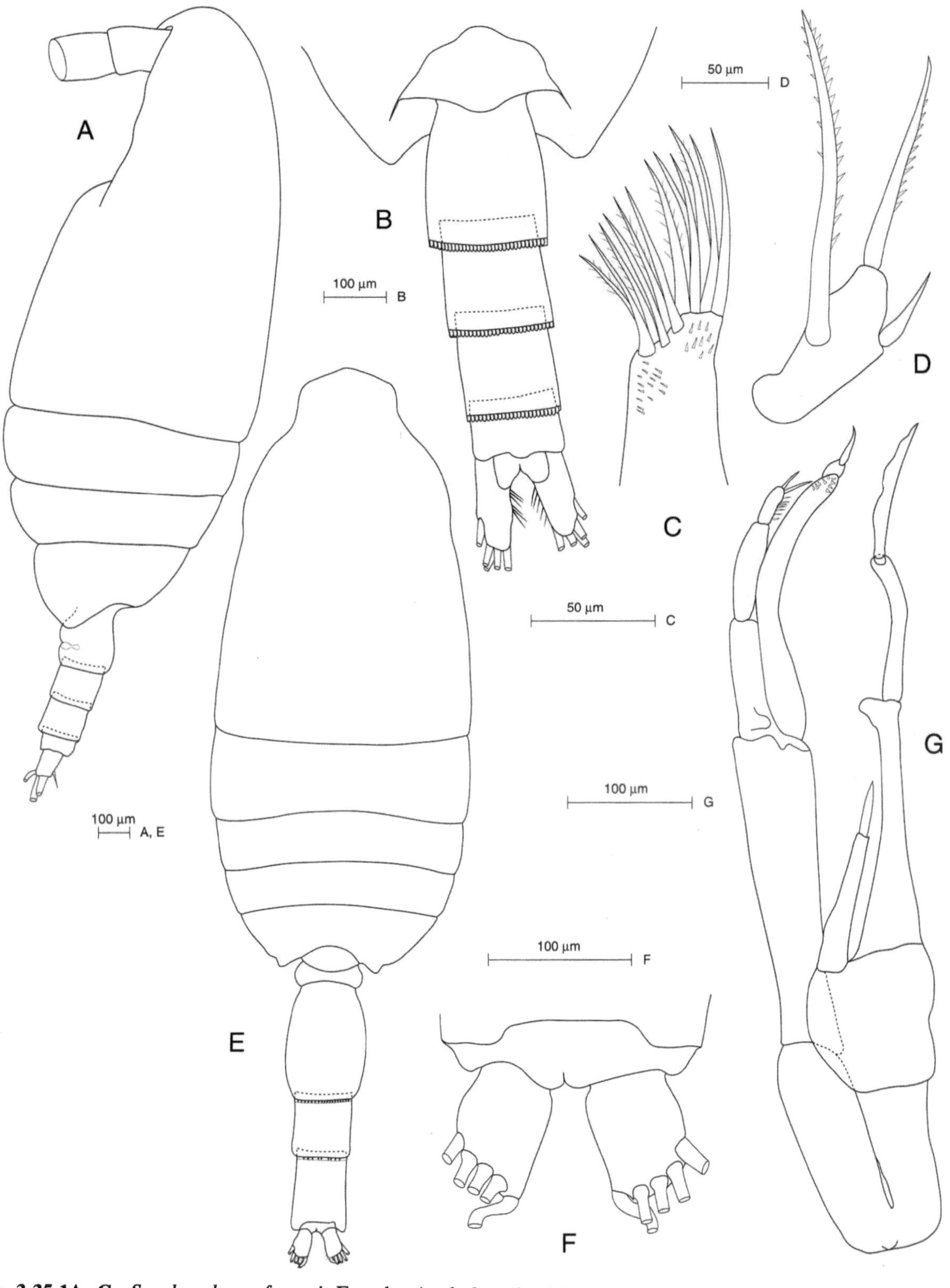

Fig. 3.35.1A–G. *Scaphocalanus farrani*. Female: **A** whole animal lateral view; **B** abdomen dorsal view; **C** first inner lobe of maxillule; **D** free joint of P5. Male: **E** whole animal dorsal view; **F** furca; **G** P5

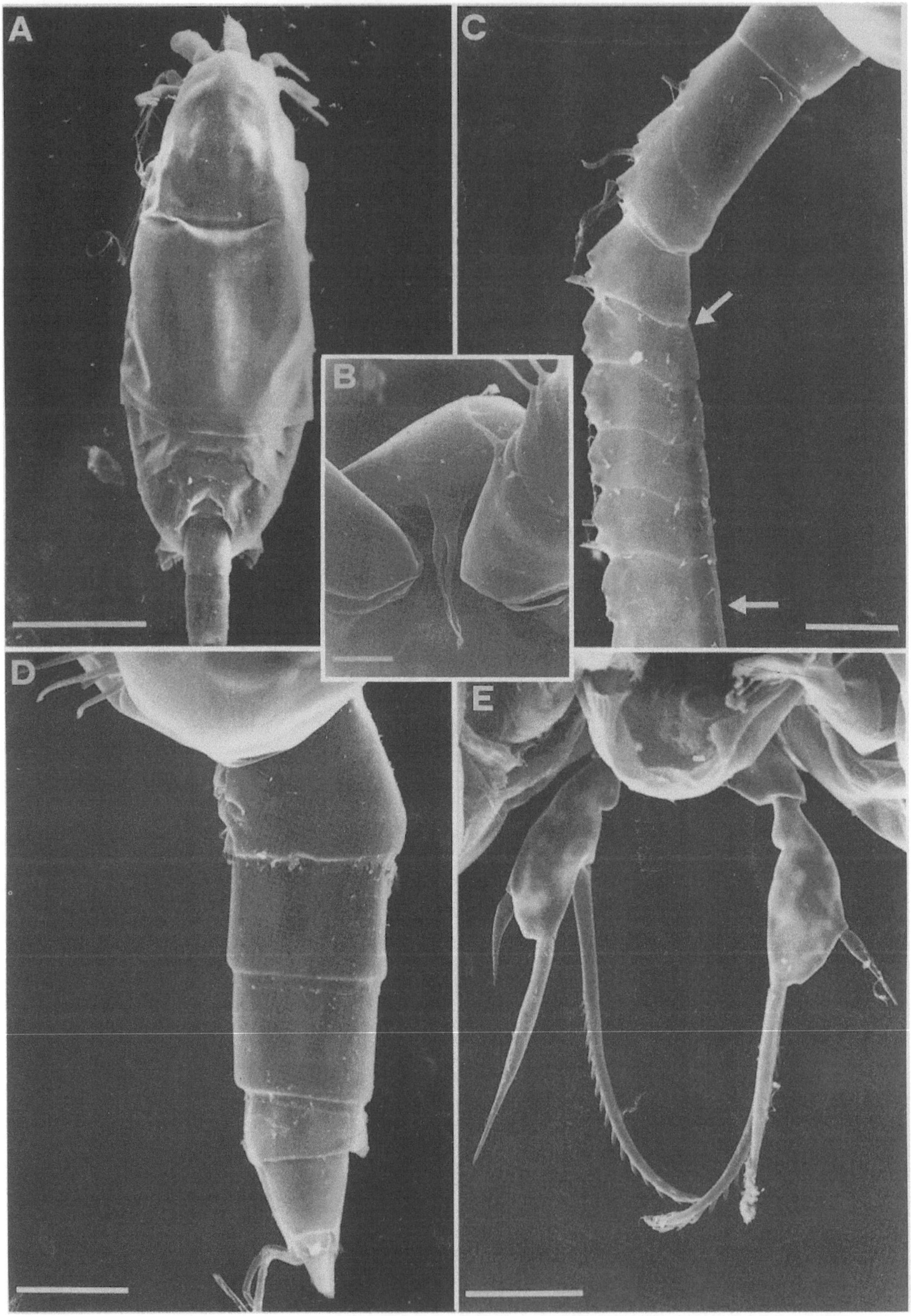

Fig. 3.35.2A–E. *Scaphocalanus farrani*. Female: **A** whole animal dorsal view; **B** rostrum; **C** A1 showing detail of transparent strip (*arrows*); **D** abdomen lateral view; **E** P5. *Bars*: **A** 500 μm; **B**, **E** 50 μm; **C**, **D** 100 μm

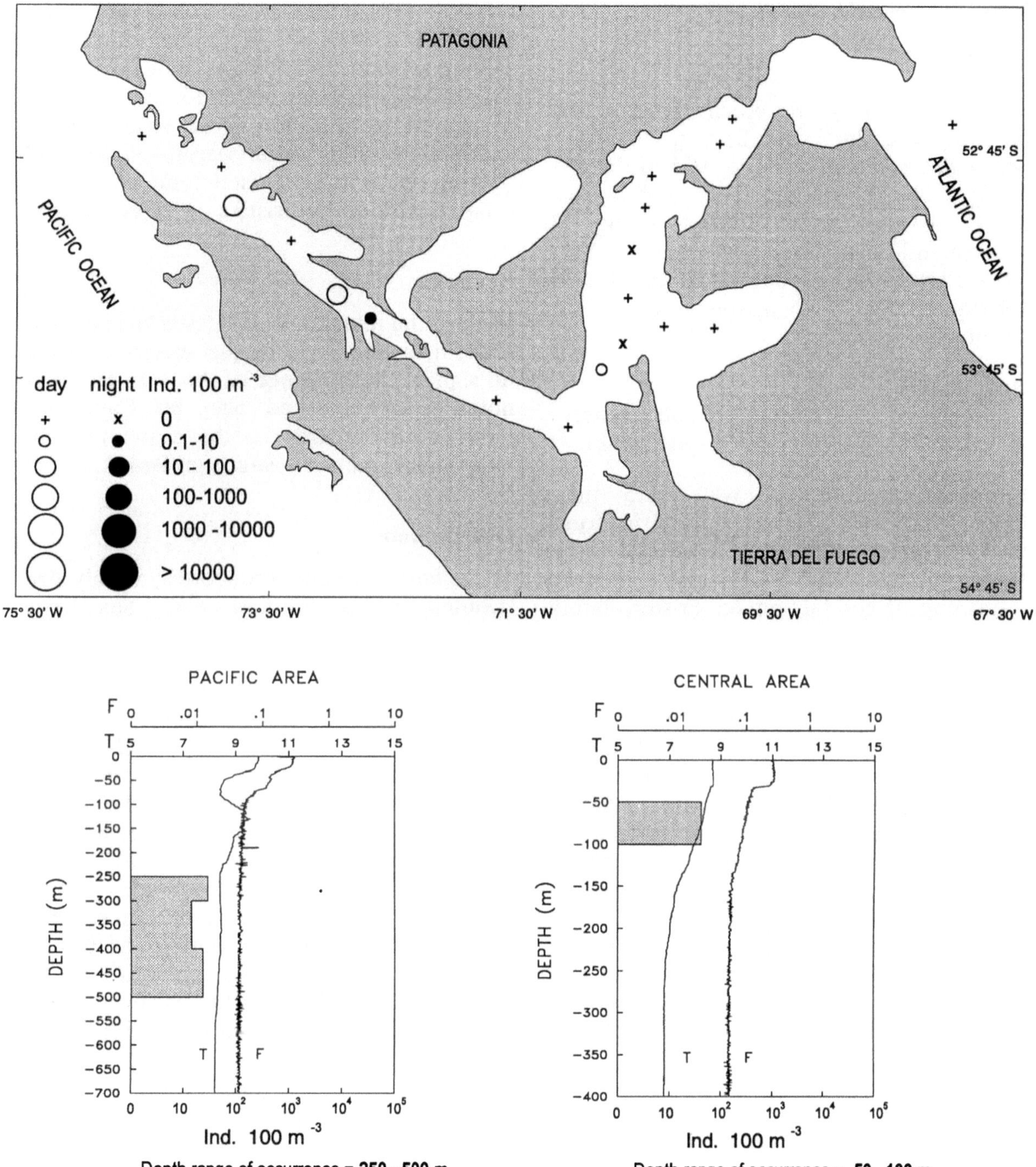

PACIFIC AREA	CENTRAL AREA
Depth range of occurrence = **250 - 500** m	Depth range of occurrence = **50 - 100** m
Population core = **250 - 300** m	Population core = **50 - 100** m
Temperature (T) = **8.20 - 9.80** °C	Temperature (T) =**6.70 - 9.01** °C
Oxygen = **348.4 - 437.2** µmol dm^{-3}	Oxygen = **346.1 - 407.3** µmol dm^{-3}
Chl fluorescence (F) = **0.04 - 0.39**	Chl fluorescence (F) = **0.05 - 0.27**

Fig. 3.35.3. *Scaphocalanus farrani.* Distribution in the Straits of Magellan in February–March 1991

3.36 *Scolecithricella dentata* Giesbrecht 1892

Scolecithrix dentata Giesbrecht, 1892: p. 266, pl. 13 figs. 12, 20, 33, pl. 37 figs. 13, 14. *Scolecithricella dentata* Sars, 1925: p. 191, pl. 52 figs. 21–23 – Rose, 1942: p. 144, figs. 38–41, 46–48 – Tanaka, 1962: p. 42, fig. 130 – Owre and Foyo, 1967: p. 61, figs. 94, 379–381 – Park, 1968: p. 555, pl. 18 figs. 13–16; 1980: p. 42, fig. 7 – Campaner, 1984: p. 176, fig. 6f–k.

Body length –
females: 1.84–1.96 mm (10 specimens)
males: 1.76–1.94 mm (4 specimens)

Female. Cephalothorax ovate, 2.5 times longer than wide. Head rounded in dorsal view. Rostrum with strong base and rather short, pointed rami. A1 almost reaching distal end of genital somite. Postero-lateral corners of last thoracic somite notched on each side. Proportional lengths of cephalothorax and abdomen 76:24. Abdomen 4-segmented, genital somite the longest; proportional lengths of abdominal somites and furca 37:22:22:8:11 = 100. P5 1 – jointed, flat, with minute outer marginal spine, one small distal spine and a relatively large inner spine. This leg may be variable in spinulation, with outer marginal spine absent in some specimens (Tanaka 1962).

Male. Cephalothorax 2.3 times longer than wide. Postero-lateral corners of last thoracic somite rounded. Proportional lengths of cephalothorax and abdomen 73:27; proportional lengths of abdominal somites and furca 15:30:21:23:3:8 = 100. Caudal rami about as long as wide. In P5, right ramus with very short endopod; both endopod and last joint of exopod bearing short spine. In left ramus, endopod reaching distal margin of Re1 and bearing spine at tip; distal joint of exopod ornated with dense bristles on inner margin.

Remarks

Most of the females we collected lacked a P5. Careful examination of the specimens gave no sign of the presence of this leg. This was noted and remarked also by Campaner (1984) with regards to the material from epipelagic waters off Southern Brazil.

Distribution

S. dentata is a common species, widely distributed in the Mediterranean Sea (type locality), and the Atlantic, Pacific and Indian Oceans. Farran (1929) reported *S. dentata* from the North temperate and tropical Atlantic and off New Zealand from the surface to 150 m. According to Campaner (1984), it was one of the most frequent and abundant *Scolecithricella* species off southern Brazil in neritic and oceanic waters, where it was first found by Björnberg (1963). Park (1980) found *S. dentata* in Antarctic and sub-Antarctic areas, north to 46°40′S, in samples collected from the surface down to 2470 m. In the present study, it was sampled from 200–400 m at three stations in the Pacific area and from 200–250 m at a Central station in the Straits of Magellan.

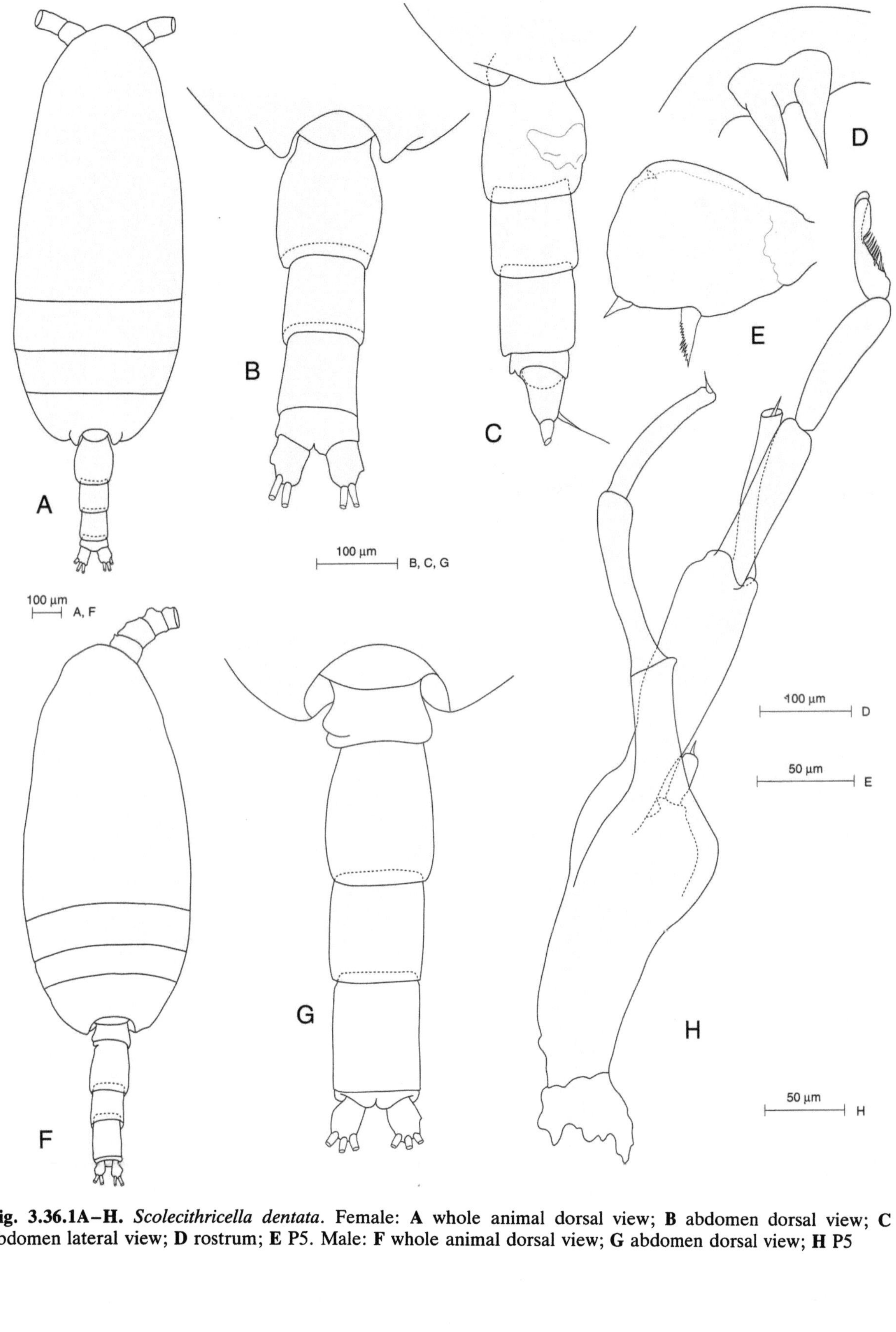

Fig. 3.36.1A–H. *Scolecithricella dentata*. Female: **A** whole animal dorsal view; **B** abdomen dorsal view; **C** abdomen lateral view; **D** rostrum; **E** P5. Male: **F** whole animal dorsal view; **G** abdomen dorsal view; **H** P5

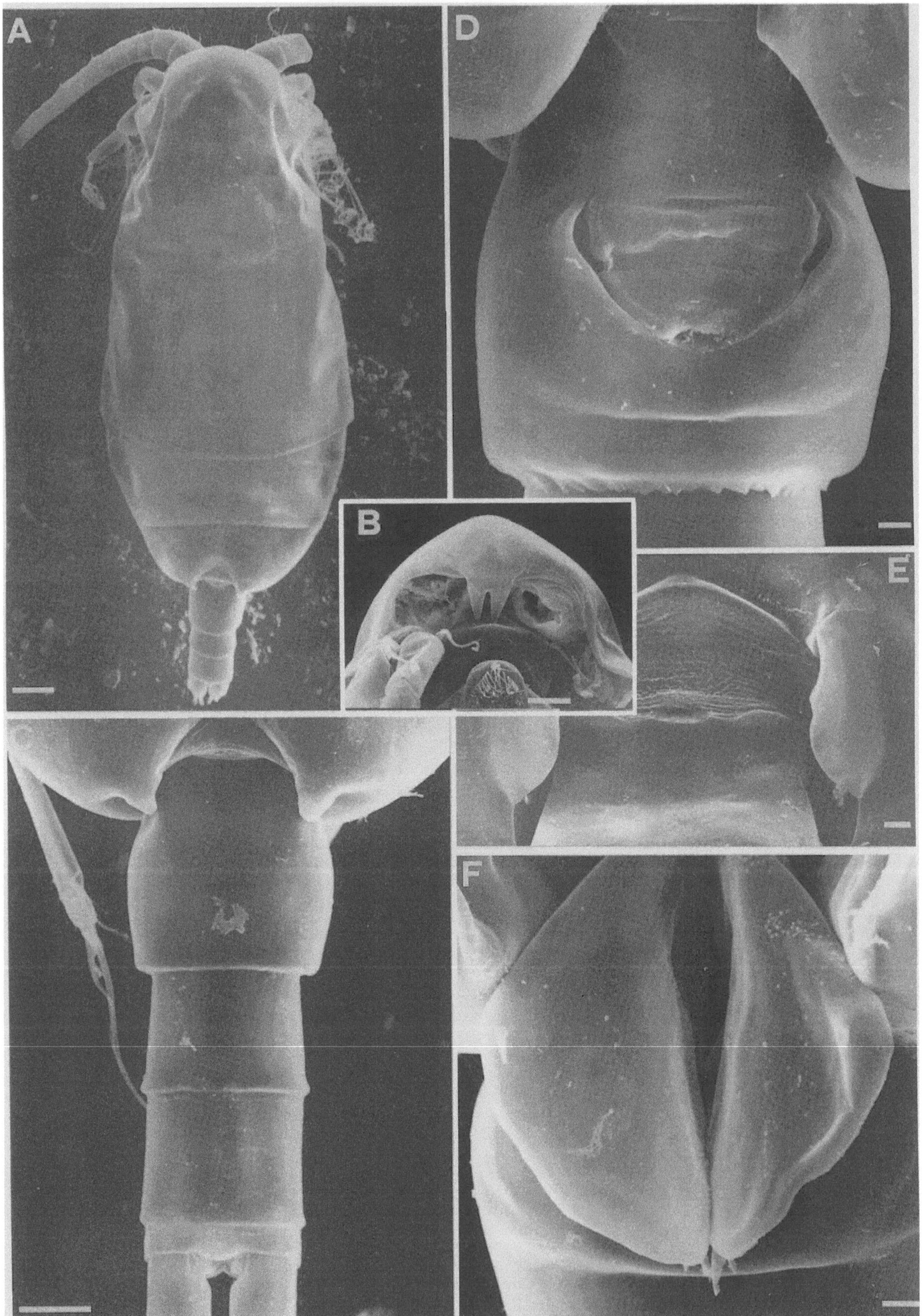

Fig. 3.36.2A–F. *Scolecithricella dentata*. Female: **A** whole animal dorsal view; **B** forehead with rostrum; **C** abdomen dorsal view; **D** genital somite ventral view; **E** inner notches on posterior margins of last thoracic somite dorsal view; **F** P5. *Bars*: **A** 100 µm; **B**, **C** 50 µm; **D**, **E**, **F** 10 µm

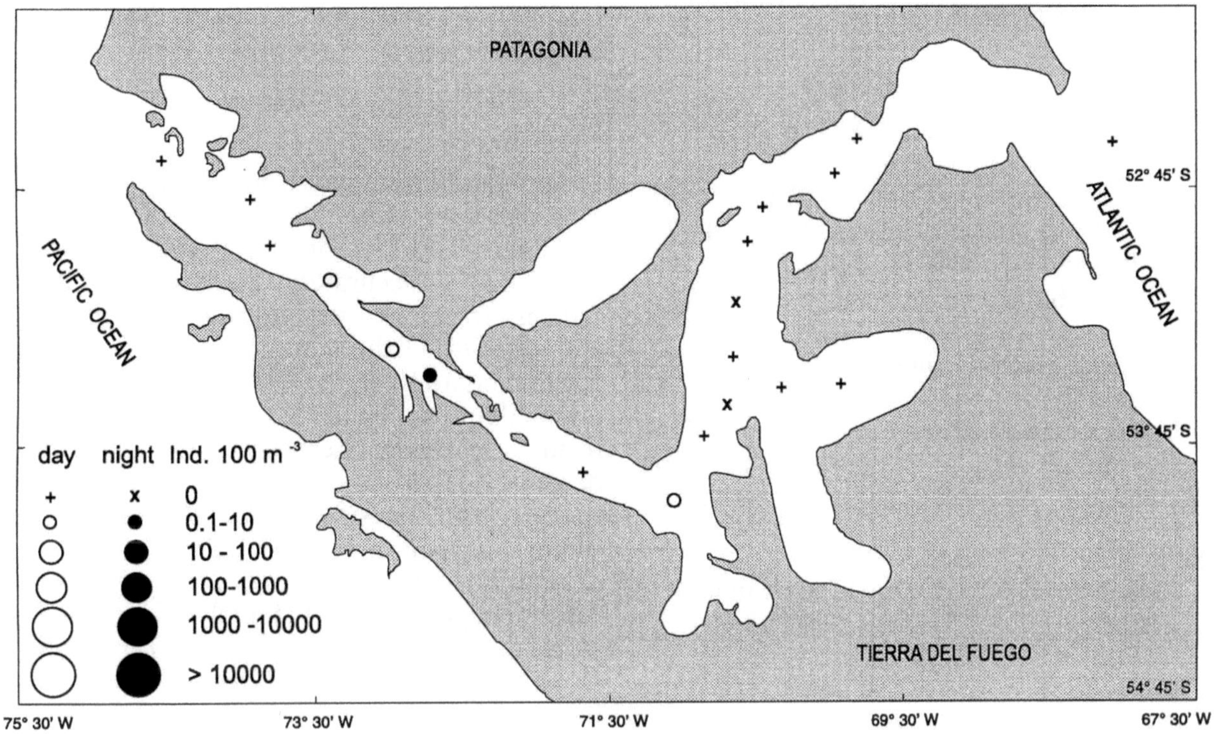

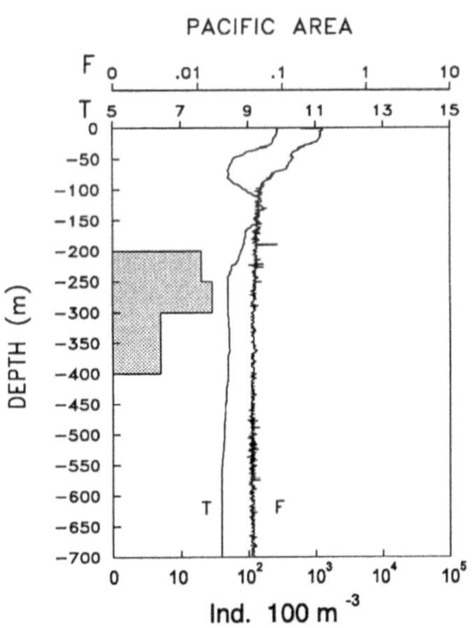

Depth range of occurrence = **200 - 400** m
Population core = **250 - 300** m
Temperature (T) = **8.20 - 9.80** °C
Oxygen = **348.4 - 437.2** μmol dm^{-3}
Chl fluorescence (F) = **0.04 - 0.39**

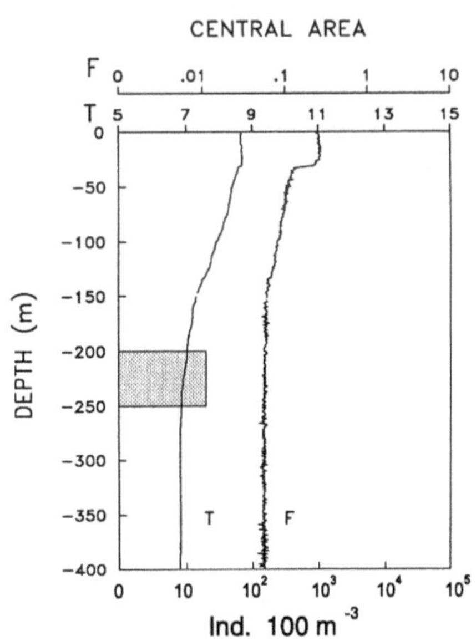

Depth range of occurrence = **200 - 250** m
Population core = **200 - 250** m
Temperature (T) = **6.70 - 9.01** °C
Oxygen = **346.1 - 407.3** μmol dm^{-3}
Chl fluorescence (F) = **0.05 - 0.27**

Fig. 3.36.3. *Scolecithricella dentata.* Distribution in the Straits of Magellan in February–March 1991

3.37 *Scolecithricella minor* Brady 1883

Scolecithrix minor Brady, 1883: p. 58, pl. 16 figs. 15, 16, pl. 18 figs. 1–5.
Scolecithricella minor Sars, 1903: p. 55, pls. 37, 38 – Mori, 1937: p. 51, pl. 25 figs. 1–7 – Brodskii, 1950: p. 269, fig. 178 – Park, 1968: p. 554, pl. 8 figs. 9–12; 1980: p. 31, figs. 2, 3.
Scolecithrix glacialis Giesbrecht, 1902, p. 25, pl. 4 figs. 1–7.

More complete lists of synonyms of *S. minor* and *S. glacialis* are given in Vervoort (1956b) and Vervoort (1957), respectively.

Body length –
females: 1.27–1.37 mm (15 specimens)
males: 1.27–1.31 mm (3 specimens)

Female. Cephalothorax slender, 2.5 times longer than wide. Head fused with thorax, narrowly rounded in dorsal view. Rostrum bifurcated with strong rami. A1 reaching posterior margin of genital somite. Posterolateral corners of last thoracic somite symmetrical, produced distally, forming an obtuse angle in lateral view. Proportional lengths of cephalothorax and abdomen 79:21; proportional lengths of abdominal somites and furca 39:18:18:8:17 = 100. P5 foliate, 1-jointed; apical spine short, inner spine longer, serrated; outer margin of joint with smaller spinule, sometimes appearing as small tubercle.

Male. Body slender; cephalothorax 2.3 times longer than wide. Rostrum similar to that of female. Proportional lengths of cephalothorax and abdomen 70:30; proportional lengths of abdominal somites and furca 16:26:20:25:4:9 = 100. P5 exceeding length of abdomen, asymmetrical; both legs biramous; right leg with vestigial endopod. Right exopod with spatula-like distal joint. Left exopod ending in "bayonet-shaped" (Brodskii 1950) joint.

Remarks

Scolecithricella minor was originally described as *Scolecithrix minor* by Brady (1883) from specimens collected in the Indian Ocean. Sars (1903) accurately re-described this species from Norwegian-water specimens and established the genus *Scolecithricella* to accommodate it. Park (1980) noted that it is impossible to distinguish *S. minor* Brady from *S. glacialis* Giesbrecht, 1902, on the basis of the original descriptions for these species, and considered *S. glacialis* as a junior synonym of *S. minor*. Brodskii (1950) distinguished two forms of *S. minor* mainly for the shape of the P5 in females: f. *orientalis* from the North Pacific and f. *occidentalis* from the North Atlantic Ocean. Our specimens are similar to the *occidentalis* form.

Distribution

Scolecithricella minor is distributed worldwide. This species is known to occur in all oceans, even though it is purely oceanic and avoids coastal waters (Vervoort 1965b). It seems to be absent in the Mediterranean Sea. Park (1980) found this species as the most common of the genus in Antarctic and sub-Antarctic waters during the Eltanin cruises, from samples collected from the surface to 3750 m. According to Park (1980), this species was more common in Antarctic than sub-Antarctic waters. It was found in Chilean areas by Wilson (1942) and Hirakawa (1989). In the present study, *S. minor* was more widely distributed in the Straits of Magellan than the congeneric species *S. dentata*. It was sampled from the surface to 500 m with maximum abundances from 150–200 m at Pacific stations.

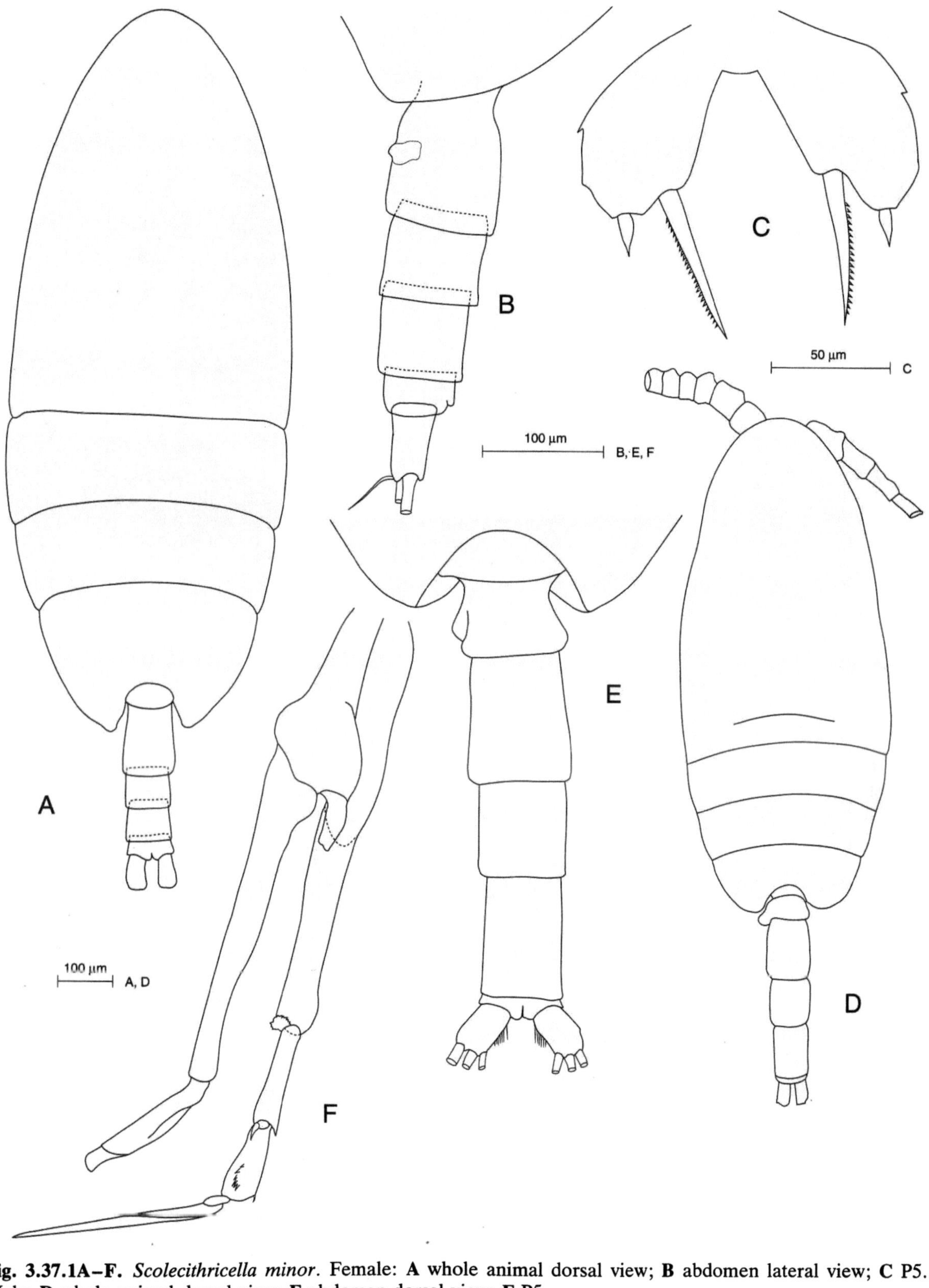

Fig. 3.37.1A–F. *Scolecithricella minor*. Female: **A** whole animal dorsal view; **B** abdomen lateral view; **C** P5. Male: **D** whole animal dorsal view; **E** abdomen dorsal view; **F** P5

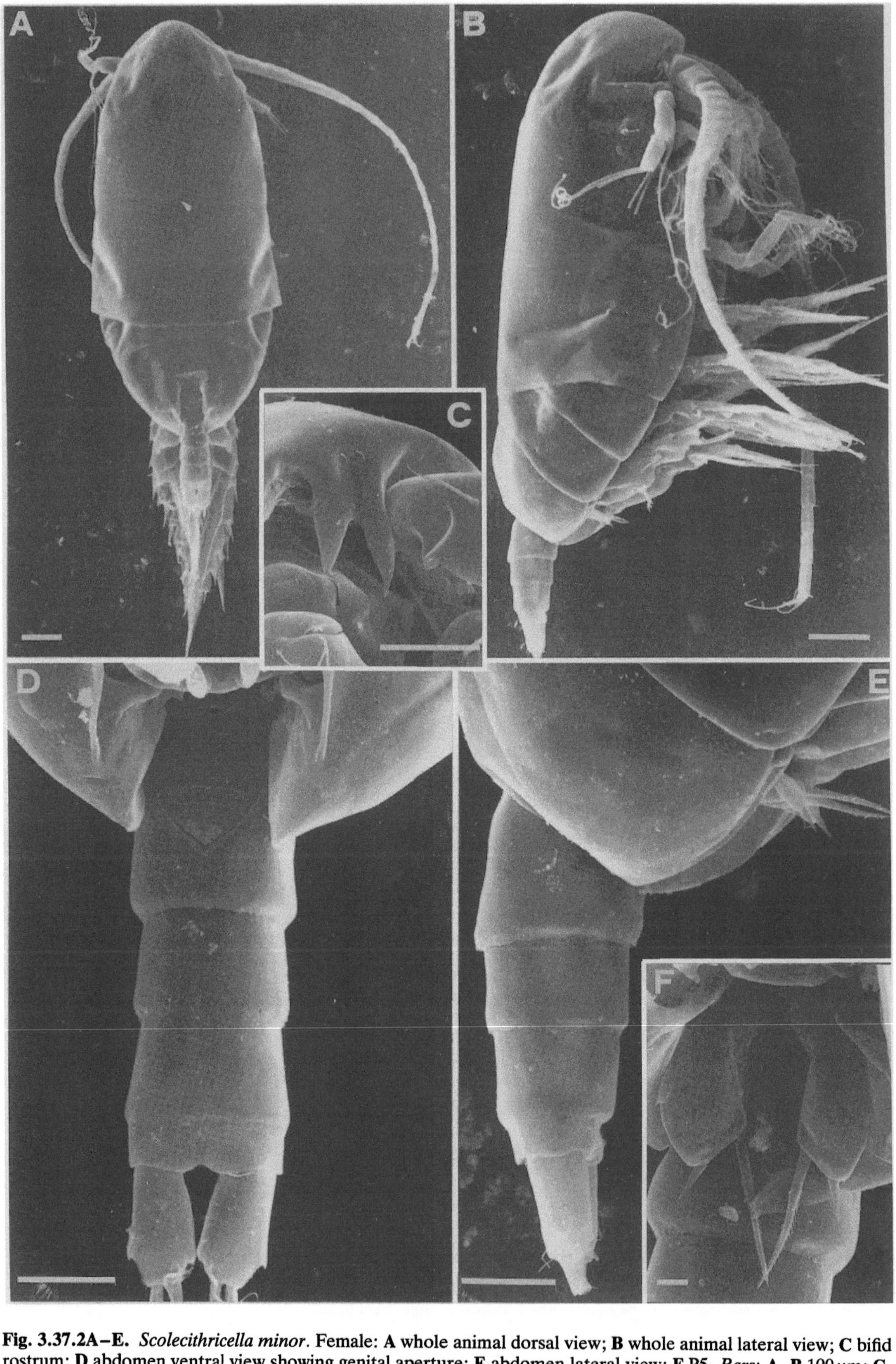

Fig. 3.37.2A–E. *Scolecithricella minor*. Female: **A** whole animal dorsal view; **B** whole animal lateral view; **C** bifid rostrum; **D** abdomen ventral view showing genital aperture; **E** abdomen lateral view; **F** P5. *Bars*: **A**, **B** 100 μm; **C**, **D**, **E** 50 μm; **F** 10 μm

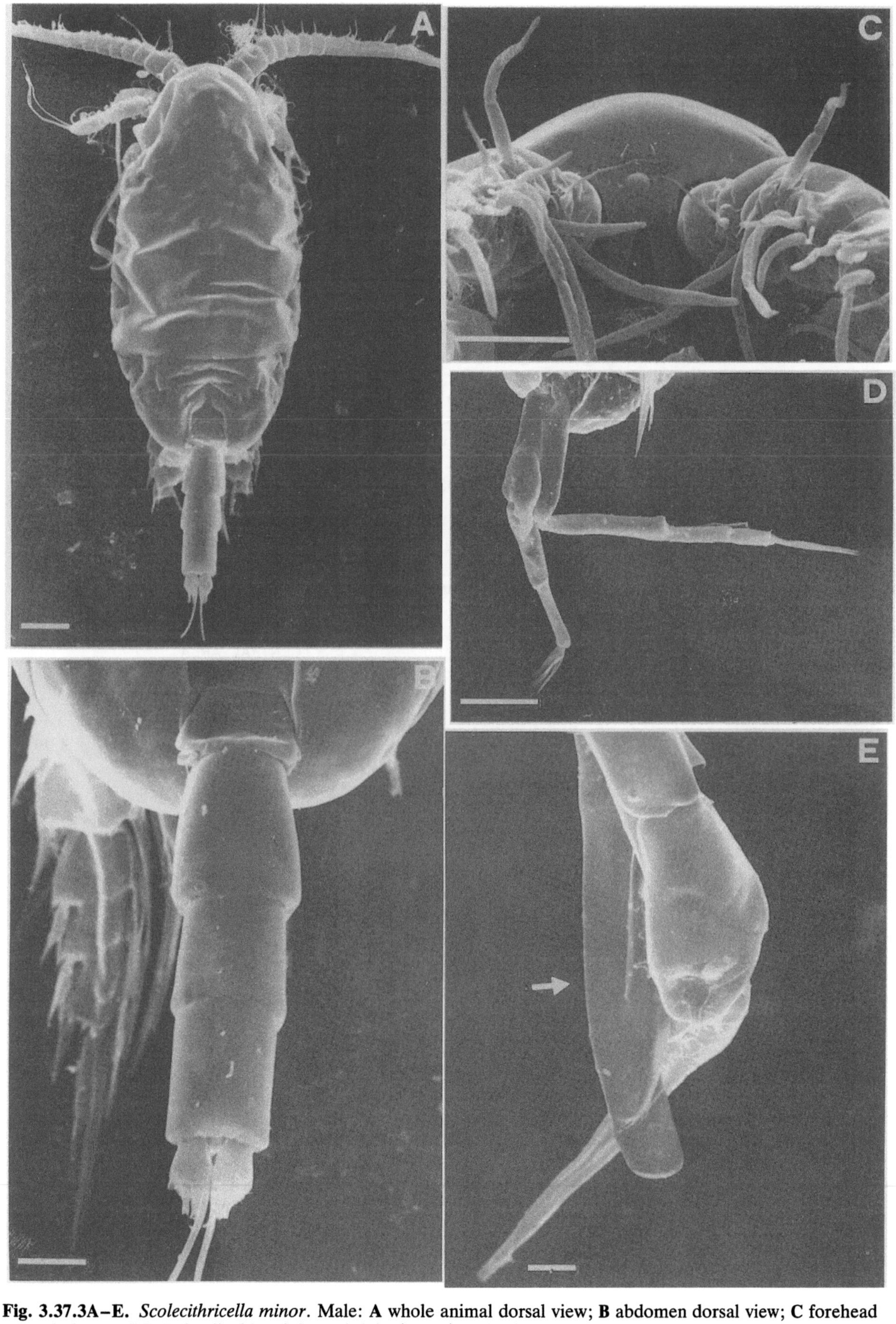

Fig. 3.37.3A–E. *Scolecithricella minor*. Male: **A** whole animal dorsal view; **B** abdomen dorsal view; **C** forehead ventral view; **D** P5; **E** detail of last joints of right (*arrow*) and left P5. *Bars*: **A**, **D** 100 µm; **B**, **C** 50 µm; **E** 10 µm

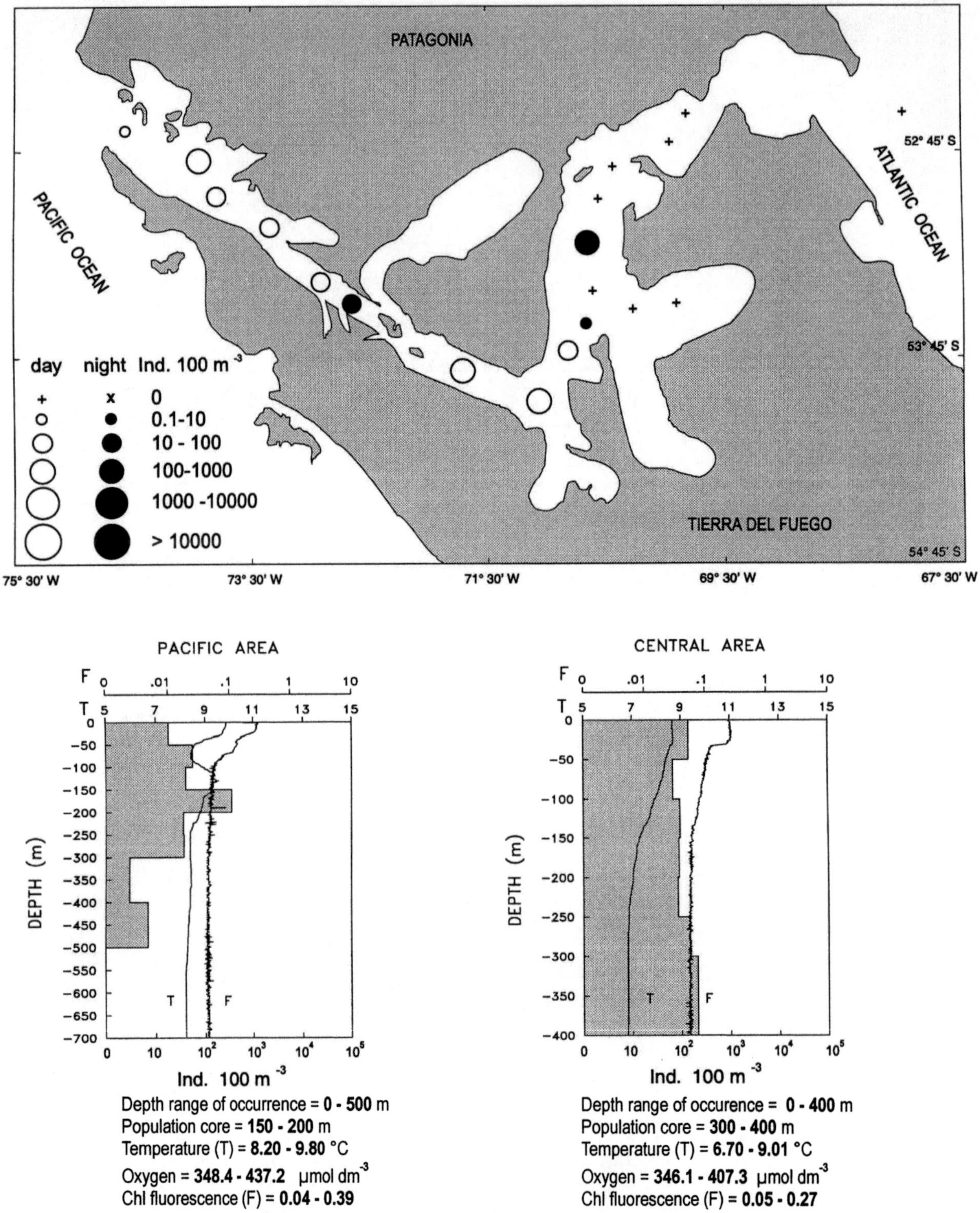

Fig. 3.37.4. *Scolecithricella minor*. Distribution in the Straits of Magellan in February–March 1991

3.38 *Spinocalanus brevicaudatus* Brodskii 1950

Spinocalanus brevicaudatus Brodskii, 1950: p. 136, fig. 55 – Semenova, 1962: p. 1571, figs. 1–4 – Damkaer, 1975: p. 53, figs. 130–136, 157.
See Damkaer (1975) for more complete synonymy.

Body length –
females: 1.55–1.80 mm (20 specimens)
male: 1.57 mm (1 specimen)

Female. Cephalothorax ovoid in dorsal view, 2.6 times longer than wide; head slightly truncated; rostrum absent. Posterior corners of last thoracic somite rounded, not prolonged. Proportional lengths of cephalothorax and abdomen 78:22; proportional lengths of abdominal somites and furca 37:17:16:19:11 = 100. Genital somite slightly longer than the following somite and protruding ventrally. Ventral surface of genital somite ornated with tuft of long hairs below genital aperture. Borders of abdominal somites ornated with fringes. Caudal rami as long as wide, symmetrical, with long hairs on dorsal side. P1 Re3 posterior surface without spines; P2 B1 inner posterior surface with setules; P3 B1 outer surface with few setules and spines, inner margins with hairs; P4 B2 posterior surface lacks ornamentation.

Male. Cephalothorax elongated, rectangular in dorsal view, 2.6 times longer than wide. Posterior corners of last thoracic somite prolonged into small rounded processes curved outwards. Proportional lengths of cephalothorax and abdomen 44:19. Abdomen 5-segmented; anal somite very short, almost hidden between the fourth somite and furca. Proportional lengths of abdominal somites and furca 11:35:20:21:3:10 = 100. Caudal rami about as long as wide, symmetrical. P5 biramous; left B1 reaching 1/3 length of right B2; left B2 reaching middle of right Re1. Right leg, including terminal blade, longer than left. Each Re3 is subtle and stylet-shaped. Endopods stylet-shaped, right Ri longer than left. Inner edge of left Re2, Re3 with long hairs.

Remarks

The systematics of the family Spinocalanidae is rather complex and confusing, particularly for the genus *Spinocalanus*, which now includes 26 species (Razouls 1993). Since they are easily damaged, it is difficult to find specimens in which the swimming legs are intact, a prerequisite for their correct taxonomic identification. To date, the best revision of this family is still the one by Damkaer (1975).

S. brevicaudatus had been known as a larger form of *S. abyssalis* until Brodskii (1950) recognized it as a separate species. Mazzocchi and Ianora (1991) reported *S. terranovae* from the previous cruise in the Straits of Magellan. Their identification was based on a single female specimen with damaged swimming legs. Therefore, it probably should be corrected to *S. brevicaudatus*.

S. brevicaudatus and *S. terranovae* have overlapping size ranges; the former species has a somewhat lower size limit (1.4–2.4 mm versus 1.7–2.2 mm). Due to the scarce iconographic information for both species, the two appear very similar in morphology.

According to the taxonomic key given by Damkaer (1975), the two differ mainly for their ornamentation of the swimming legs; *S. brevicaudatus* lacks strong spines on P1 Re3, P2 B1, P4 B2 which are characteristic of *S. terranovae*. The absence of this ornamentation on the swimming legs of our specimens is clearly shown in the SEM figures. The only male we found concords perfectly with the description given by Damkaer (1975) both in general shape and for the details of the P5. Unfortunately, the left and right P5 Re3 were broken in our specimen.

Distribution

Damkaer (1975) re-orders, on the basis of the complex and confusing synonymy, the records of this species and its areas of distribution. It is widely distributed in the North and Central Pacific, West Indian and Atlantic Oceans. It has also been reported in the SE Atlantic Ocean by Unterüberbacher (1964). The present record is the first for sub-Antarctic waters. *S. brevicaudatus* is considered an oceanic bathypelagic species, recorded down to 4000 m. In the present study, it was sampled only in the Pacific area of the Straits of Magellan, where it occurred at all depths but with maximum abundances from 400–500 m.

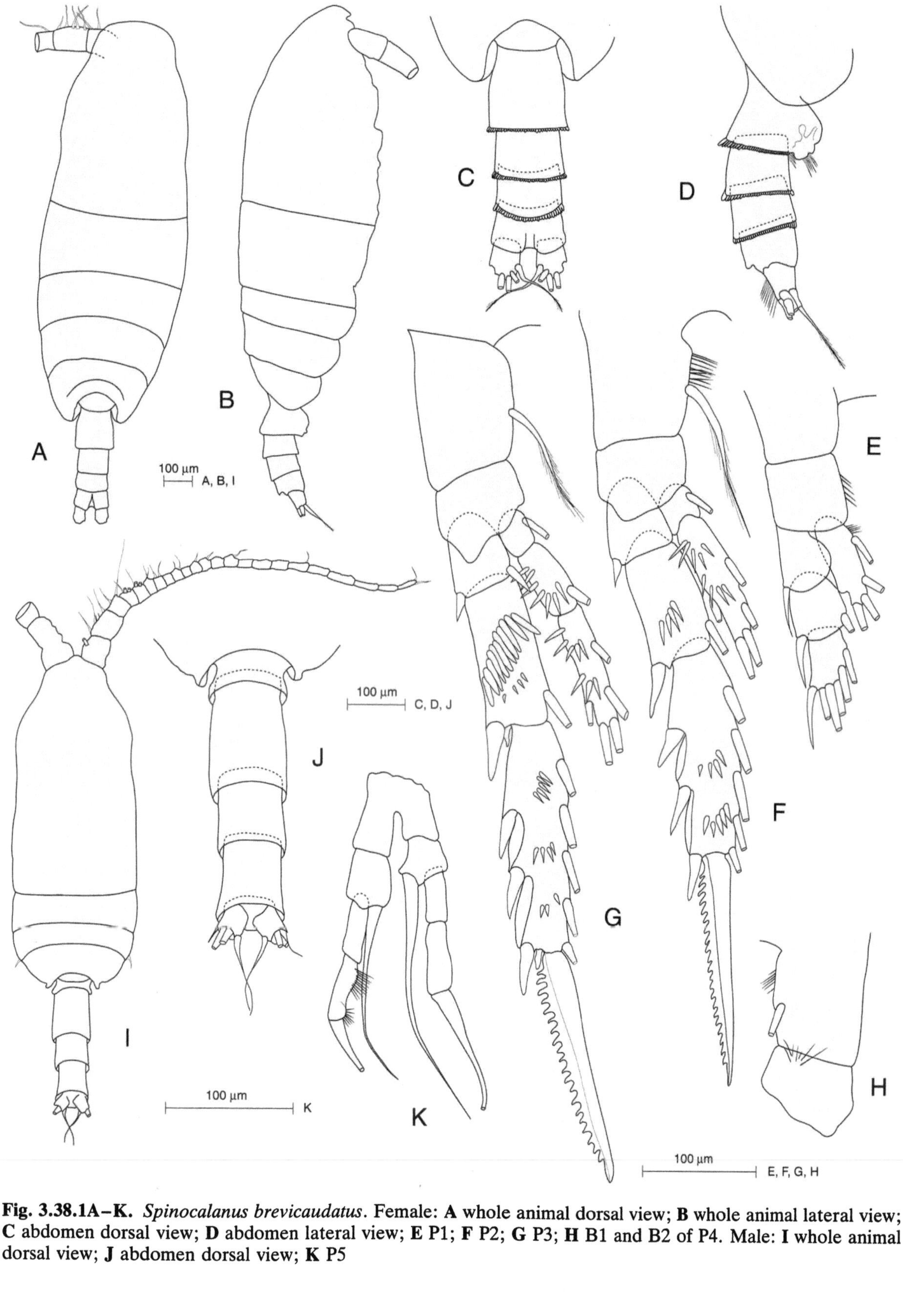

Fig. 3.38.1A–K. *Spinocalanus brevicaudatus*. Female: **A** whole animal dorsal view; **B** whole animal lateral view; **C** abdomen dorsal view; **D** abdomen lateral view; **E** P1; **F** P2; **G** P3; **H** B1 and B2 of P4. Male: **I** whole animal dorsal view; **J** abdomen dorsal view; **K** P5

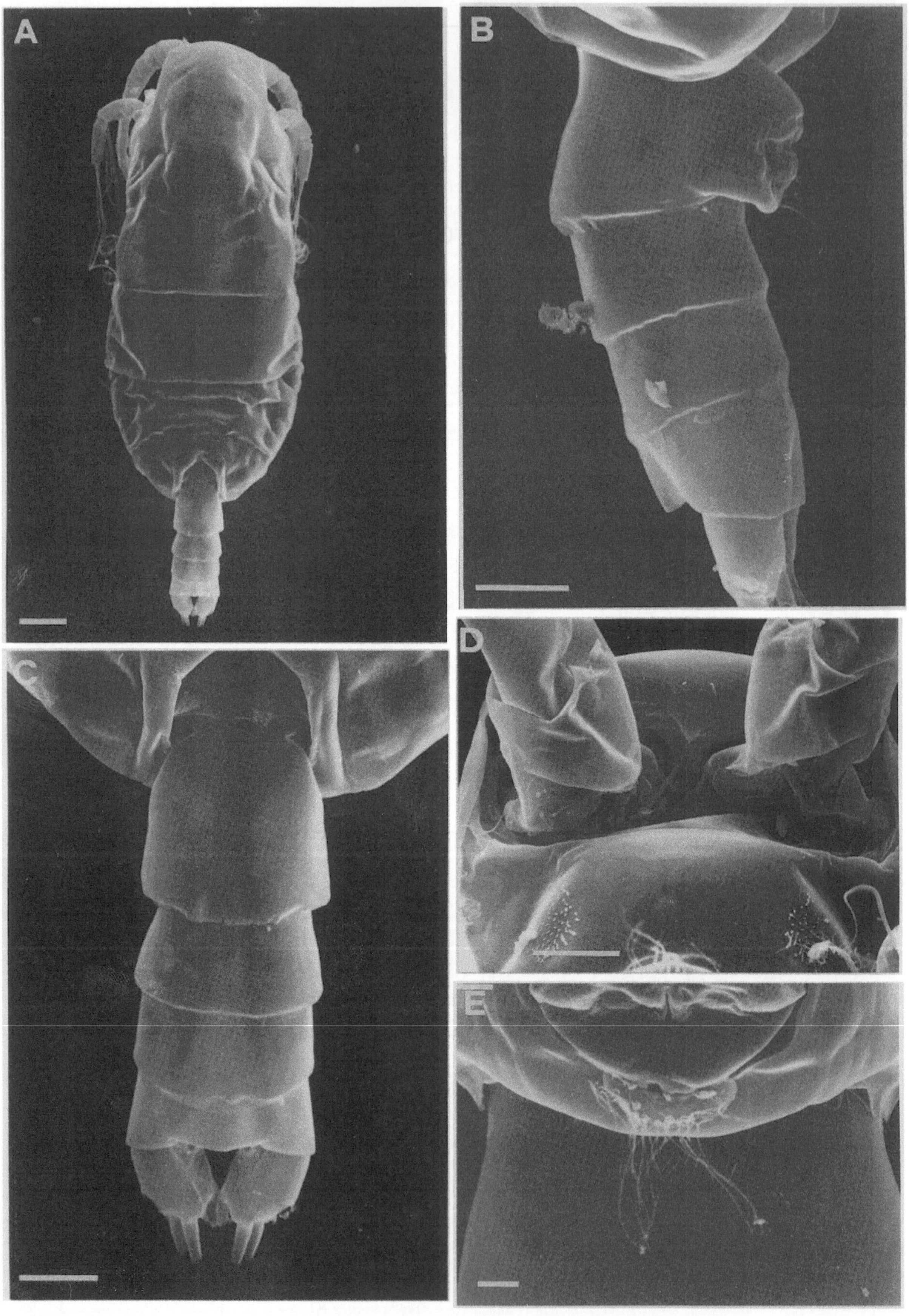

Fig. 3.38.2A–E. *Spinocalanus brevicaudatus*. Female: **A** whole animal dorsal view; **B** abdomen lateral view; **C** abdomen dorsal view; **D** forehead ventral view; **E** detail of genital aperture. *Bars*: **A** 100 µm; **B**, **C**, **D** 50 µm; **E** 10 µm

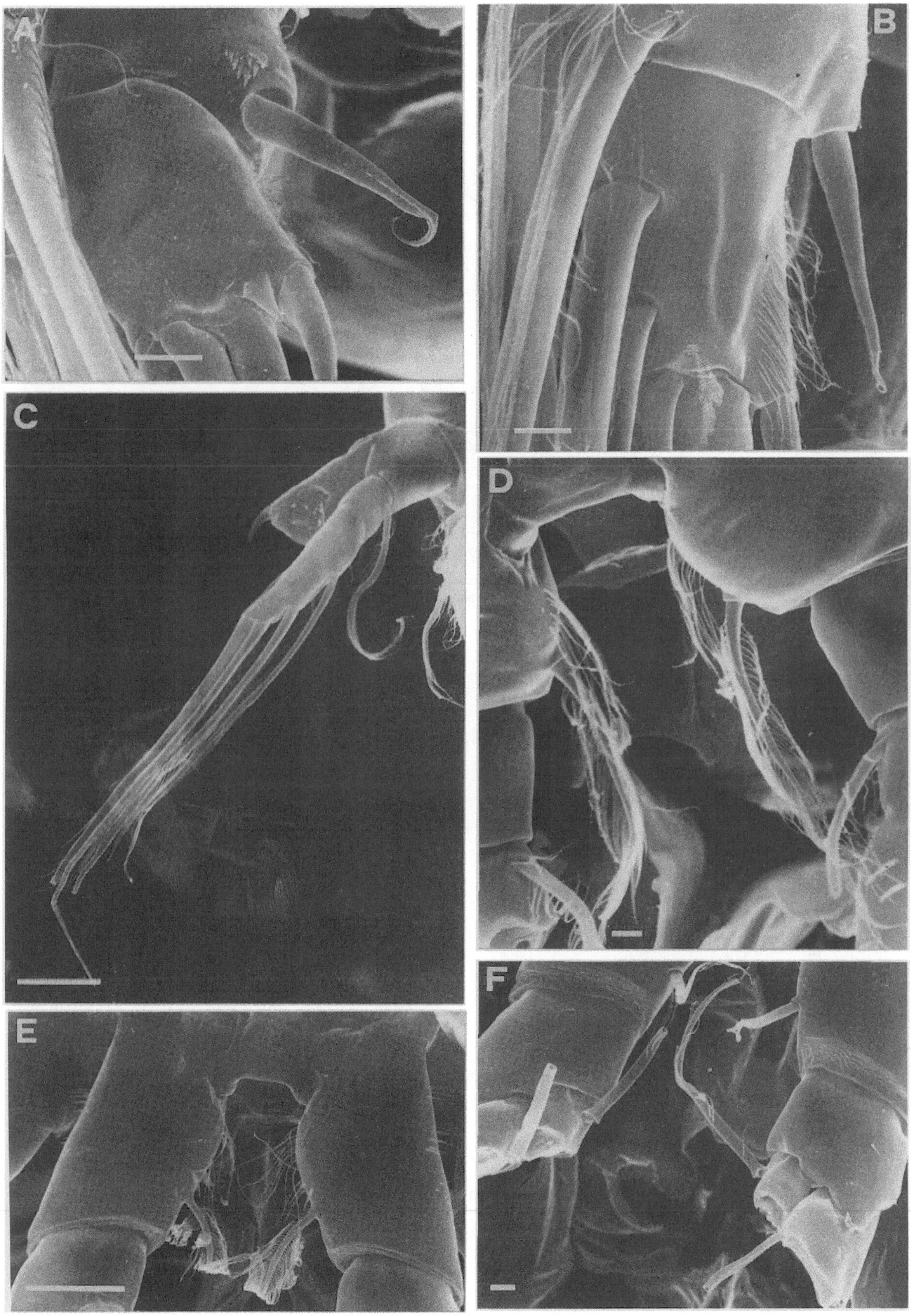

Fig. 3.38.3A–F. *Spinocalanus brevicaudatus*. Female: **A** P1 Re3 posterior view; **B** P1 Re3 anterior view; **C** P2 Ri anterior view; **D** P2 B1 posterior view; **E** P3 B1 posterior view; **F** P4 B2 posterior view. *Bars* **A**, **B**, **D** 10 μm; **C**, **E**, **F** 50 μm

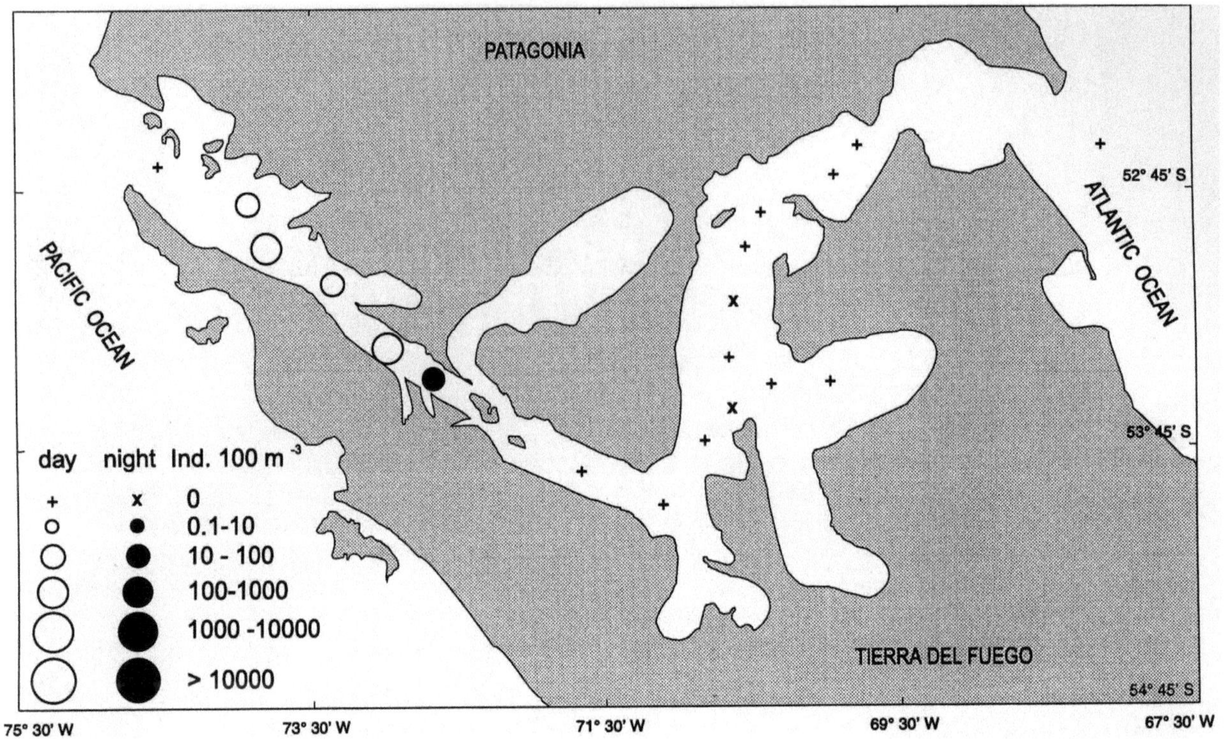

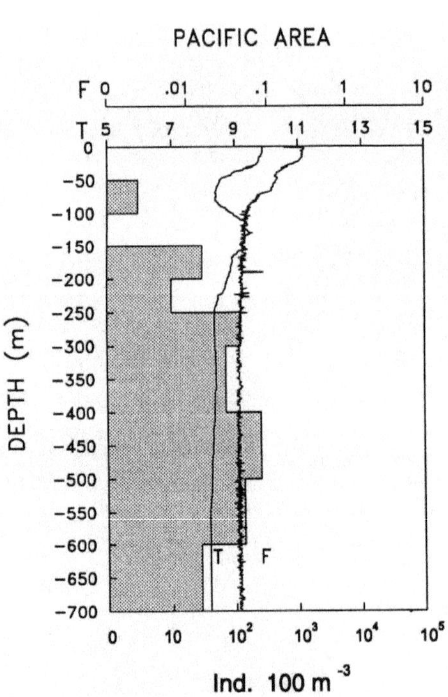

Depth range of occurrence = **50 - 700** m
Population core = **400 - 500** m
Temperature (T) = **8.20 - 9.80** °C
Oxygen = **348.4 - 437.2** μmol dm^{-3}
Chl fluorescence (F) = **0.04 - 0.39**

Fig. 3.38.4. *Spinocalanus brevicaudatus*. Distribution in the Straits of Magellan in February–March 1991

Order CYCLOPOIDA

Family Oithonidae

3.39 *Oithona atlantica* Farran 1908

Oithona atlantica Farran, 1908: p. 500, pl. 16 figs. 124–128, 131, pl. 17 fig. 137 – Rosendorn, 1917: p. 12, fig. 2a–f – Nishida et al., 1977: p. 126, fig. 2a–h – Nishida, 1985: p. 17, figs. 3, 4.
Oithona spinirostris Sars, 1918: p. 6, pls. 1, 2.
For more complete synonymy see Nishida (1985).

Body length –
females: 1.16–1.39 mm (47 specimens)

Female. Prosome fusiform in dorsal view, 2.5 times longer than wide, 1.1 times the urosome. In dorsal view, head ending anteriorly in pointed rostrum. Proportional lengths of urosomites and furca 11:32:15:14:17:11 = 100. P1–P4 Re with 1,1,2; 1,0,2; 1,0,1; 0,0,1 outer marginal spines and 1,1,4; 0,1,5; 0,1,5; 0,1,5 inner marginal setae. Outer marginal spine of Re3 rather straight, curved inwards at about 1/2 its length, and about half as long as the terminal spine. P5 with one long seta attached directly to the lateral process of Th5 and one moderately long free segment with one long seta.

Remarks

O. atlantica is closely related to *O. longispina* but this latter species has not been found in sub-Arctic and sub-Antarctic waters (Nishida 1985). Nishida (1985) reported that clear differences distinguished *O. atlantica* and *O. longispina* in the North Pacific; however, many specimens with intermediate features occurred in the South Pacific. The specimens collected in the Straits of Magellan concord with *O. atlantica* as reported in the detailed description by Nishida (1985).

Distribution

O. atlantica occurs in sub-Antarctic and Antarctic waters of the Pacific Ocean (Nishida 1985). In the Atlantic Ocean it has been recorded from cold-water areas (Sars 1918; Farran 1926; Jespersen 1940) as well as from the Mediterranean Sea (Razouls and Durand 1991). Ramirez (1969) reported *O. atlantica* as a poorly dominant species in the Argentinian sector of the SW Atlantic. Marin and Antezana (1985) and Hirakawa (1989) reported *O. atlantica* in Chilean fjords. In the present study, it was frequently sampled throughout the Pacific and Central areas of the Straits of Magellan; maximum abundances were recorded from 50–100 m.

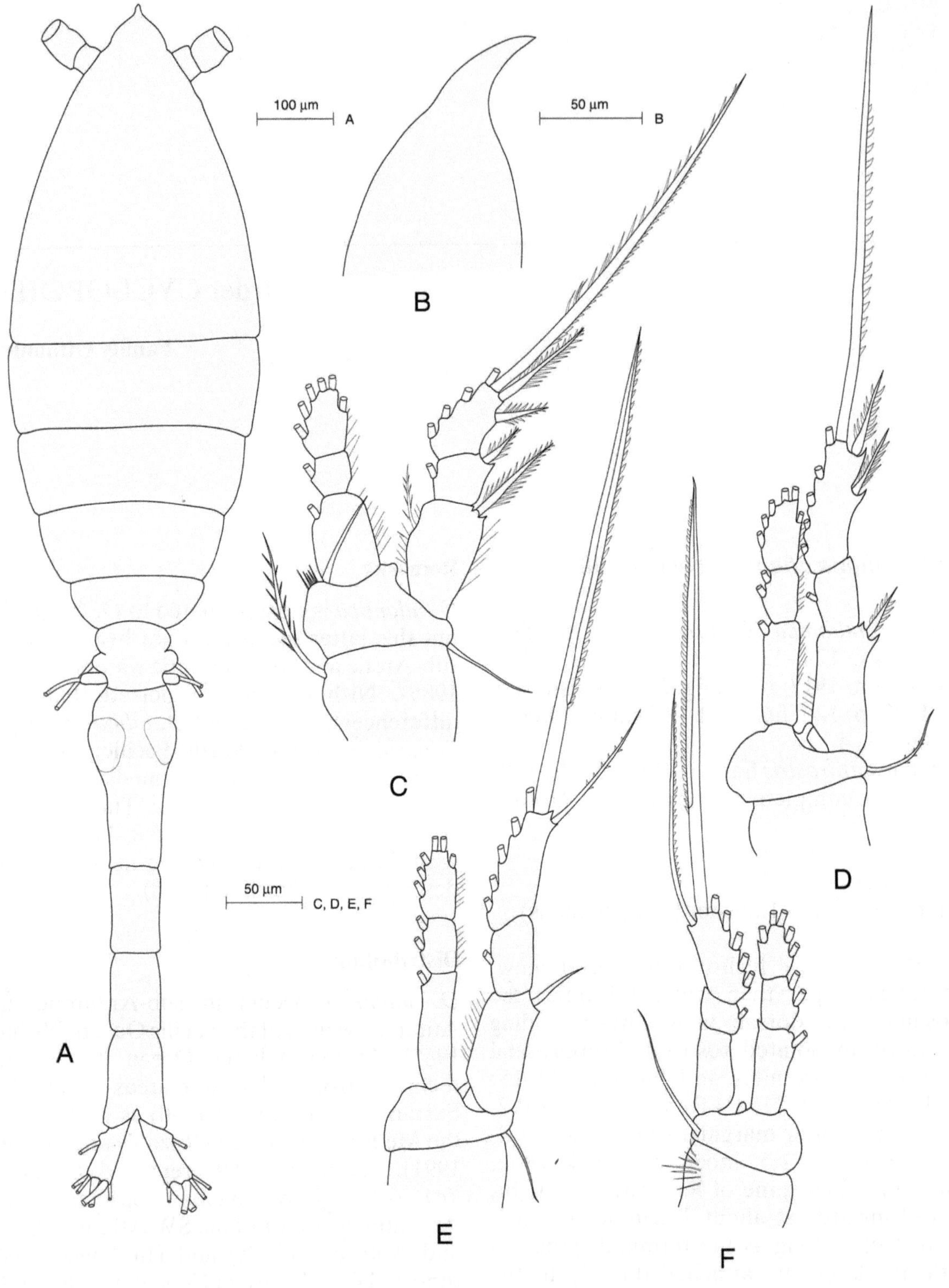

Fig. 3.39.1A–F. *Oithona atlantica*. Female: **A** whole animal dorsal view; **B** forehead lateral view; **C** P1; **D** P2; **E** P3; **F** P4

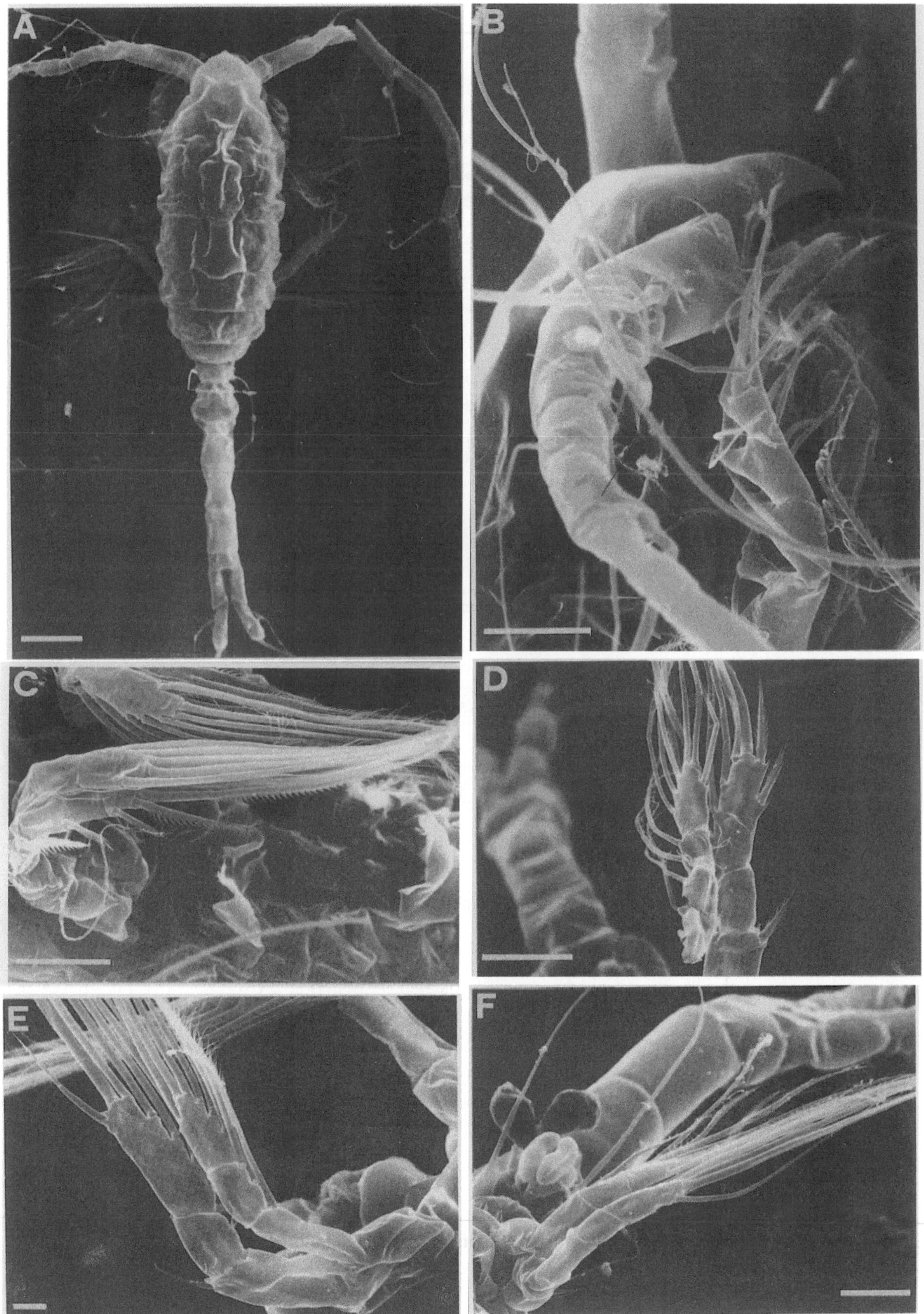

Fig. 3.39.2A–F. *Oithona atlantica*. Female: **A** whole animal dorsal view; **B** beak-like forehead in lateral view; **C** P1; **D** P2; **E** P3; **F** P4. *Bars*: **A** 100 μm; **B**, **C**, **D**, **F** 50 μm; **E** 10 μm

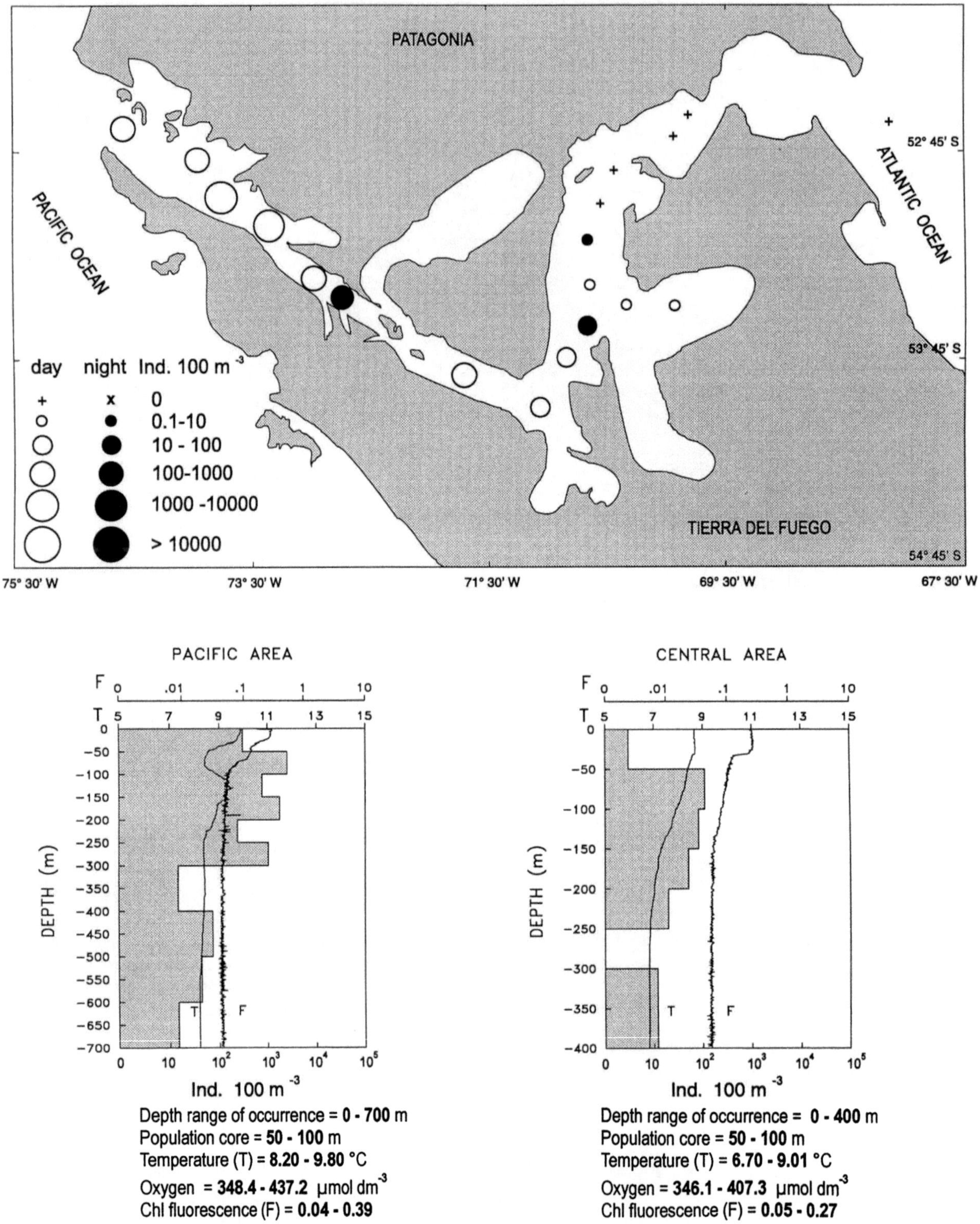

Fig. 3.39.3. *Oithona atlantica*. Distribution in the Straits of Magellan in February–March 1991

3.40 *Oithona similis* Claus 1866

Oithona similis Claus, 1966: p. 14 – Giesbrecht, 1892: p. 537, pl. 34 figs. 18, 19, 21, 36–39, pl. 44 figs. 3, 5, 8–11 – Rosendorn, 1917: p. 24, fig. 13a–e – Mori, 1937: p. 112, pl. 62 figs. 1–12 – Tanaka, 1960a: p. 62, pl. 27 figs. 1-9 – Nishida et al., 1977: p. 149, figs. 21a–h, 22a–c – Nishida, 1985: p. 88, figs. 50, 51.
Oithona helgolandica Claus, 1863: p. 105, pl. 1 fig. 2 – Sars, 1918: p. 8, pl. 3.

Body length –
females: 0.80–0.92 mm (39 specimens)

Female. Prosome elongated, ellipsoid in dorsal view, 2.6 times longer than wide, 1.2 times urosome. Head truncated anteriorly in dorsal view; in lateral view produced ventrally into sharply pointed rostrum. Proportional lengths of urosomites and furca 15:36:14: 12:12:11 = 100. Anterior 1/3 of genital somite swollen laterally. P1–P4 Re with 1,1,2; 1,0,1; 1,0,1; 0,0,1 outer marginal spines and 0,1,4; 0,1,5; 0,1,5; 0,1,5 inner marginal setae. Outer marginal spine on P4 Re3 minute.

Remarks

Our specimens agree in general shape, size range and swimming legs with the typical *O. similis*. There seems to be some confusion about the name of this species, as reported by Crisafi (1959), Ramirez (1969) and Nishida et al. (1977), who widely discussed the relationships between *O. similis* and *O. helgolandica*. Nishida et al. (1977) are of the opinion that the species described as *O. helgolandica* by Claus (1863) should be rejected.

Distribution

O. similis is considered a cosmopolitan species with a very wide geographical distribution. Farran (1929) reported *O. similis* common in the Pacific sector of the Antarctic and frequent in hauls made under the ice. It has been reported in the Atlantic Sector of the Antarctic by Vervoort (1951). Its distribution in the Pacific and Indian Oceans is discussed by Nishida (1985), who reported this species as very abundant in sub-Arctic, sub-Antarctic and Antarctic waters. Bradford (1971b) reported *O. similis* as abundant in the Ross Sea (Antarctica) and in the warm deep waters of the Antarctic. North of the Antarctic Convergence, it occurs in surface waters (Vervoort 1957). Ramirez (1969, 1977) reported *O. helgolandica* as common in Argentine waters. Hirakawa (1989) reported *O. similis* for southern Chilean fjords. According to Nishida (1985), it is still unknown if there are morphologically distinct populations within the wide geographical range of *O. similis*. In the present study, it was one of the most frequently sampled species. It was recorded at all depths but with maximum abundances in the upper 50 m.

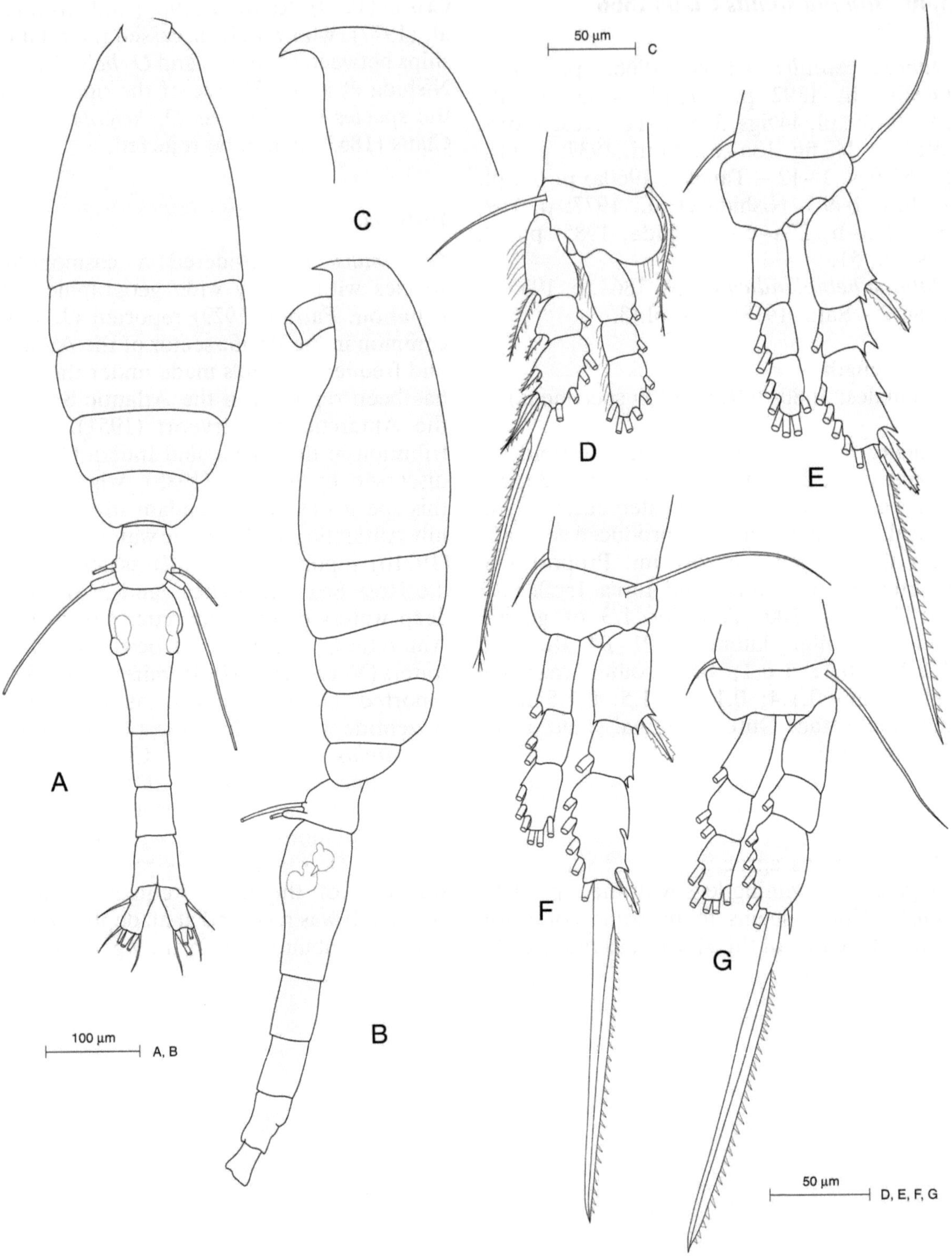

Fig. 3.40.1A–G. *Oithona similis*. Female: **A** whole animal dorsal view; **B** whole animal lateral view; **C** rostrum lateral view; **D** P1; **E** P2; **F** P3; **G** P4

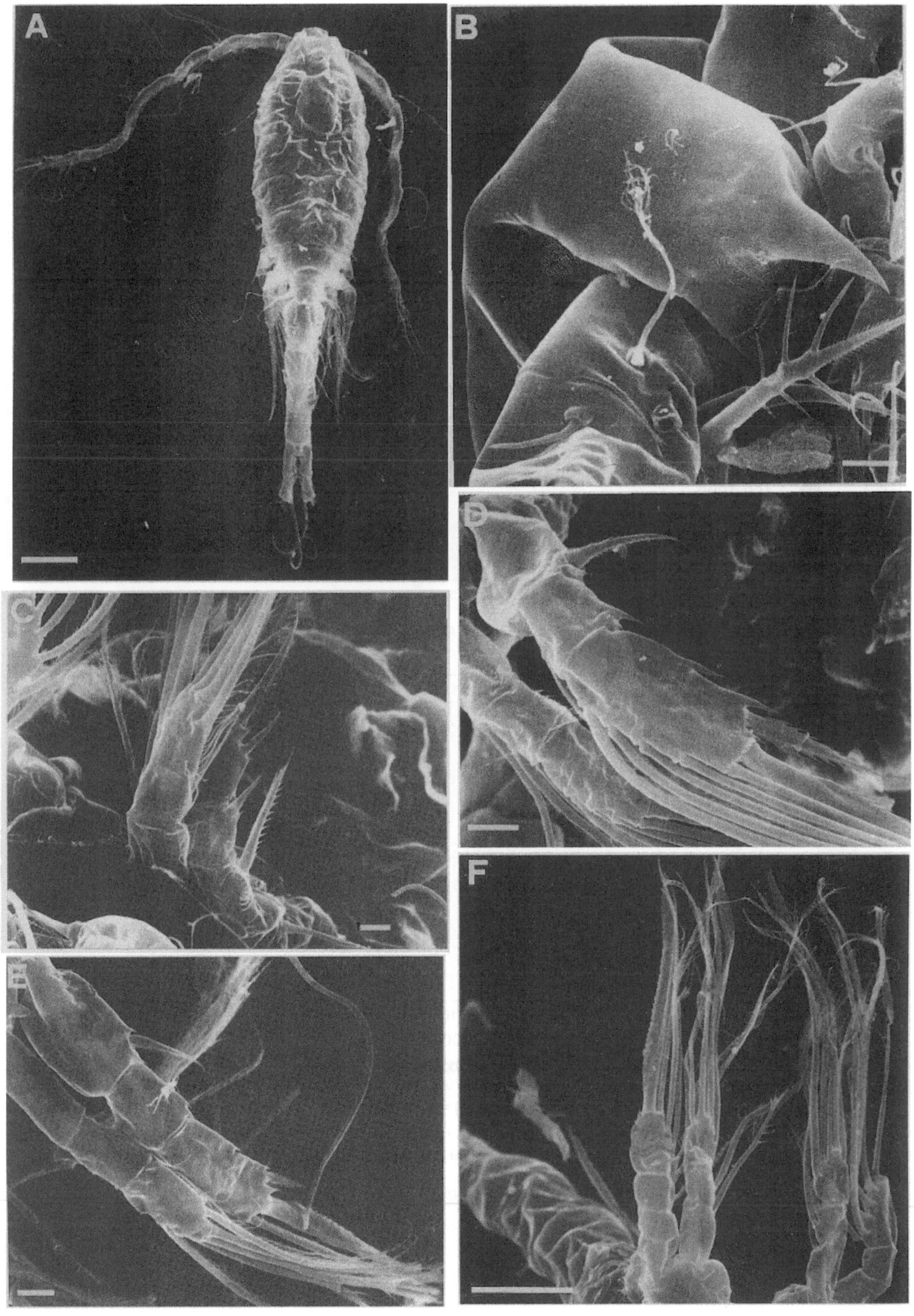

Fig. 3.40.2A–F. *Oithona similis*. Female: **A** whole animal dorsal view; **B** rostrum ventro-lateral view; **C** P1; **D** P2; **E** P3; **F** P4. *Bars*: **A** 100 µm; **B**, **C**, **D**, **E** 10 µm; **F** 50 µm

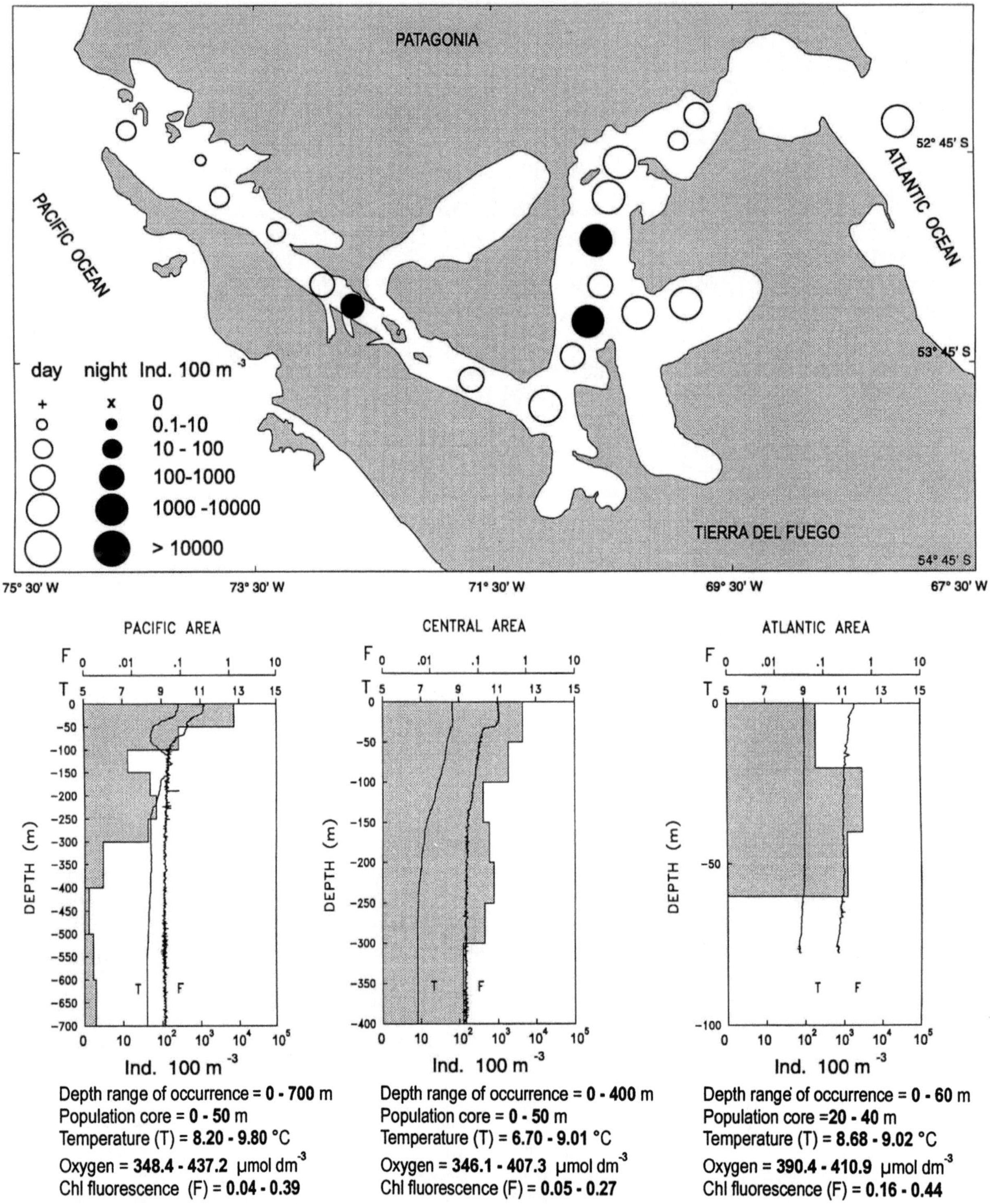

Fig. 3.40.3. *Oithona similis*. Distribution in the Straits of Magellan in February–March 1991

Order POECILOSTOMATOIDA

Family Oncaeidae

3.41 *Lubbockia aculeata* Giesbrecht 1891

Lubbockia aculeata Giesbrecht, 1891: p. 477 – Giesbrecht, 1892: p. 611, pl. 48 figs. 3, 9, 11, 13, 16, 20 – Boxshall, 1977: p. 110, figs. 4, 5 – Heron and Damkaer, 1978: p. 15, figs. 10, 11 – Stephen, 1988: p. 36, fig. 1a,b.

Body length –
females: 2.55–2.84 mm (4 specimens)

Female. Body elongated. Prosome 2.3 times longer than wide. Postero-lateral corners of last thoracic somite ending in slender processes. Proportional lengths of prosome and urosome 55:45; proportional lengths of urosomites and furca 11:27:17:16:16:13 = 100. Posterior margins of urosomites with notched hyaline fringe. Maxilliped basal segment with one inner denticle; second segment with four denticles on inner surface; third segment very short and unarmed; terminal segment clawshaped with distal row of small denticles on concave margin and small spinule at end of row. Re3 of P1–P4 with two spines on external margin. P5 with two setae of different length; longer seta not reaching posterior margin of genital complex.

Distribution

L. aculeata has been reported from all oceans including the Mediterranean Sea. This species has a tropical distribution, although a few specimens have been collected outside of these waters (Heron and Damkaer 1978). It has been reported also for the Antarctic (Ramirez and Dinofrio 1976) and in Argentine (Ramirez 1969) and Chilean (Wilson 1942) waters. In the present study, the species was sporadically present from 100–500 m at two stations in the Pacific area of the Straits of Magellan.

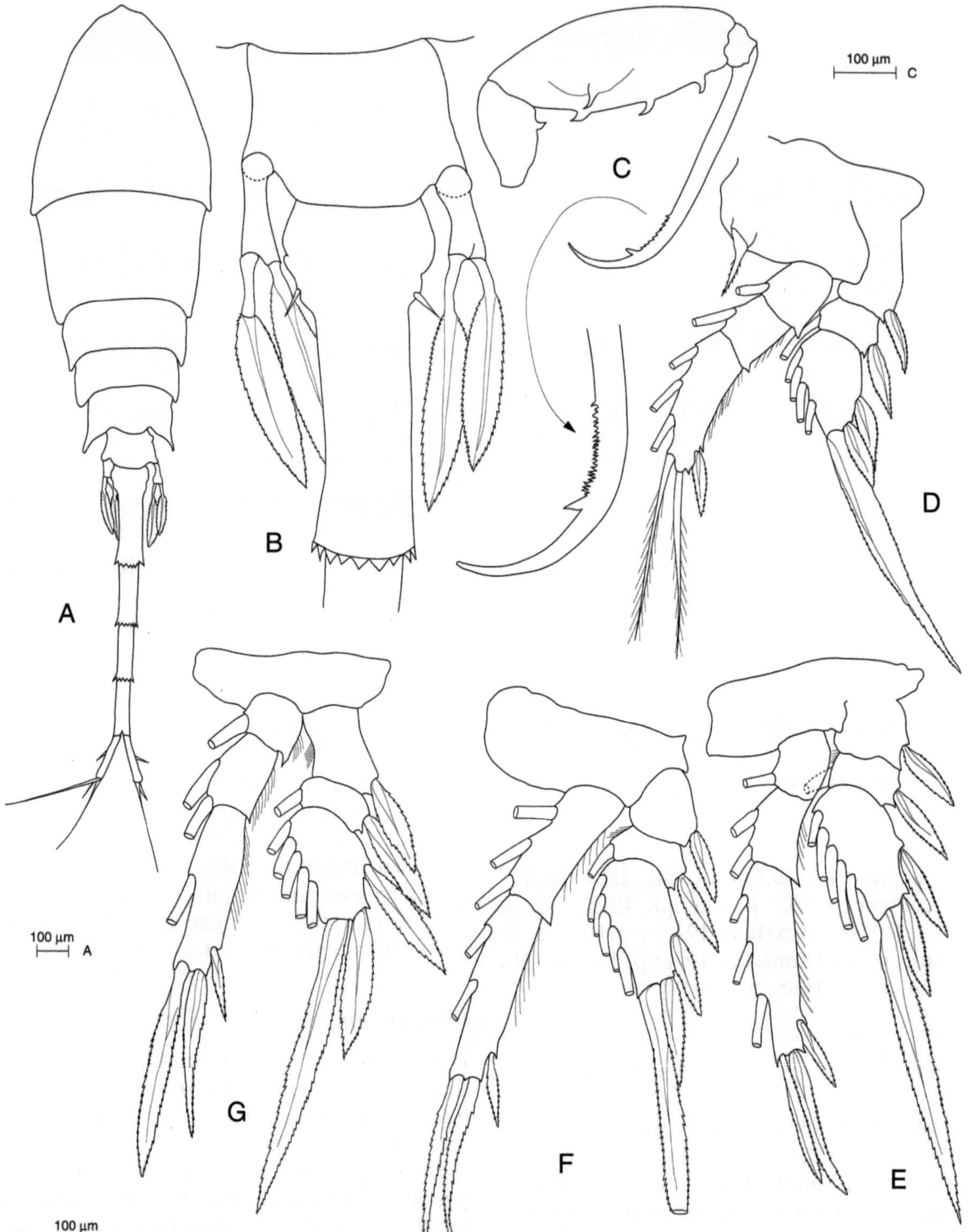

Fig. 3.41.1A–G. *Lubbockia aculeata*. Female: **A** whole animal dorsal view; **B** last thoracic somite and genital somite dorsal view; **C** maxilliped; **D** P1; **E** P2; **F** P3; **G** P4

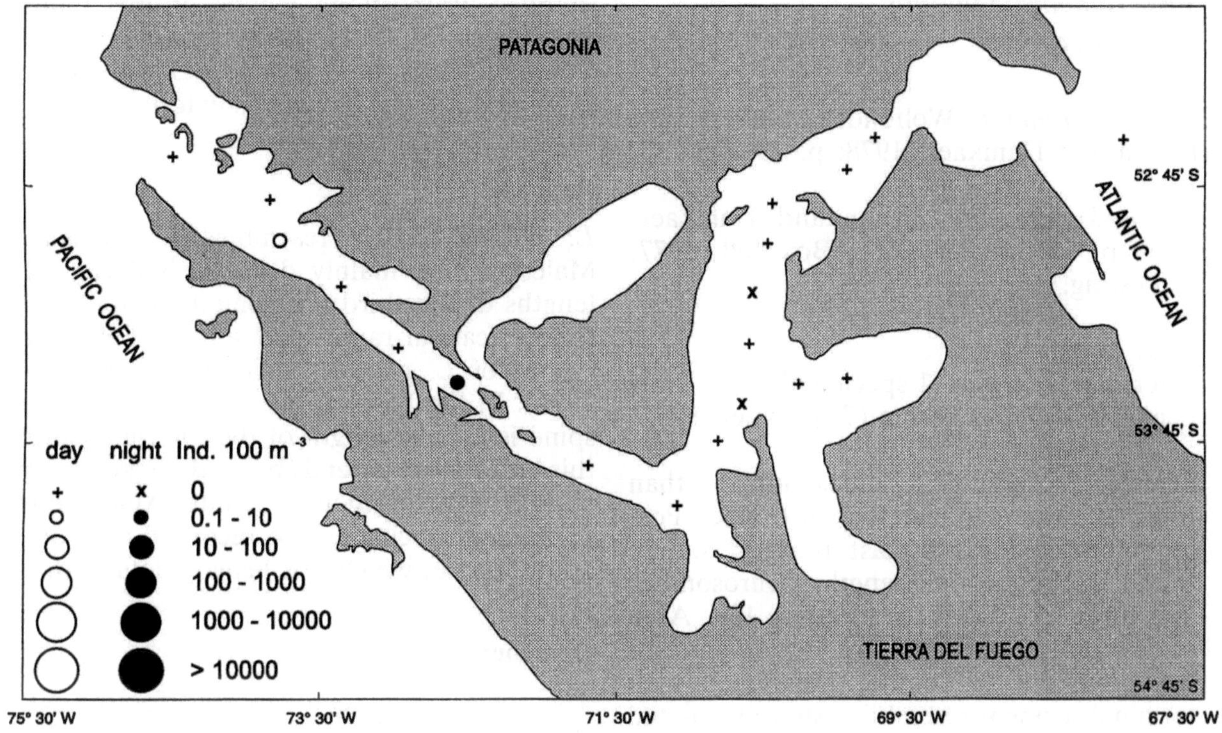

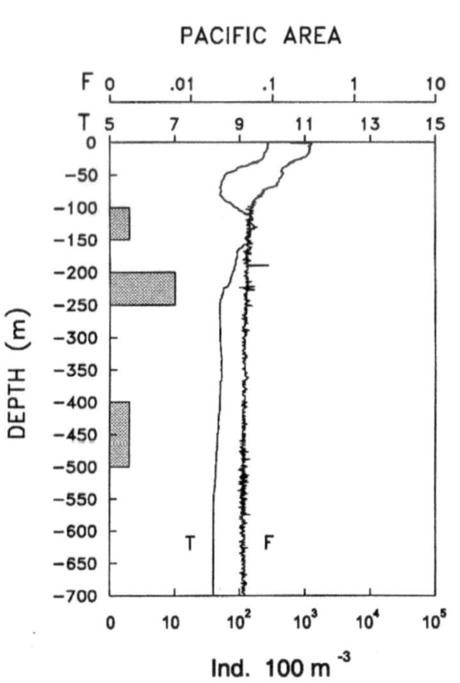

Depth range of occurrence = **100 - 500** m
Population core = **200 - 250** m
Temperature (T) = **8.20 - 9.80** °C
Oxygen = **348.4 - 437.2** µmol dm^{-3}
Chl fluorescence (F) = **0.04 - 0.39**

Fig. 3.41.2. *Lubbockia aculeata*. Distribution in the Straits of Magellan in February–March 1991

3.42 *Lubbockia minuta* Wolfenden 1905

Lubbockia minuta Wolfenden, 1905: p. 15 – Heron and Damkaer, 1978: p. 19, figs. 13, 14, 15.
Lubbockia glacialis Heron and Damkaer, 1969: p. 16, figs. 20, 21 – Boxshall, 1977: p. 118, fig. 8.

Body length –
female: 1.30 mm (1 specimen)
males: 1.90–1.98 mm (7 specimens)

Female. Prosome 2.8 times longer than wide, 1.2 times longer than urosome. Posterolateral corners of last thoracic somite rounded. Proportional lengths of urosomites and furca 15:25:17:20:10:13 = 100. Area of external genital apparatus located anterior to mid-length of somite. Maxilliped with terminal segment resembling sharply pointed claw. P1–P2 Re3 with three short outer, lanceolated spines. P5 with slender segment not reaching area of external genital apparatus; two terminal spines, the inner longer and reaching mid-length of genital somite.

Male. Urosome 1.5 longer than prosome. Postero-lateral corners of last thoracic somite rounded. Proportional lengths of urosomites and furca 9:12:16:19:29:5:10 = 100. Swimming legs as in female. P5 with inner spine about twice as long as external spine.

Remarks

L. minuta closely resembles *L. glacialis*. Males can be mainly distinguished for the lengths of the third postgenital somite, anal somite, caudal ramus and spines of P5 (see figures of *L. glacialis* male in Heron et al. 1984). In the P5 of *L. glacialis*, the external spine is 1/4 the length of the internal spine; this latter spine extends beyond the posterior margin of the postgenital somite. Among our *L. minuta* specimens, we recorded only one adult female, which was heavily damaged.

Distribution

The species has been reported for the NE Atlantic (Wolfenden 1905), tropical Atlantic and NE Pacific (Heron and Damkaer 1978) Oceans. To our knowledge, *L. minuta* has never been reported in Argentinian or Chilean waters. In the present study, it was sampled from 400–500 m at one Pacific station in the Straits of Magellan and from 100–150 m at the entrance to Baia Inutil.

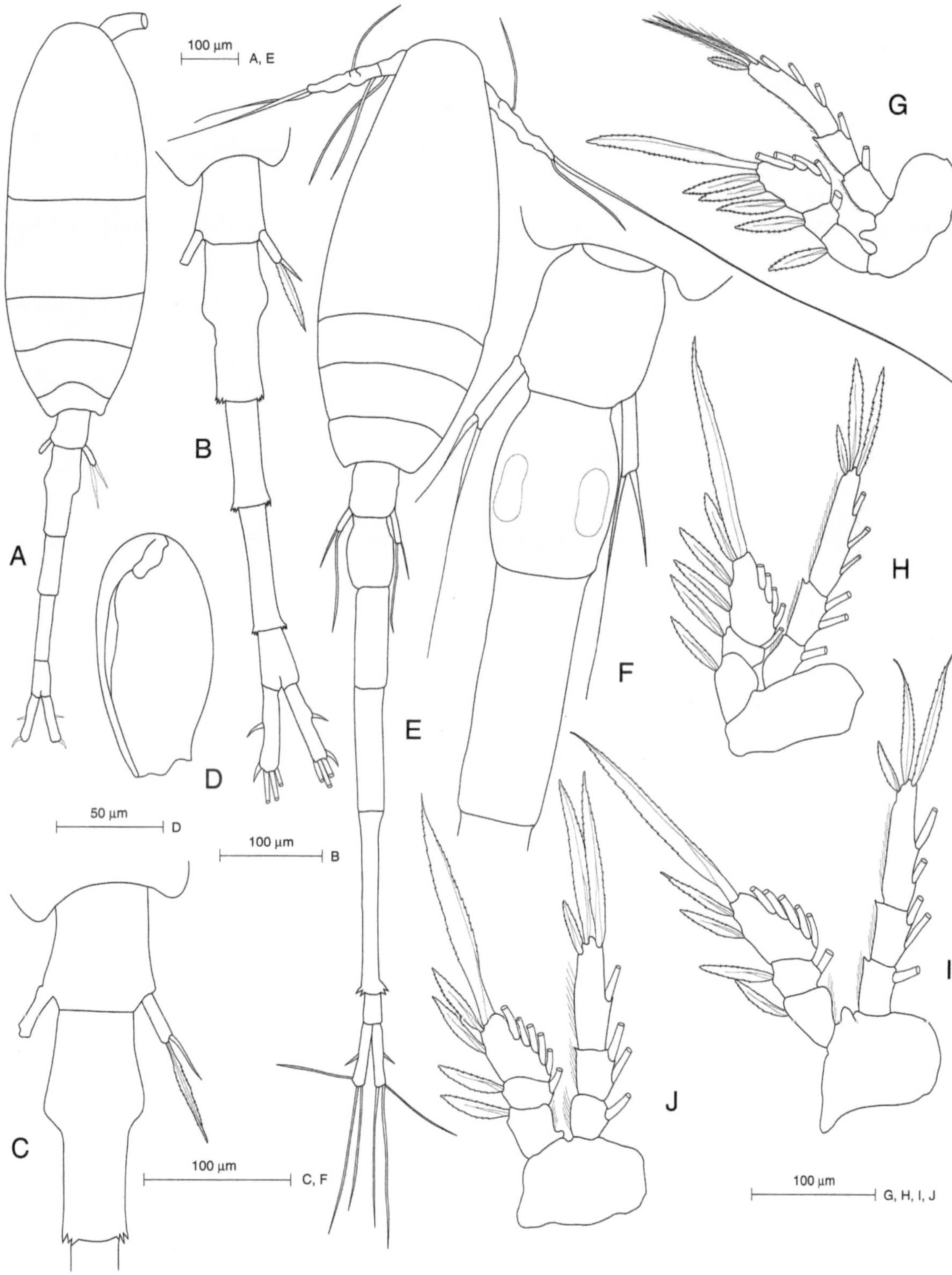

Fig. 3.42.1A–J. *Lubbockia minuta*. Female: **A** whole animal dorsal view; **B** urosome dorsal view; **C** last thoracic somite and genital somite dorsal view; **D** maxilliped. Male: **E** whole animal dorsal view; **F** last thoracic somite, and 1st and 2nd urosomal somites in dorsal view; **G** P1; **H** P2; **I** P3; **J** P4

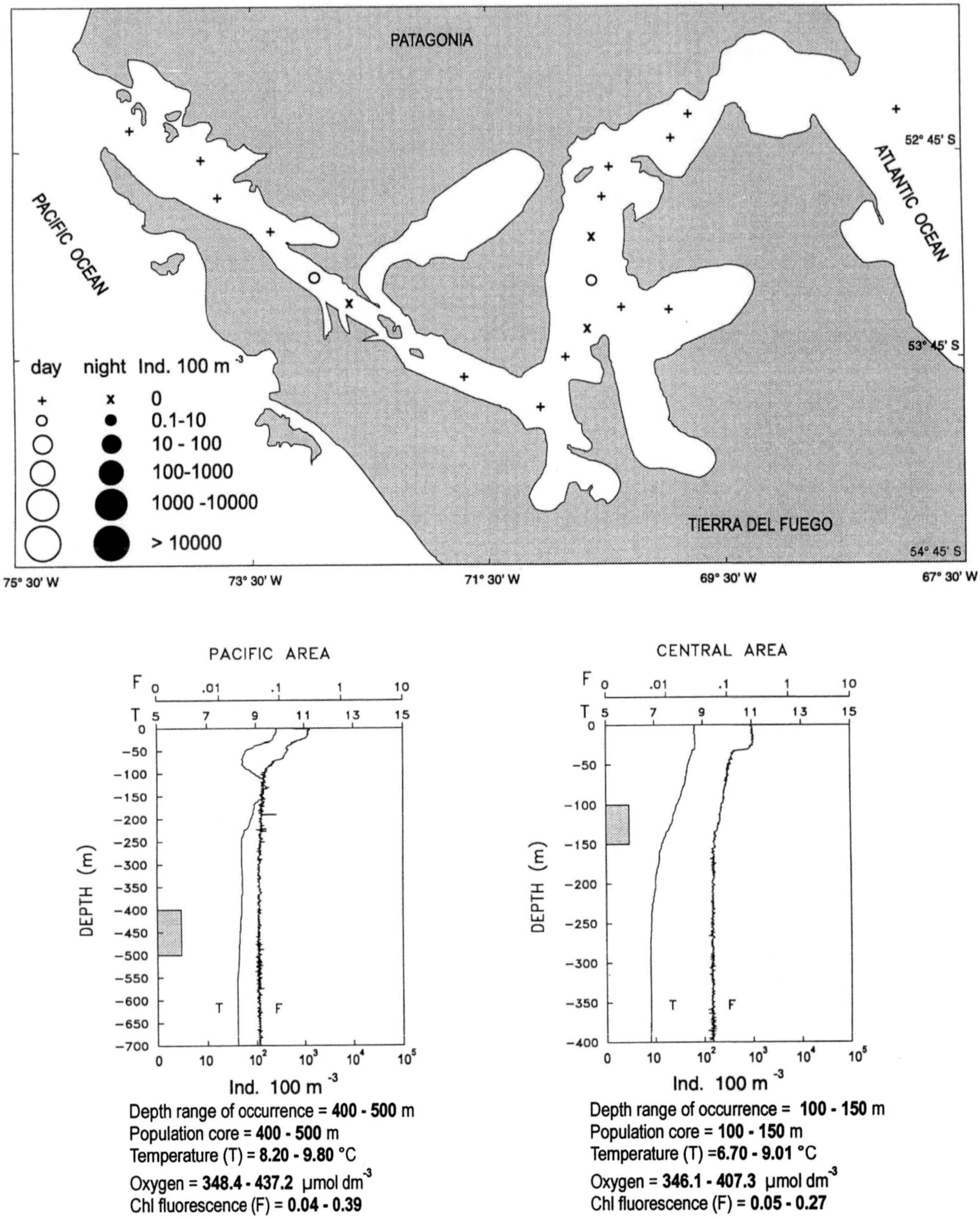

Fig. 3.42.2. *Lubbockia minuta*. Distribution in the Straits of Magellan in February–March 1991

3.43 *Oncaea conifera* Giesbrecht 1891

Oncaea conifera Giesbrecht, 1891: p. 471; 1892: p. 591, pl. 2 fig. 10, pl. 47 figs. 4, 16, 21, 23, 28, 34–38, 42, 55, 56 – Mori, 1937: p. 120, pl. 66 figs. 10–13 – Tanaka, 1960a: p. 66, pl. 29 figs. 1–8 – Owre and Foyo, 1967: p. 111, figs. 807–811 – Ramirez, 1969: p. 89, pl. 17 figs. 139, 140, 142, 147, pl. 18 fig. 150.

Body length –
females: 1.12–1.33 mm (20 specimens)
males: 0.88–1.00 mm (20 specimens)

Female. Prosome twice as long as wide. Rostral area with thickened, rounded ventral margin, like a fold. Second thoracic somite bearing conspicuous dorsal projection, characteristic in lateral view. Proportional lengths of prosome and urosome 64:36; proportional lengths of urosomites and furca 16:48:5:5:10:16 = 100. Genital somite with two setules in mid-region of dorsal surface. Caudal rami almost three times longer than wide. Second segment of maxilliped with two spines on inner surface and long row of setules. Terminal segment bearing long claw with row of denticules on concave surface. P4 endopod terminates in conical projection with patch of hairs and two spines of about the same length, with serrated, hyaline flange. P5 free segment elongated, with one seta almost twice as long as the other.

Male. Prosome more slender than female, 2.2 times longer than wide, lacking dorsal protrusion on third thoracic somite. Proportional lengths of prosome and urosome 66:34; proportional lengths of urosomites and furca 8:51:14:2:2:11:12 = 100. Genital somite with postero-lateral flap on ventral surface, ending with pointed posterior corners. Flap bearing denticulated margin on ventral side and minute, hyaline denticules on both ventral and dorsal surfaces.

Remarks

Giesbrecht (1902) distinguished this species as three varieties: "variety I", the specimens he originally described from the Mediterranean Sea and Pacific Ocean; "variety II", specimens from the Arctic Ocean; and "variety III", specimens from the Antarctic-Pacific Ocean area. Giesbrecht's Antarctic "variety III" has been elevated to full species status as *O. antarctica* by Heron (1977). This author discussed all morphological details distinguishing *O. conifera* from *O. antarctica*. Malt (1983) summarized in a table (p. 458) the characters distinguishing *O. conifera* females from specimens of other *Oncaea* species having the same dorsal projection. The same author reported and widely discussed polymorphic forms of *O. conifera* found in the North Atlantic (Malt 1983). Our specimens do not show the characteristics of *O. antarctica* and agree in main features with Giesbrecht's "variety I". However, in our specimens, the terminal lanceolate spines in P4 Ri3 are almost of the same length, whereas in Giesbrecht's (1892) fig. 38 (pl. 47) the terminal spine on the external side is longer. A more detailed morphological study is necessary to verify if polymorphism characterizes also co-occurring populations in the Straits of Magellan.

Distribution

This species is reported as cosmopolitan. It has an almost world-wide distribution, ranging from Arctic to Antarctic waters and is also distributed over the Atlantic, Pacific and Indian Oceans as well as the Mediterranean Sea. Farran (1929) found it in samples collected from New Zealand, southwards to the Antarctic continent, also in hauls beneath the ice. Vervoort (1951) reported *O. conifera* in the Atlantic sector of the Antarctic, Bradford (1971b) in the Ross Sea (Antarctica). Takana (1960a) reported *O. conifera* from the Japanese Antarctic Expedition, in regions south of 66°S. This species has been reported from Argentine waters (Ramirez 1969) and from Chilean fjords (Marin and Antezana

1985). In the present study, it was quite common in the Pacific area but rare in the Central and Atlantic areas of the Straits of Megellan up to the entrance of the Primera Angostura. It was sampled throughout the water column but had maximum abundances from 100–200 m.

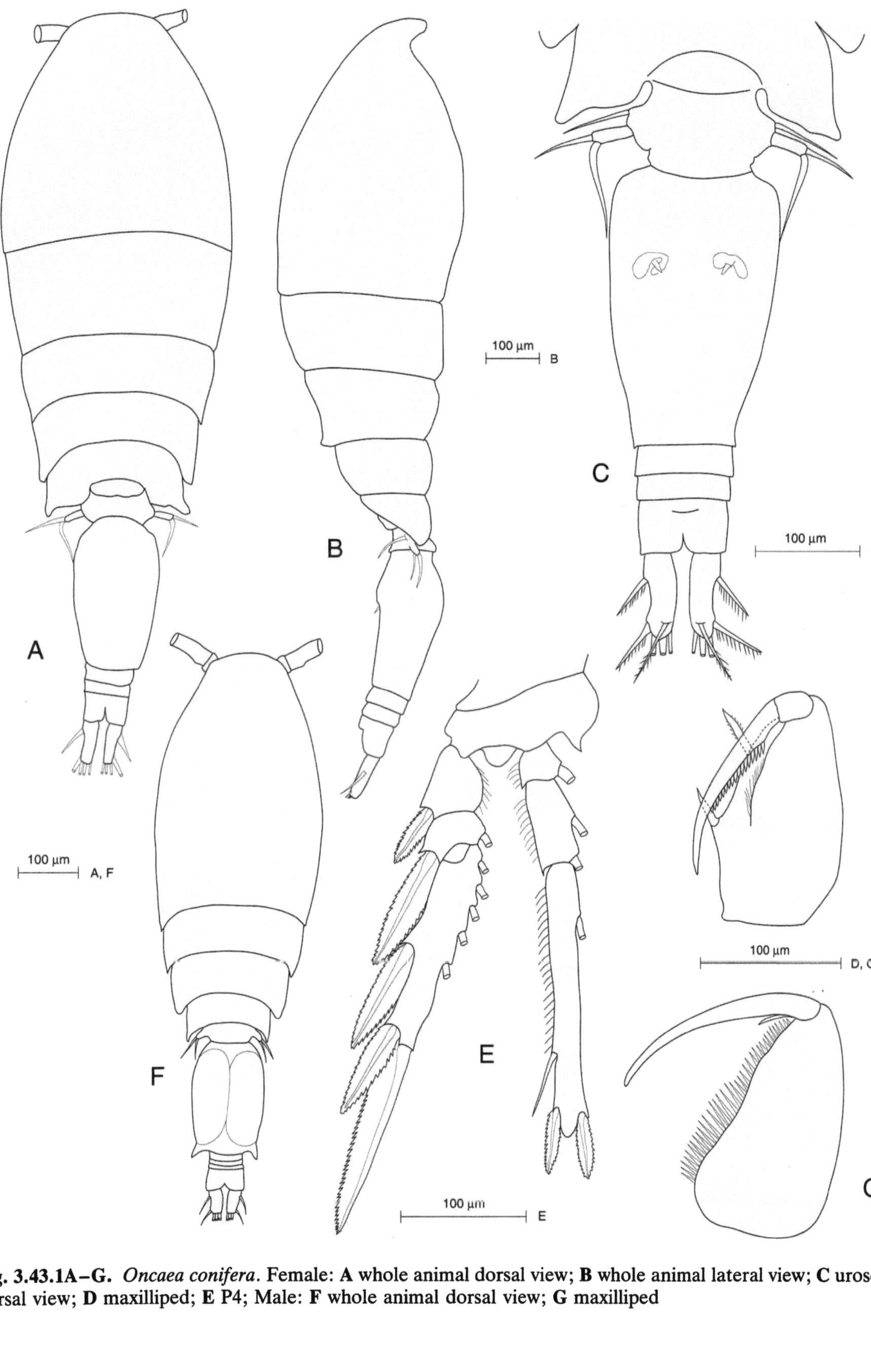

Fig. 3.43.1A–G. *Oncaea conifera*. Female: **A** whole animal dorsal view; **B** whole animal lateral view; **C** urosome dorsal view; **D** maxilliped; **E** P4; Male: **F** whole animal dorsal view; **G** maxilliped

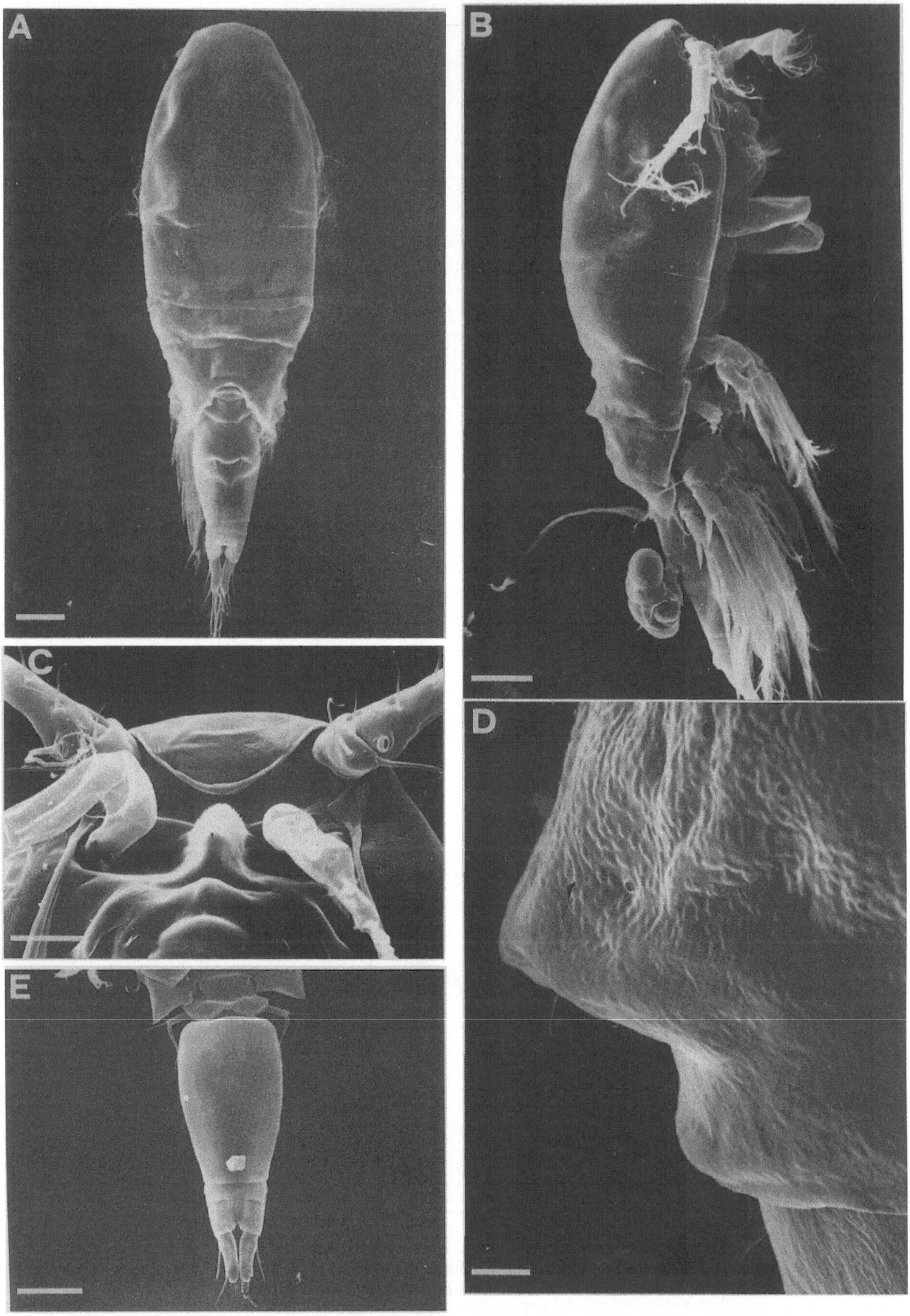

Fig. 3.43.2A–E. *Oncaea conifera*. Female: **A** whole animal dorsal view; **B** whole animal lateral view with ovigerous sacs; **C** forehead ventral view; **D** characteristic dorsal hump on 2nd thoracic somite; **E** urosome ventral view. *Bars*: **A**, **B**, **E** 100 μm; **C** 50 μm; **D** 10 μm

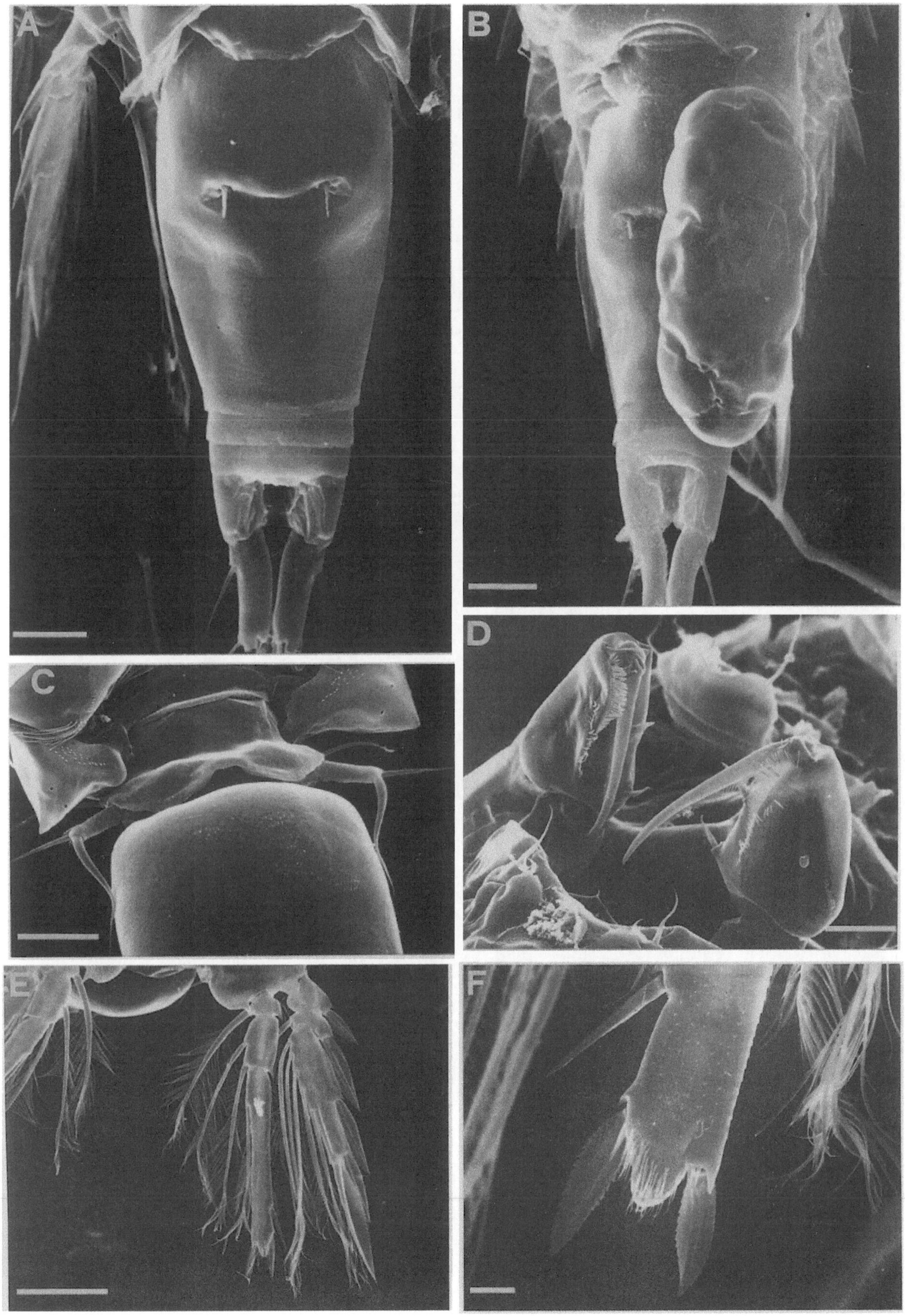

Fig. 3.43.3A–F. *Oncaea conifera*. Female: **A** urosome dorsal view; **B** urosome dorsal view with attached ovigerous sac; **C** P5; **D** maxilliped; **E** P4; **F** detail of extremity of last joint of P4 endopod. *Bars*: **A**, **B**, **C**, **D** 50 µm; **E** 100 µm; **F** 10 µm

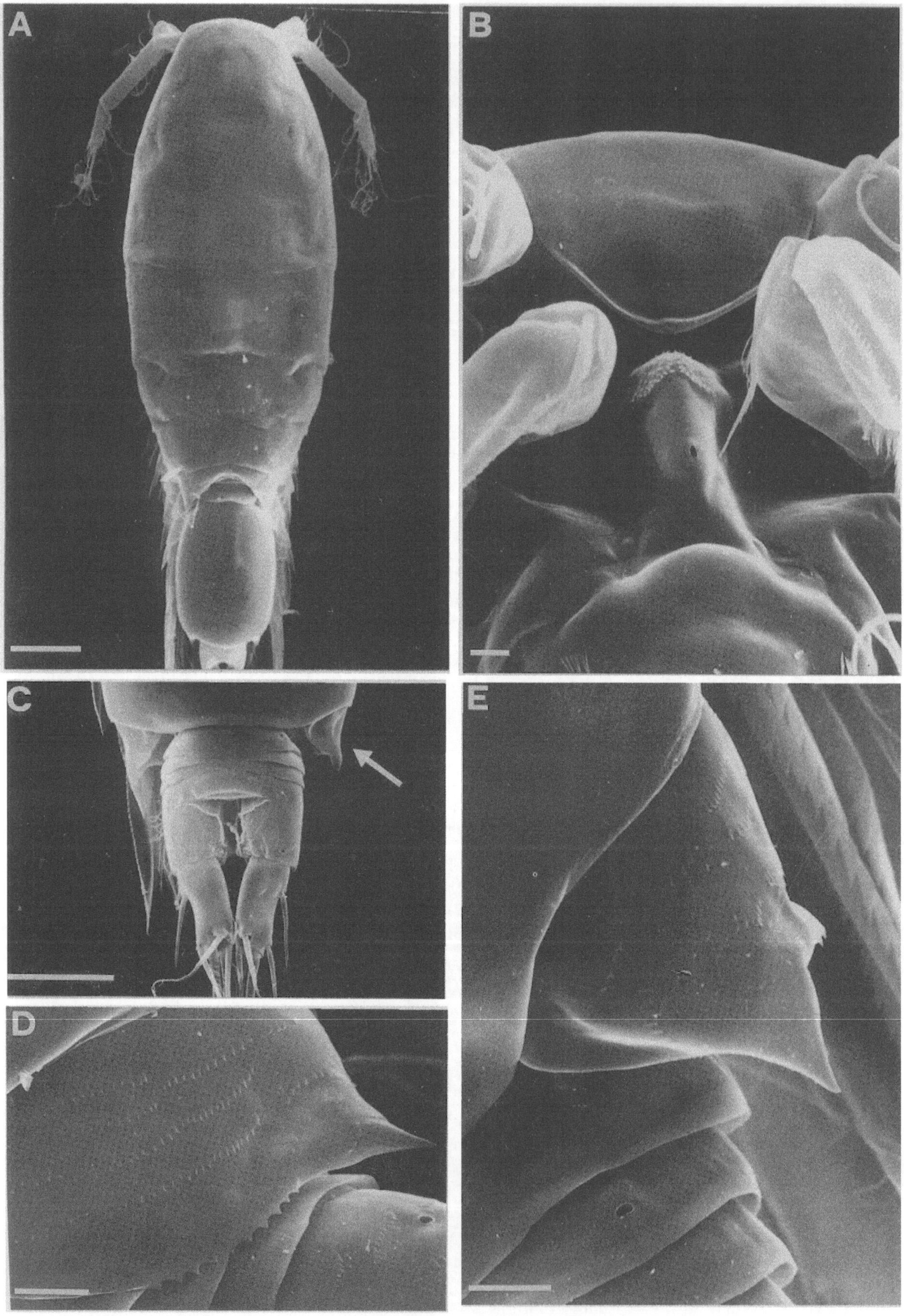

Fig. 3.43.4A–E. *Oncaea conifera*. Male: **A** whole animal dorsal view; **B** forehead ventral view; **C** urosome dorsal view; **D** postero-lateral flap on genital somite ventral view (*arrow* in **C**); **E** same flap dorsal view. *Bars*: **A** 100 μm; **B**, **D**, **E** 10 μm; **C** 50 μm

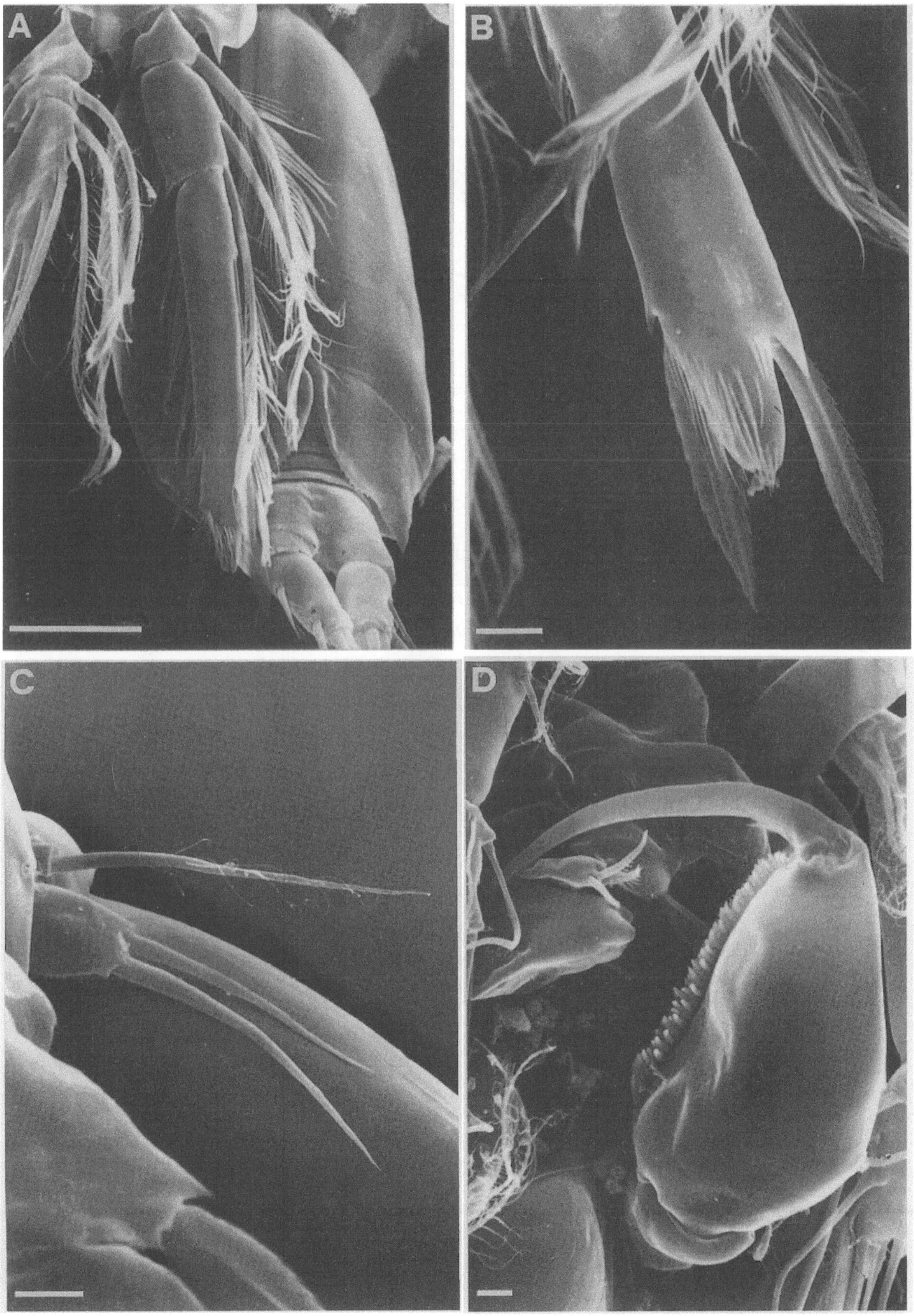

Fig. 3.43.5A–D. *Oncaea conifera*. Male: **A** P4 and urosome ventral view; **B** extremity of last joint of P4 endopod; **C** P5; **D** maxilliped. *Bars*: **A** 50 µm; **B**, **C**, **D** 10 µm

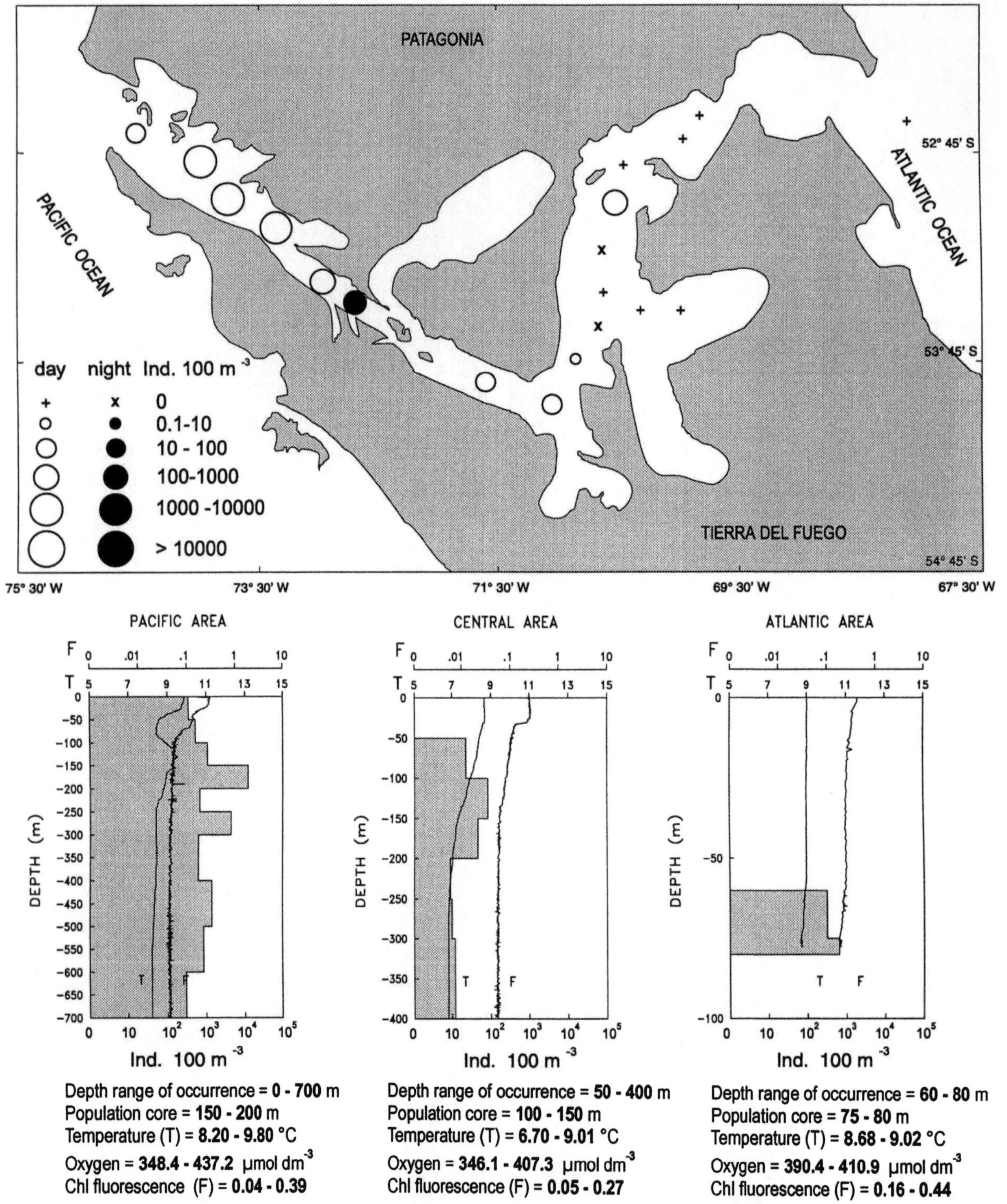

Fig. 3.43.6. *Oncaea conifera*. Distribution in the Straits of Magellan in February–March 1991

3.44 *Oncaea curvata* Giesbrecht 1902

Oncaea curvata Giesbrecht, 1902: p. 42, pl. 13 figs. 12–17 – Tanaka, 1960a: p. 68, pl. 30 figs. 1–11, pl. 31 figs. 1–3 – Heron, 1977: p. 70, fig. 21d–q.

Body length –
females: 0.57–0.69 mm (10 specimens)
male: 0.53 mm (1 specimen)

Female. Prosome 2.1 times longer than wide. Proportional lengths of prosome and urosome 63:37; proportional lengths of urosomites and furca 5:45:10:9:15:16 = 100. Genital somite about 1/2 as long as wide, swollen laterally about mid-length in dorsal view. Caudal rami about three times longer than wide. Inner surface of second segment of maxilliped with two spines, proximal spine with long and short spinules, distal spine with two rows of denticles; anterior surface with long hairs. Endopods of P2, P3 with terminal conical projection; P4 lacking projection. In P4 endopod, distal spine about as long as last joint. P5 small; segment not delimited from Th5 somite, with two terminal setae.

Male. General body shape as female, with prosome and urosome having the same proportional lengths. Prosome 2.2 times longer than wide. Proportional lengths of urosomites and furca 7:51:5:4:4:13:16 = 100. Genital somite ovate, about 1.6 times longer than wide.

Distribution

O. curvata is characteristic of the Antarctic (Farran 1929; Vervoort 1957; Tanaka 1960a; Heron 1977; Bradford 1971b). It has been reported for Chilean (Fagetti 1962) and Argentinian waters (Ramirez 1969; Ramirez and Dinofrio 1976), and was previously recorded in high numbers in the Straits of Magellan by Mazzocchi and Ianora (1991). In the present study, the species was very rare. A few specimens were sampled from 300–400 m at a single Pacific station and from 100–250 m at two stations in the Central area of the Straits of Magellan.

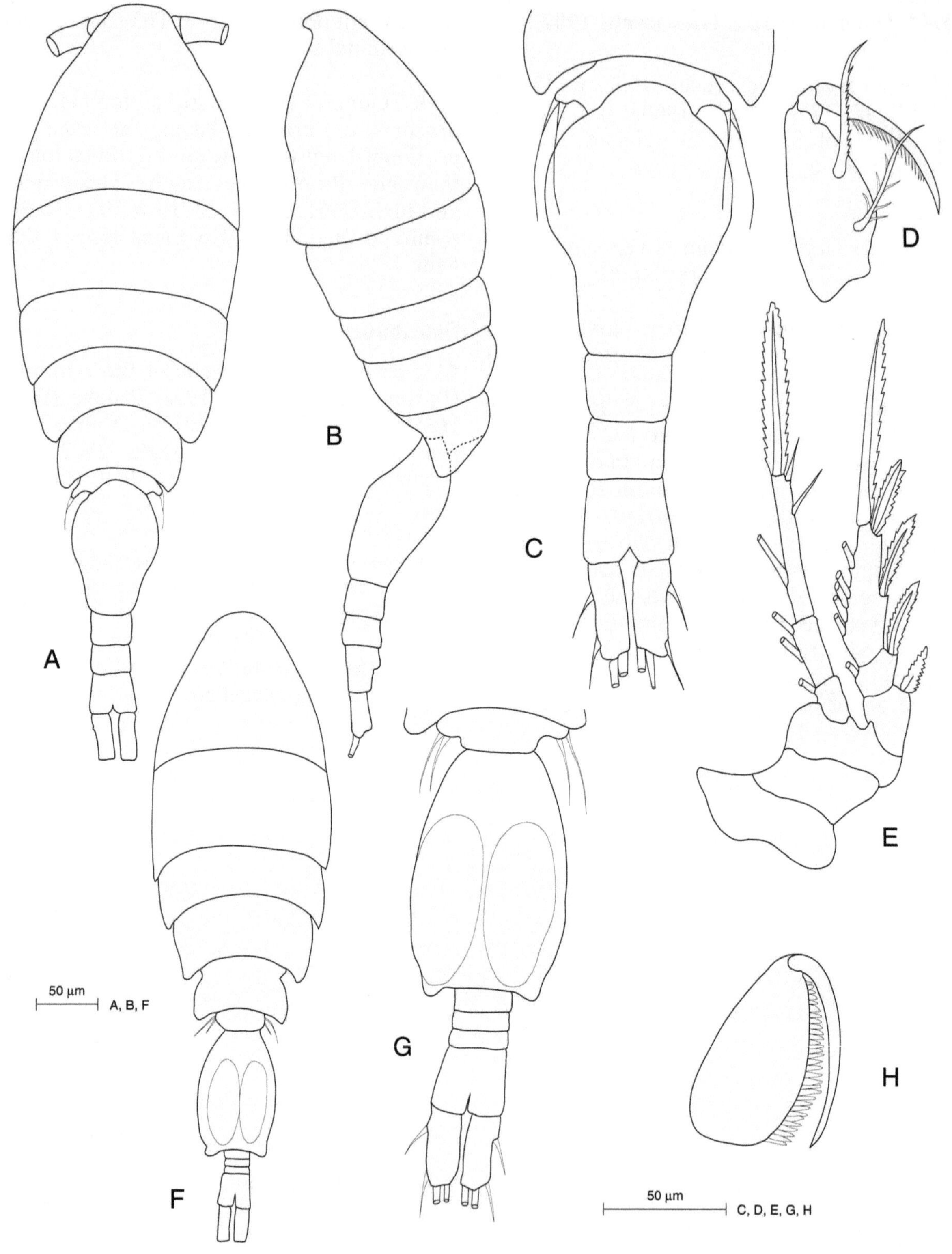

Fig. 3.44.1A–H. *Oncaea curvata*. Female: **A** whole animal dorsal view; **B** whole animal lateral view; **C** abdomen dorsal view; **D** maxilliped; **E** P4. Male: **F** whole animal dorsal view; **G** abdomen dorsal view; **H** maxilliped

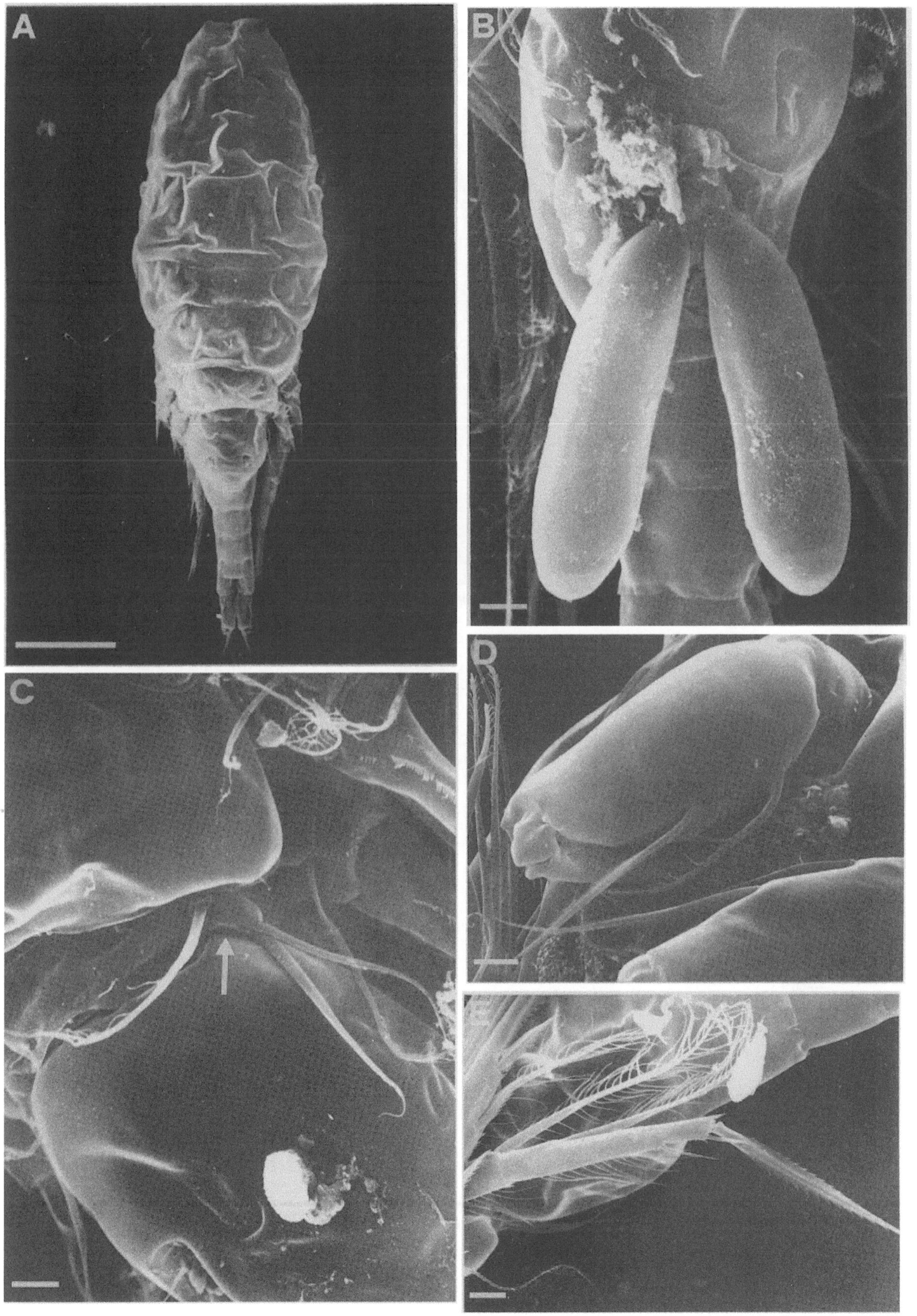

Fig. 3.44.2A–E. *Oncaea curvata*. Female: **A** whole animal dorsal view; **B** pair of spermatophores attached to genital pores on dorsal side of genital somite; **C** right P5 (*arrow*); **D** maxilliped; **E** last joint of P4 endopod. *Bars*: **A** 100 μm; **B**, **C**, **D**, **E** 10 μm

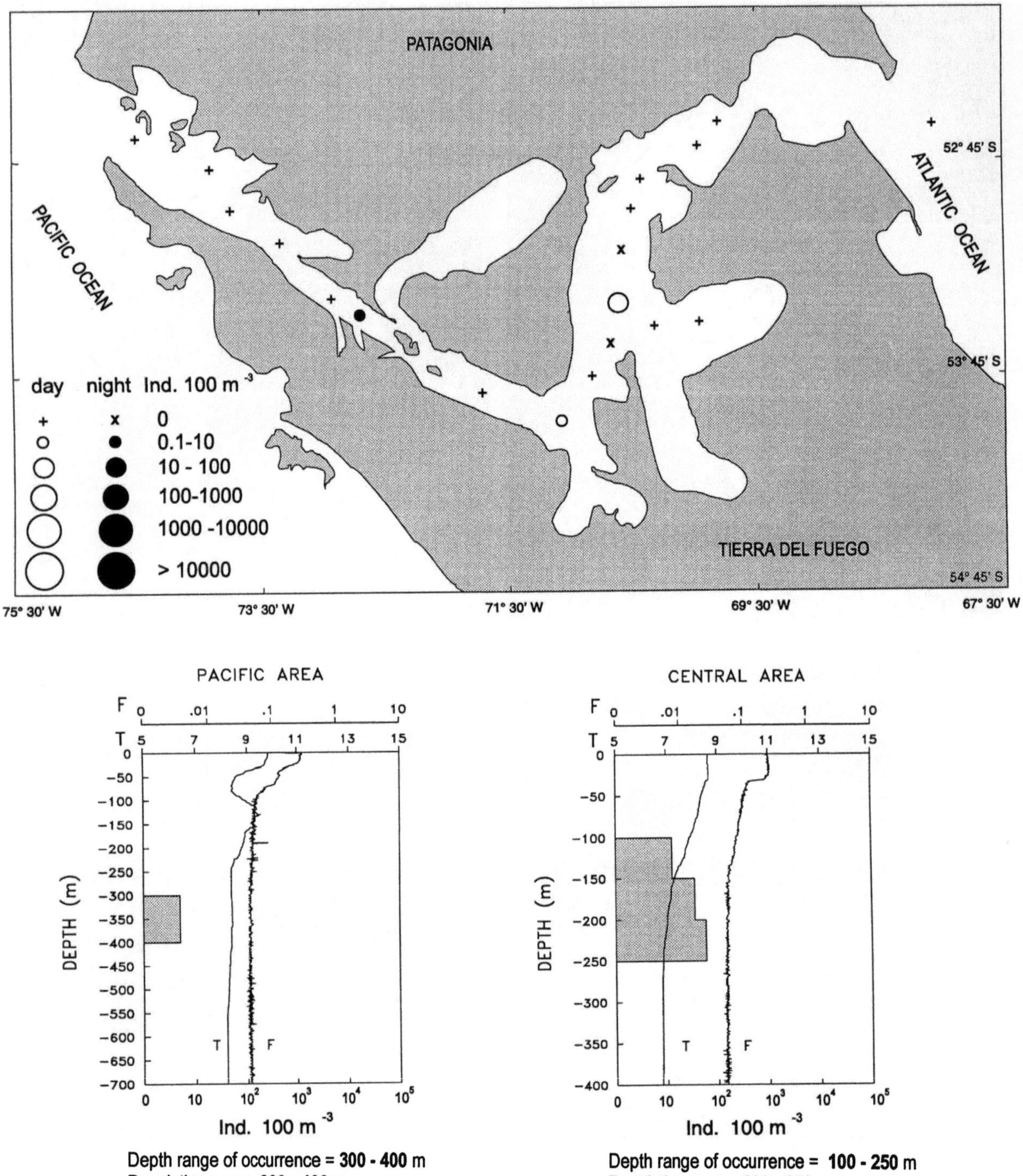

Depth range of occurrence = **300 - 400** m
Population core = **300 - 400** m
Temperature (T) = **8.20 - 9.80** °C
Oxygen = **348.4 - 437.2** μmol dm^{-3}
Chl fluorescence (F) = **0.04 - 0.39**

Depth range of occurrence = **100 - 250** m
Population core = **200 - 250** m
Temperature (T) =**6.70 - 9.01** °C
Oxygen = **346.1 - 407.3** μmol dm^{-3}
Chl fluorescence (F) = **0.05 - 0.27**

Fig. 3.44.3. *Oncaea curvata*. Distribution in the Straits of Magellan in February–March 1991

3.45 *Oncaea englishi* (Heron 1977)

Oncaea englishi Heron, 1977: p. 79, figs. 25n, 25o, 26, 27.

Body length –
females: 0.78–0.88 mm (22 specimens)
male: 0.76 mm (1 specimen)

Female. Prosome 1.9 times longer than wide, dilated at first somite, about 1.7 times as long as urosome. Head narrow and truncated in dorsal view. Rostral area with thickened, rounded ventral margin. Posterior points of fifth thoracic somite with slightly curved tip. Proportional lengths of urosomites and furca 11:53:9:7:9:11 = 100. Anal somite with patch of spinules on dorsal distal side; caudal ramus with very small spinules on dorsal side. Second segment of maxilliped with two spines on inner surface; proximal spine short, distal spine elongated, armed with two longitudinal rows of denticles. P4 Re3 with two spines; the long terminal spine longer than third segment. P5 with small free segment, terminally bearing one seta and one short setule.

Male. General shape resembling female, with head narrow and truncated in dorsal view. Proportional lengths of urosomites and furca 8:61:4:4:4:9:10 = 100. Genital somite ovoid in dorsal view. Anal somite and caudal rami with spinules. Maxilliped lacking small segment proximal to claw on female; second segment with several rows of setules and denticles. Claw sinuous, with membranous flange on inner curvature, bearing small setules.

Remarks

As discussed by Heron et al. (1984), *O. englishi* closely resembles *O. ornata* Giesbrecht, 1891, differing in the relative lengths of the female urosomites, the terminal endopod segments and the spinulation on P2 and P4. Malt (1983) discussed the differences between the two species also on the basis of the pore signature pattern. Our specimens closely agree with Herons' description except for the size range, being smaller than those reported by Heron (females 0.95–1.15 mm; males 0.85–1.00 mm).

Distribution

O. englishi was described from specimens sampled in the SW Pacific-Antarctic area (Heron 1977); it has also been recorded in Arctic waters, the NE Pacific Ocean and Norvegian Sea (Heron et al. 1984). In the present study, the species was widely distributed in the Pacific area. It was sampled at all depths but with maximum abundances from 400–500 m. A few specimens were also found from 300–400 m off Cape Froward.

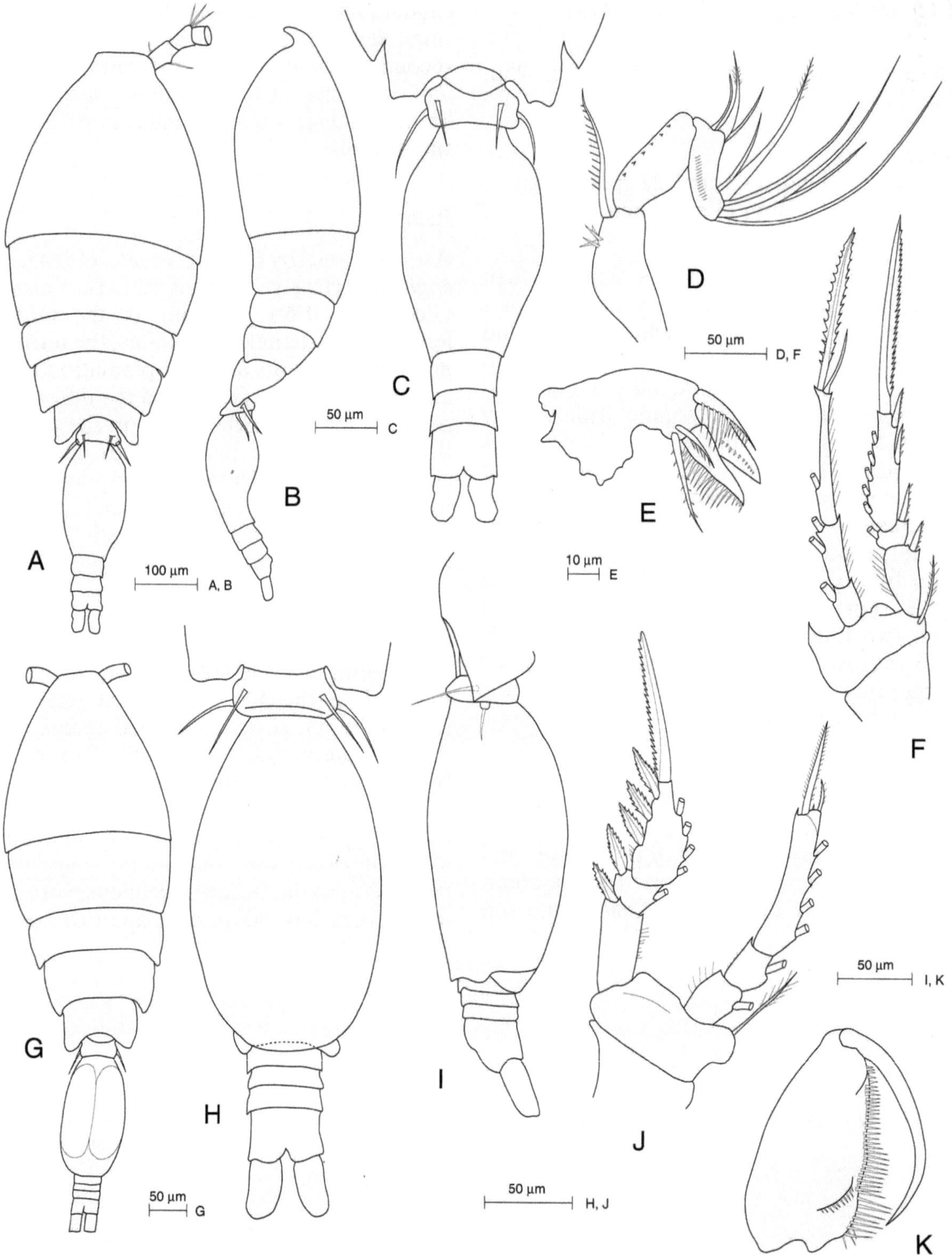

Fig. 3.45.1A–K. *Oncaea englishi*. Female: **A** whole animal dorsal view; **B** whole animal lateral view; **C** abdomen dorsal view; **D** A2; **E** mandible; **F** P4. Male: **G** whole animal dorsal view; **H** abdomen dorsal view; **I** abdomen lateral view; **J** P1; **K** maxilliped

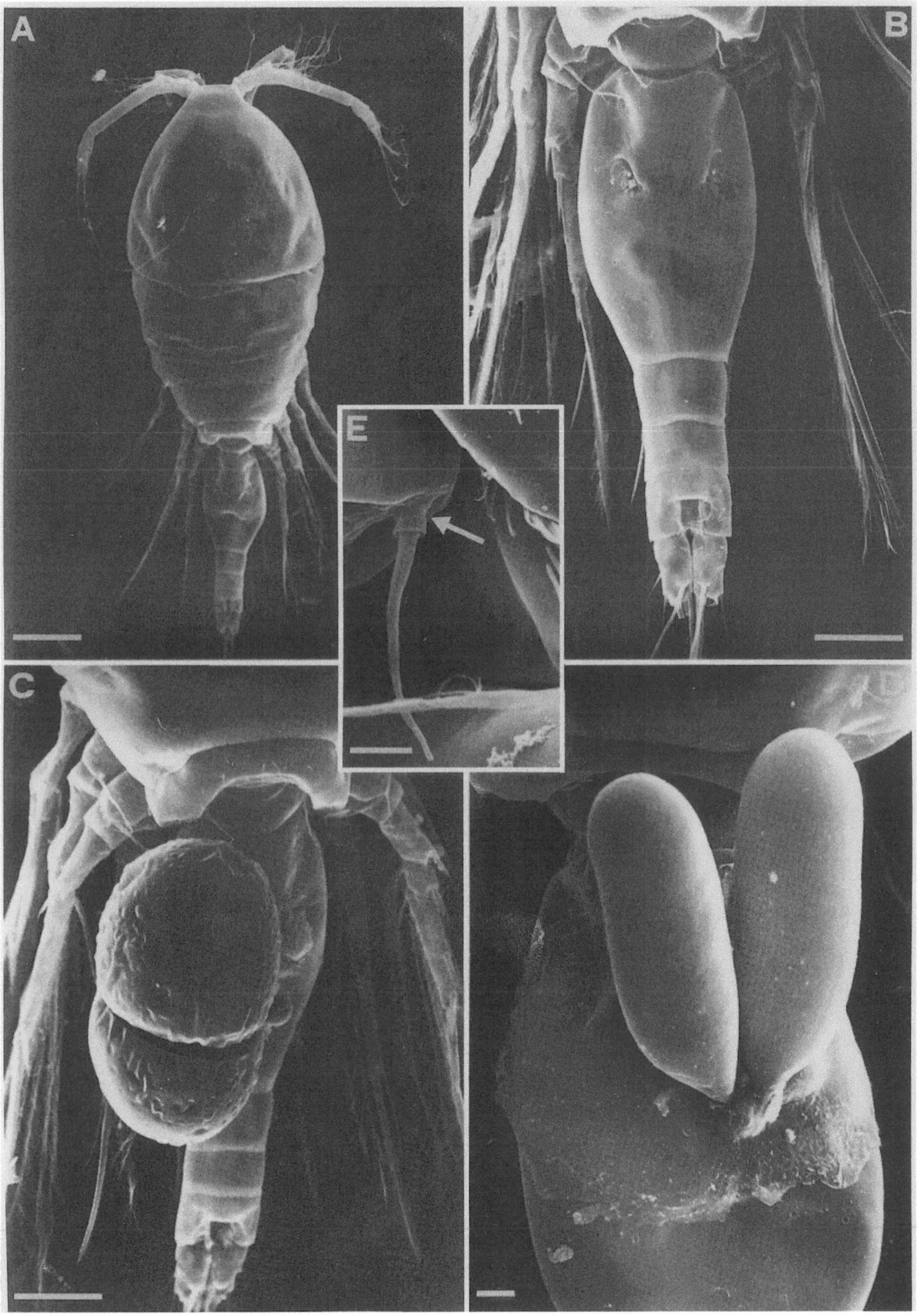

Fig. 3.45.2A–E. *Oncaea englishi*. Female: **A** whole animal dorsal view; **B** abdomen dorsal view; **C** abdomen dorsal view bearing one of two ovigerous sacs; **D** pair of spermatophores attached to genital pores on dorsal side of genital somite; **E** P5 (*arrow*). *Bars*: **A** 100 µm; **B**, **C** 50 µm; **D** 10 µm; **E** 5 µm

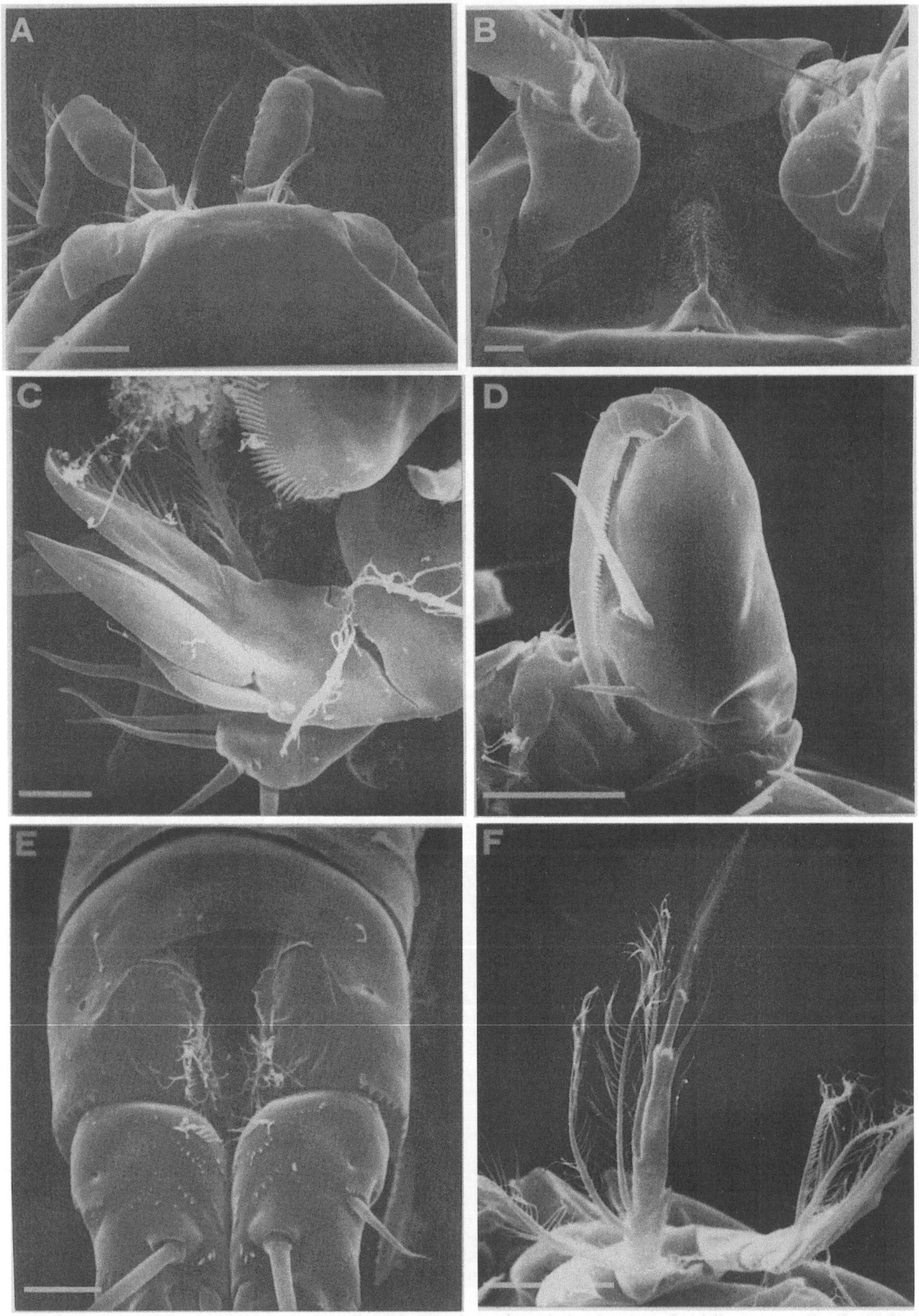

Fig. 3.45.3A–F. *Oncaea englishi*. Female: **A** truncated forehead in dorsal view; **B** forehead ventral view; **C** mandible; **D** maxilliped; **E** anal somite and caudal rami dorsal view; **F** P4. *Bars*: **A**, **D**, **F** 50 µm; **B**, **C**, **E** 10 µm

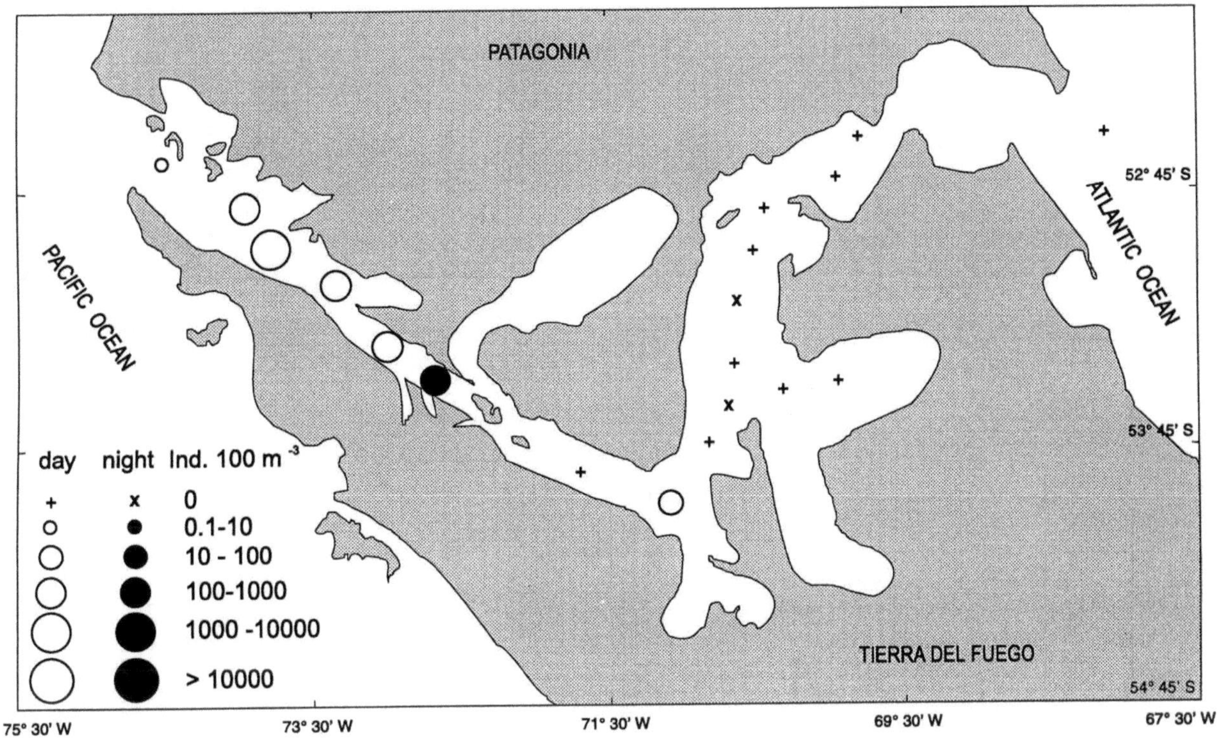

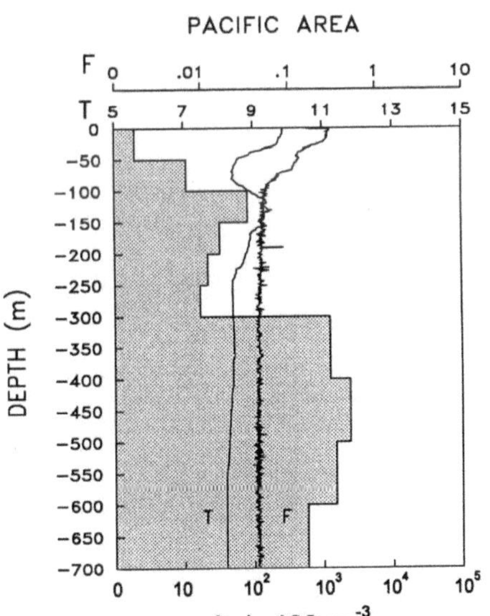

Depth range of occurrence = **0 - 700** m
Population core = **400 - 500** m
Temperature (T) = **8.20 - 9.80** °C
Oxygen = **348.4 - 437.2** μmol dm^{-3}
Chl fluorescence (F) = **0.04 - 0.39**

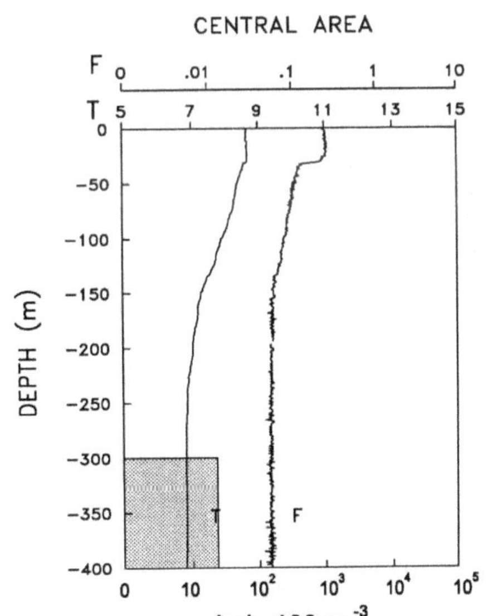

Depth range of occurrence = **300 - 400** m
Population core = **300 - 400** m
Temperature (T) = **6.70 - 9.01**°C
Oxygen = **346.1 - 407.3** μmol dm^{-3}
Chl fluorescence (F) = **0.05 - 0.27**

Fig. 3.45.4. *Oncaea englishi*. Distribution in the Straits of Magellan in February–March 1991

References

Ali-Khan S, Ali-Khan J (1982) Seven new records of family Lucicutiidae from Pakistan (Copepoda, Calanoida). Crustaceana 43: 265–270

Antezana T, Guglielmo L, Ghirardelli E (1992) Microbasins within the Strait of Magellan affecting zooplankton distribution. In: Gallardo VA, Ferretti O, Moyano HI (eds) Oceanografia in Antartide. ENEA-PNRA, Rome, pp 453–458

Arcos DF (1974) Los Copépodos Calanoideos colectados en la region Magallanica par la expedicion Hero 72–4b. Bol Soc Biol Concepcion 47: 215–225

Arcos DF (1975) Copepodos Calanoideos de la bahia de Concepcion, Chile. Conoscimento sistematico y variacion estacional. Gayana Zool 32: 1–31

Arcos DF (1976) Los Copepodos Calanoideos de la Region Magallanica. Expedicion Hero 72–4. Rev Com Perm Pacifico Sur 5: 85–100

Artegiani A, Paschini E, Andueza Calderon J (1991) Physical oceanography of the Straits of Magellan. Nat Sc Com Ant, Magellan Cruise, Feb–March 1991, Data Rep I: 11–52

Atkinson A, Peck JM (1988) A summer-winter comparison of zooplankton in the oceanic area around South Georgia. Polar Biol 8: 463–473

Benassi G, Ferrari I, Gentile G, Menozzi P, McKenzie KG (1994) Ostracoda in zooplankton samples from the Straits of Magellan: 1991 cruise. Nat Sc Com Ant, Magellan Cruise, Feb–March 1991, Data Rep III: 197–213

Björnberg TKS (1963) On the marine free-living copepods off Brazil. Bol Inst Oceanogr S Paulo 13: 3–142

Björnberg TKS (1973) The planktonic copepods of the Marchile I Expedition and of the Eltanin cruises 3–6 taken in the SE Pacific. Bol Zool Biol Mar (Nova Ser) 30: 245–394

Boeck A (1864) Oversight over de ved Norges Kyster iagttagne Copepoder henhörende til Calanidernes, Cyclopidernes og Harpactidernes Familier. Forh VidenskSelsk Krist: 1–57

Boltovskoy E (1970) Masas de agua (caracteristica, distribucion, movimentos) en la superficie del Atlantico sudoeste, segun indicadores biologicos. Foraminiferos. Servicio de Hidrografia Naval, Argentina H 643, pp 1–99

Bowman TE (1971) The distribution of calanoid copepods off the southeastern United States between Cape Hatteras and Southern Florida. Smithson Contrib Zool 96: 1–58

Boxshall GA (1977) The planktonic copepods of the northeastern Atlantic Ocean: some taxonomic observations on the Oncaeidae (Cyclopoida). Bull Br Mus (Nat Hist) Zool 31: 101–155

Boysen-Ennen E, Piatkowski U (1988) Meso- and macrozooplankton communities in the Weddell Sea, Antarctica. Polar Biol 9: 17–35

Bradford JM (1971a) *Aetideus* and *Euaetideus* (Copepoda: Calanoida) from the Atlantic and Pacific Oceans. NZ J Mar Freshwater Res 5: 12–40

Bradford JM (1971b) The fauna of the Ross Sea. 8. Pelagic Copepoda. Bull NZ Dep Sci Ind Res 206: 9–31

Bradford JM (1974) *Euchaeta marina* (Prestandrea) (Copepoda, Calanoida) and two closely related species from the Pacific. Pac Sci 28: 159–169

Bradford JM (1981) Records of *Paraeuchaeta* (Copepoda: Calanoida) from Mc Murdo Sound, Antarctica, with a description of three hitherto unknown males. NZ J Mar Freshwater Res 15: 391–402

Bradford JM (1988) Review of the taxonomy of the Calanidae (Copepoda) and the limits to the genus *Calanus*. Hydrobiologia 167/168: 73–81

Bradford JM, Jillett JB (1980) The marine fauna of New Zealand: pelagic calanoid copepods: family Aetideidae. Mem NZ Oceanogr Inst 86: 1–102

Bradford JM, Kaakonssen L, Jillett JB (1983) The marine fauna of New Zealand: pelagic calanoid copepods: families Euchaetidae, Phaennidae, Scolecithricidae, Diaixidae, and Tharybidae. Mem NZ Oceanogr Inst 90: 1–150

Brady GS (1883) Report on the Copepoda collected by HMS *Challenger* during the years 1873–76. Zool Chall Exp 23: 1–142, 55 pls

Brady G, Robertson D (1873) Contribution to the study of the Entomostraca. VIII. On marine Copepoda taken in the West of Ireland. Ann Mag Nat Hist 12: 126–142

Brodskii KA (1948) The free-swimming Copepoda of the Japan Sea. Izv Tikhookean Nauchno-issled Inst Rybn Khoz Okeanogr 26: 28–32 (in Russian)

Brodskii KA (1950) Calanoida of the far eastern seas and polar basin of the USSR Zool Inst Acad Sci

USSR Fauna 35: 1–440 (Translation: Israel Program for Scientific Translations, Jerusalem, 1967)
Brodskii KA (1959) On phylogenetic relations of some *Calanus* (Copepoda) species of northern and southern hemispheres. Zool Zh 38: 1537–1553
Brodskii KA, Vyshkvarzeva NV, Kos MS, Marchaseva EL (1983) Crustaceans (Copepoda: Calanoida) from SSSR seas and adjacent waters, vol 1. Academy of Sciences, Nauka, Leningrad, pp 1–358
Cabrini M, Fonda Umani S (1991) Phytoplankton populations in the Strait of Magellan. Boll Oceanol Teor Appl 9 (2–3 spec issue): 137–144
Campaner AF (1984) *Scaphocalanus* and *Scolecithricella* (Copepoda, Calanoida, Scolecithricidae) from the epipelagial off southern Brazil: a taxonomic and distributional survey. Bol Zool Univ S Paulo 8: 165–187
Campodonico IG, Guzman LM, Lembeye GV (1975) Una discoloracion causada por el ciliado *Mesodinium rubrum* (Lohmann) en ensenada Wilson, Magallanes. An Inst Patagonia 6: 225–239
Catalano G, Goffart A (1991) Dissolved oxygen and nutrients in the Straits of Magellan. Nat Sc Com Ant, Magellan Cruise, Feb–March 1991, Data Rep I: 53–66
Claus C (1863) Die freilebenden Copepoden mit besonderer Berücksichtigung der Fauna Deutschlands, der Nordsee und des Mittelmeeres. Engelmann, Leipzig, 230 pp, 37 pls
Claus C (1866) Die Copepoden-Fauna von Nizza. Ein Beitrag zur Characteristik der Formen und deren Abänderungen "im Sinne Darwin's". Schr Ges Naturwiss, Marburg, Suppl-Heft, 34 pp, 5 pls
Cleve PT (1904) The plankton of the South African seas. I. Copepoda. Mar Invest S Afr 3: 177–210
Corral J (1972) Copepoda. Sub-order: Calanoida. Family: Calocalanidae (Paracalanidae part.) Genera: *Calocalanus*, *Ischnocalanus*, *Mecynocera*. Cons Int Explor Mer Zooplankton Sheet 138: 1–7
Crisafi P (1959) Sulla *Oithona helgolandica* Claus (Copepoda, Cyclopoida) dello Stretto di Messina. Boll Zool 26: 49–57
Dahl F (1893) *Pleuromamma*, ein Krebs mit Leuchtorgan. Zool Anz 16: 104–109
Damkaer DM (1975) Calanoid copepods of the genera *Spinocalanus* and *Mimocalanus* from the central Arctic Ocean, with a review of the Spinocalanidae. NOAA Tech Rep NMFS CIRC 391: 1–88
Dana JD (1849) Conspectus Crustaceorum quae in orbis terrarum circumnavigatione, C. Wilkes e classe Reibublicae Foederatae duce, lexit et descripsit J.D. Dana. Proc Am Acad Arts Sci, vol 1, 1847: 149–154; vol 2, 1849: 9–61
Dana JD (1853–1855) United States Exploring Expedition during the years 1838–42, under the command of C. Wilkes. Crustacea, vol 14(2), text (1853), pp 691–1618; Atlas (1855), 96 pls
de Decker A, Mombeck FJ (1965) South African contribution to the International Indian Ocean Expedition. 4. A preliminary report on the planktonic Copepoda. Invest Rep Div Sea Fish Rep S Afr 51: 1–67
Deevey GB, Brooks AL (1977) Copepods of the Sargasso Sea off Bermuda: species composition, and vertical and seasonal distribution between the surface and 2000 m. Bull Mar Sci 27: 256–291
Esterly CO (1905) The pelagic Copepoda of the San Diego Region. Univ Calif Publ Zool 2: 113–233
Esterly CO (1906) Additions to the copepod fauna of the San Diego region. Univ Calif Publ Zool 6: 313–352
Fagetti EG (1962) Catalogo de los Copepodos planctonicos chilenos. Gayana Zool 4: 1–59
Farran GP (1908) Second report on the Copepoda of the Irish Atlantic slope. Sci Invest Fish Brch Ire 1906 2: 1–104
Farran GP (1926) Biscayan plankton collected during a cruise of HMS *Research*, 1900, Part XIV. The Copepoda. J Linn Soc Lond Zool 36: 219–310
Farran GP (1929) Copepoda. Nat Hist Rep Br Antarct Terra Nova Exped (Zool) 8: 203–306, 4 pls
Farran GP (1936) Copepoda. Sci Rep Gt Barrier Reef Exped 5: 73–142
Farran GP (1948) Copepoda. Sub-order Calanoida. Family Candaciidae. Genus *Candacia*. Cons Int Explor Mer Zooplankton Sheet 13: 1–4
Farran GP, Vervoort W (1951a) Copepoda. Sub-order: Calanoida. Family: Paracalanidae. Genus: *Paracalanus*. Cons Int Explor Mer Zooplankton Sheet 35: 1–4
Farran GP, Vervoort W (1951b) Copepoda. Sub-order: Calanoida. Family: Eucalanidae. Cons Int Explor Mer Zooplankton Sheet 11: 1–4
Farran GP, Vervoort W (1951c) Copepoda. Sub-order: Calanoida. Family: Pseudocalanidae. Genera: *Pseudocalanus*, *Microcalanus*. Cons Int Explor Mer Zooplankton Sheet 37: 1–4
Ferrari F, Dojiri M (1987) The calanoid copepod *Euchaeta antarctica* from Southern Ocean Atlantic sector midwater trawls, with observations on spermatophore dimorphism. J Crustacean Biol 7: 458–480
Fish AG (1962) Pelagic Copepods from Barbados. Bull Mar Sci 12: 1–38
Fleminger A (1973) Pattern, number, variability, and taxonomic significance of integumental organs (sensilla and glandular pores) in the genus *Eucalanus* (Copepoda, Calanoida). Fish Bull 71: 965–1010
Fontaine M (1988) Taxonomy and distribution of the *antarctica* species group of the genus *Euchaeta* (Copepoda, Calanoida). Antarct Res Ser 47: 27–5
Frost B, Fleminger A (1968) A revision of the genus *Clausocalanus* (Copepoda: Calanoida) with remarks on distributional patterns in diagnostic characters. Bull Scripps Inst Oceanogr 12: 1–235
Gaudy R, Viñas MD (1985) Premiere signalization en Méditerranée du copepode pelagique *Acartia tonsa*. Rapp Comm Int Mer Médit 29: 227–229

Ghirardelli E, Guglielmo L, Zagami G (1991) Chaetognaths in the Straits of Magellan (a preliminary note). Mem Biol Mar Oceanogr 19: 167–171

Ghirardelli E, Guglielmo L, Zagami G (1993) Chaetognaths in the Straits of Magellan. In: Isabel Moreno (ed) Proceedings of the II International Workshop of the Chaetognatha, Palma 1–16 September 1992, pp 79–83

Giesbrecht W (1888) Elenco dei Copepodi pelagici raccolti dal tenente di vascello Gaetano Chierchia durante il viaggio della R. Corvetta Vettor Pisani negli anni 1882–1885 e dal tenente di vascello Francesco Orsini nel Mar Rosso nel 1884. Atti Accad Naz Lincei Rend 4, 2° sem: 284–287, 330–338

Giesbrecht W (1889) Elenco dei Copepodi pelagici raccolti dal tenente di vascello Gaetano Chierchia durante il viaggio della R. Corvetta Vettor Pisani negli anni 1882–1885 e dal tenente di vascello Francesco Orsini nel Mar Rosso nel 1884. Atti Accad Naz Lincei Rend 5, 1° sem: 811–815

Giesbrecht W (1891) Elenco dei Copepodi pelagici raccolti dal tenente di vascello Gaetano Chierchia durante il viaggio della R. Corvetta Vettor Pisani negli anni 1882–1885 e dal tenente di vascello Francesco Orsini nel Mar Rosso nel 1884. Atti Accad Naz Lincei Rend 7, 1° sem: 474–481

Giesbrecht W (1892) Systematik und Faunistik der pelagischen Copepoden des Golfes von Neapel. Fauna Flora Golf Neapel 19: 1–831, 54 pls

Giesbrecht W (1902) Copepoden. Result Voy Belg Rapp Sci Zool: 1–49

Giesbrecht W, Schmeil O (1898) Copepoda. I. Gymnoplea. Tierreich 6: 1–169, 6 pls

Glorioso PD (1987) Temperature distribution related to shelf-sea fronts on the Patagonian Shelf. Continental Shelf Res 7: 27–34

Gonzalez JG, Bowman TE (1965) Planktonic copepods from Bahia Fosforescente, Puerto Rico, and adjacent waters. Proc US Nat Mus 117: 241–304

Grice GD (1962) Calanoid copepods from equatorial waters of Pacific Ocean. Fish Bull Fish Wildl Serv US 186: 167–246

Grice GD (1963) A revision of the genus *Candacia* (Copepoda: Calanoida) with an annotated list of the species and a key for their identification. Zool Meded (Leiden) 10: 171–194

Grice GD, Hart AD (1962) The abundance, seasonal occurrence and distribution of the epizooplankton between New York and Bermuda. Ecol Monogr 32: 287–309

Grice GD, Hülsemann K (1965) Abundance, vertical distribution and taxonomy of calanoid copepods at selected stations in the northeast Atlantic. J Zool 146: 213–262

Grice GD, Hülsemann K (1967) Bathypelagic calanoid copepods of the western Indian Ocean. Proc US Nat Mus 122: 1–67

Grice GD, Jones EC (1960) Two new species of *Candacia* (Copepoda: Calanoida) from the central Pacific, with notes on two other species. Pac Sci 14: 280–291

Guglielmo L, Antezana T, Costanzo G, Zagami G (1991) Zooplankton communities in the Straits of Magellan. Mem Biol Mar Oceanogr 19: 157–162

Guglielmo L, Arena G, Costanzo G, Crescenti N, Vanucci S, Zagami G (1993) Zooplankton ecology in the Straits of Magellan. Nat Sc Com Ant, Magellan Cruise, Feb–March 1991, Data Report II: 137–250

Guzman LM, Campodonico IG, Hermosilla JS (1975) Estudios sobre un florecimiento toxico causado pro *Gonyaulax catenella* en Magallanes. I. Distribucion espacial y temporal de *G. catenella*. An Inst Patagonia 6: 173–183

Hardy AC, Gunther ER (1935) The plankton of the South Georgia whaling ground and adjacent waters, 1926–1927. "Discovery" Rep 11: 1–456

Heptner MV (1971) On the copepod fauna (Copepoda, Calanoida) of the Kurile-Kamchatka Trench. Families Euchaetidae, Lucicutiidae, Heterorhabdidae. Tr Inst Okeanol 92: 73–161 (in Russian)

Heron GA (1977) Twenty-six species of Oncaeidae (Copepoda, Cyclopoida) from the southwest Pacific Antarctic area. Antarct Res Ser 26: 37–96

Heron GA, Bowman TE (1971) Postnaupliar developmental stages of the copepod crustaceans *Clausocalanus laticeps*, *C. brevipes*, and *Ctenocalanus citer* (Calanoida: Pseudocalanidae). Antarct Res Ser 17: 141–165

Heron GA, Damkaer DM (1969) Five species of deep-water cyclopoid copepods from the plankton of the Gulf of Alaska. Smithson Contrib Zool 20: 1–24

Heron GA, Damkaer DM (1978) Seven *Lubbockia* species (Copepoda: Cyclopoida) from the plankton of the Northeast Pacific, with a review of the genus. Smithson Contrib Zool 267: 1–36

Heron GA, English TS, Damkaer DM (1984) Arctic ocean copepoda of the genera *Lubbockia*, *Oncaea*, and *Epicalymna* (Poecilostomatoida: Oncaeidae), with remarks on distributions. J Crustacean Biol 4: 448–490

Hirakawa K (1989) Planktonic copepods from Aysen fjord and adjacent waters, Southern Chile. Proc NIPR Symp Polar Biol 2: 46–50

Hirakawa K, Zama A (1985) Zooplankton investigations in Lagoon San Rafael and Elefantes fjord. In: Nakajima C (ed) Glaciological studies in Patagonian northern icefield 1983–1984. Nagoya Univ Data Center for Glacier Research and Japanese Society of Snow and Ice, Nagoya, pp 115–120

Hopkins TL (1985) The Zooplankton community of Croker Passage, Antarctic Peninsula. Polar Biol 4: 161–170

Hülsemann K (1966) A revision of the genus *Lucicutia* (Copepoda: Calanoida) with a key to its species. Bull Mar Sci 16: 703–747

Hülsemann K (1985) Two species of *Drepanopus* Brady (Copepoda: Calanoida) with discrete ranges in the Southern Hemisphere. J Plankton Res 7: 909–925

Hülsemann K (1991) The copepodid stages of *Drepanopus forcipatus* Giesbrecht, with notes on the genus and a comparison with other members of the family Clausocalanidae (Copepoda Calanoida). Helgol Meeresunters 45: 199–224

Hure J (1955) Distribution annuelle verticale du zooplancton sur une station de l'Adriatique meridionale. Acta Adriat 7: 1–69

Hure J, Scotto di Carlo B (1968) Two new species of *Scaphocalanus* (Copepoda: Calanoida) from the Mediterranean Sea. Pubbl Staz Zool Napoli 36: 152–166

Huys R, Boxshall GA (1991) Copepod evolution. Gresham Press, Old Woking, Surrey (The Ray Society, vol 159) 468 pp

Iriarte JL, Uribe JC, Valladares C (1993) Biomass of size-fractionated phytoplankton during the spring-summer season in Southern Chile. Bot Mar 36: 443–450

Jespersen P (1934) Copepoda. The Godthaab expedition 1928. Medd Grϕnl 79: 1–166

Jespersen P (1940) Non-parasitic Copepoda. Zool Iceland 3: 1–116

Jillett JB (1968) *Calanus tonsus* (Copepoda, Calanoida) in southern New Zealand waters with notes on the male. Aust J Mar Freshwater Res 19: 19–30

Johnson MW (1938) Concerning the copepod *Eucalanus elongatus* Dana and its varieties in the Northeast Pacific. Bull Scripps Inst Oceanogr Tech Ser 4: 165–180

Johnson MW (1963) Zooplankton collections from the high Polar Basin with special reference to the Copepoda. Limnol Oceanogr 8: 89–102

Lembeye GV (1981) Estructura del fitoplancton asociado a la presencia del veneno paralitoco del los mariscos en seno Uninion y areas adyacentes (Magallanes, Chile, 1981). An Inst Patagonia 12: 277–288

Lembeye GV, Guzman LM, Campodonico IG (1975) Estudios sobre un florecimiento toxico causado por *Gonyaulax catenella* en Magallanes. III. Fitoplancton asociado. An Inst Patagonia 6: 197–208

Lembeye GV, Guzman LM, Campodonico IG (1978) Fitoplancton del sector oriental del estrecho de Magallanes, Chile (5 al 13 abril del 1976). An Inst Patagonia 9: 221–228

Lusquinos A, Valdez AJ (1971) Aportes al conocimiento de las masas de agua del Atlantico sudoccidental. Servicio de Hidrografia Naval, Argentina H 659, pp 1–48

Mackintosh WA (1934) Distribution of the macroplankton in the Atlantic sector of the Antarctic. "Discovery" Rep 9: 65–160

Magazzù G, Saggiomo V, Decembrini F (1991) Primary production and biomass of pico-, nano-, and microphytoplankton in the Straits of Magellan (February–March 1991). Mem Biol Mar Oceanogr 19: 127–130

Malt SJ (1983) Polymorphism and pore signature patterns in the copepod genus *Oncaea* (Cyclopoida). J Mar Biol Assoc UK 63: 449–466

Marin V, Antezana T (1985) Species composition and relative abundance of copepods in Chilean fjords. J Plankton Res 7: 961–966

Marino D, Sarno D, Zingone A (1991) Distribution of phytoplankton populations in the Straits of Magellan (February–March 1991). Mem Biol Mar Oceanogr 19: 147–150

Marques E (1956) Copepodes marinhos de Angola. Trabalhos de Missao de Biologia maritima. Campanhas em Angola. Ann Jta Invest Ultramar 8: 1–30

Marques E (1959) Copepodes marinhos de Angola (2a Campanha, 1952–1953). Mem Jta Invest Ultramar 4: 197–222

Matthews LH (1925) A new antarctic copepod, belonging to the genus *Eucalanus*. Ann Mag Nat Hist (Ser 9) 15: 127–129

Mazza J (1965) Les copépodes du genre *Heterorhabdus* Giesbrecht en Méditerranée. Morphologie et écologie. Rev Trav Inst Pêches Marit 29: 401–412

Mazzocchi MG, Ianora A (1991) A faunistic study of the copepod assemblages in the Strait of Magellan. Boll Oceanol Teor Appl 9 (2–3 spec issue): 163–177

Medeiros C, Kjerfve B (1988) Tidal characteristics of the Strait of Magellan. Continental Shelf Res 8: 947–960

Michelato A, Accerboni E, Berger P (1991) Current meter observations in the eastern and central sectors of the Strait of Magellan. Boll Oceanol Teor Appl 9 (2–3 spec issue): 261–271

Mori T (1937) The pelagic Copepoda from the neighbouring waters of Japan. 2nd edn 1964 Soyo Tokyo, 150 pp

Nishida S (1985) Taxonomy and distribution of the family Oithonidae (Copepoda, Cyclopoida) in the Pacific and Indian Oceans. Bull Ocean Res Inst Univ Tokyo 20: 1–167

Nishida S, Tanaka O, Omori M (1977) Cyclopoid copepods of the family Oithonidae in Suruga Bay and adjacent waters. Bull Plankton Soc JPN 24: 43–82

Ohman MD (1987) Energy sources for recruitment of the subantarctic copepod *Neocalanus tonsus*. Limnol Oceanogr: 1317–1330

Owre HB, Foyo M (1967) Copepods of the Florida current. Fauna Caribaea 1. Crustacea, Part 1. Copepoda. Institute of Marine Science University of Miami, Miami, 137 pp

Panella S, Magazzù G, Decembrini F (1991a) Photosynthetic pigments in the strait of Magellan (November 1989, April 1991). Mem Biol Har Oceanogr 19: 135–140

Panella S, Michelato A, Perdicaro R, Magazzù G, Decembrini F, Scarazzato P (1991b) A preliminary contribution to understanding the hydrological characteristics of the Strait of Magellan: austral spring 1989. Mem Biol Mar Oceanogr 19: 65–75

Park T (1968) Calanoid copepods from the central North Pacific Ocean. Fish Bull Fish Wildl Serv US 66: 527–572, 13 pls

Park T (1970) Calanoid copepods from the Caribbean Sea and Gulf of Mexico. 2. New species and new records from plankton samples. Bull Mar Sci 20: 472–546

Park T (1975) Calanoid copepods of the genera *Aetideopsis*, *Pseudoaetideus*, and *Chiridius* from the Gulf of Mexico. Bull Mar Sci 25: 272–290

Park T (1978) Calanoid copepods belonging to the families Aetideidae and Euchaetidae from antarctic and subantarctic waters. Antarct Res Ser 27: 91–290

Park T (1980) Calanoid copepods of the genus *Scolecithricella* from antarctic and subantarctic waters. Antarct Res Ser 31: 25–79

Park T (1982) Calanoid copepods of the genus *Scaphocalanus* from antarctic and subantarctic waters. Antarct Res Ser 34: 75–127

Park T (1983) Calanoid copepods of the family Phaennidae from antarctic and subantarctic waters. Antarct Res Ser 39: 317–368

Park T (1988) Calanoid copepods of the genus *Haloptilus* from Antarctic and subantarctic waters. Antarct Res Ser 47: 1–25

Peterson WT, Arcos DF, McManus GB, Dam H, Bellantoni D, Johnson T, Tiselius P (1988) The nearshore zone during coastal upwelling: daily variability and coupling between primary and secondary production off Central Chile. Prog Oceanogr 20: 1–40

Povero P, Danovaro R, Fabiano M (1991) Observations on particulate organic matter in Magellan Strait waters (February–March 1991). Mem Biol Mar Oceanogr 19: 119–123

Ramirez FC (1966) Copepodos Calanoidos marinos del area de Mar del Plata. Bol Inst Biol Mar (Mar del Plata) 11: 1–24

Ramirez FC (1969) Copepodos planctonicos del sector bonaerense del Atlantico suroccidental. Inst Biol Mar (Mar del Plata) Ser Contrib 98: 1–116

Ramirez FC (1970) Copepodos planctonicos del sector patagonico. Resultados de la campaña *Pesqueria XI*. Physis (B Aires) 24: 473–476

Ramirez FC (1971) Copepodos planctonicos de lo sectores bonaerense y norpatagonico. Resultados de la campaña *Pesqueria III*. Rev Mus La Plata 11: 73–94

Ramirez FC (1977) Planktonic indicators of Argentine shelf and adjacent areas. Proc Symp Warm Water Zoopl Spec Publ UNESCO/NIO: 65–68

Ramirez FC, Dinofrio EO (1976) Resultados planctonologicos de la Campaña *Oceantar I*. III. Copepodos. Contrib Inst Antarct Argent 196: 3–49

Razouls C (1993) Bilan taxinomique actuel des copépodes planctoniques marins et des eaux saumâtres. Crustaceana 64: 300–313

Razouls C, Durand J (1991) Inventaire des copépodes planctoniques méditerranéens. Vie Milieu 41: 73–77

Razouls C (1994) Manuel d' identification des principales espèces de copépodes pélagiques antarctiques et subantarctiques. Ann Inst Océanogr 70: 3–204

Roe HSJ (1972) The vertical distribution and diurnal migrations of calanoid copepods collected on the Sond Cruise, 1965. IV. Systematic account of the families Lucicutiidae to Candaciidae. The relative abundance of the numerically most important genera. J Mar Biol Assoc UK 52: 1021–1044

Roe HSJ (1975) Some new and rare species of calanoid copepods from the northeastern Atlantic. Bull Br Mus (Nat Hist) Zool 28: 295–375

Rose M (1933) Copépodes pélagiques. Faune Fr 26: 1–374

Rose M (1942) Les Scolecithricidae (Copépodes pélagiques) de la Baie d'Alger. Ann Inst Océanogr 21: 113–170

Rosendorn I (1917) Die Gattung *Oithona*. Wiss Ergeb Dtsch Tiefsee-Exped "Valdivia" 23: 1–58

Sabatini ME (1990) The developmental stages (copepodids I to VI) of *Acartia tonsa* Dana, 1849 (Copepoda, Calanoida). Crustaceana 59: 53–61

Saggiomo V, Goffart A, Carrada GC, Hecq JH (1994) Spatial patterns of phytoplanktonic pigments and primary production in a semi-enclosed periantarctic ecosystem: the Strait of Magellan. J Mar Syst 5: 119–142

Sameoto DD, Jaroszynski LO, Fraser WB (1980) BIONESS, a new design in multiple net zooplankton samples. Can J Fish Aquat Sci 37: 722–724

Sars GO (1900) Crustacea. The Norvegian North Polar Expedition 1893–1896. Sci Res 1: 1–137, 36 pls

Sars GO (1903) Copepoda Calanoida. An account of the Crustacea of Norway, vol 4. Bergen Museum, Alb Cammermeyer's Forlag, Christiana, pp 1–171, 106 pls

Sars GO (1918) Copepoda Cyclopoida. An account of the Crustacea of Norway, vol 6. Bergen Museum, Alb Cammermeyer's Forlag, Christiana, pp 1–225, 118 pls

Sars GO (1924–1925) Copépodes particulierement bathypélagiques provenant des campagnes scientifiques du Prince Albert Ier de Monaco. Résult Camp Sci Prince Albert I, vol 69. Imprimerie de Monaco, Monaco, text (1925), pp 1–408; Atlas (1924), 127 pls

Scott A (1909) The Copepoda of the Siboga Expedition 1. Free-swimming, littoral and semiparasitic Copepoda. Siboga Exped 17 (Mon 29a): 1–323, 69 pls

Scott T (1894) Report on Entomostraca from the Gulf of Guinea, collected by John Rattray. B Sc Trans Linn Soc Lond (Zool) 6: 1–161

Scott T (1914) Remarks on some Copepoda from the Falkland Islands collected by Mr. Rupert Vallentin, FLS Ann Mag Nat Hist (Ser 8) 13: 1–11

Scotto di Carlo B, Ianora A, Fresi E, Hure H (1984) Vertical zonation patterns for Mediterranean copepods from the surface to 3000 m at a fixed station in the Tyrrhenian Sea. J Plankton Res 6: 1031–1056

Semenova TN (1962) On the diagnostics of the species *Spinocalanus brevicaudatus* Brodsky, 1950 (Copepoda, Calanoida). Zool Zh 41: 1571–1574 (in Russian)

Senô J, Komaki Y, Takeda A (1963) Reports on the biology of the *Umitaka-Maru* expedition. Plankton collected by the *Umitaka-Maru* in the Antarctic and adjacent waters, with special references to Copepoda. J Tokyo Univ Fish 49: 55–62

Sewell RBS (1932) The Copepoda of Indian Seas. Calanoida. Mem Indian Mus 10: 223–407

Sewell RBS (1947) The free-swimming planktonic Copepoda. Systematic Account. Sci Rep John Murray Exped 8: 1–303

Silva N, Neshyba S (1977) Corrientes superficiales frente a la Costa de Chile. Ciencia y Tec del Mar, CONA 3: 37–42

Stephen R (1988) Oncaeidae (Copepoda: Poecilostomatoida) in the Indian Ocean with comments on the species of *Lubbockia* and *Conaea*. Mahasagar 21: 35–43

Steuer A (1915) Revision der Gattung *Acartia* Dana. Zool Anz 45: 392–397

Steuer A (1931) Grössen- und Formvariation der Planktoncopepoden. Sber Akad Wiss Vien Kl Math Naturw 140: 1–22

Steuer A (1933) Bericht über die Bearbeitung der Copepodengattung *Pleuromamma* Giesbr., 1898 der Deutschen Tiefsee Expedition *Valdivia*. Thalassia 1: 1–48

Tanaka O (1953) The pelagic copepods of the Izu Region. Rec Oceanogr Works JPN 1: 126–137

Tanaka O (1956a) Further note on *Calanus tonsus* Brady in Japanese waters. J Oceanogr Soc JPN 12: 49–52

Tanaka O (1956b) The pelagic copepods of the Izu Region, middle Japan. Systematic account II. Families Paracalanidae and Pseudocalanidae. Publ Seto Mar Biol Lab 5: 367–406

Tanaka O (1957) The pelagic copepods of the Izu region, middle Japan. Systematic account III. Family Aetideidae (Part 1). Publ Seto Mar Biol Lab 6: 31–68

Tanaka O (1960a) Pelagic Copepoda. Spec Publ Seto Mar Biol Lab Biol Results Jap Antarct Res Exped 10: 1–95

Tanaka O (1960b) The pelagic copepods of the Izu Region, middle Japan. Systematic account VI. Families Phaennidae and Tharybidae. Publ Seto Mar Biol Lab 8: 85–135

Tanaka O (1961) The pelagic copepods of the Izu Region, middle Japan. Systematic account VII. Family Scolecithricidae (Part 1). Publ Seto Mar Biol Lab 9: 139–190

Tanaka O (1962) The pelagic copepods of the Izu Region, middle Japan. Systematic account VIII. Family Scolecithricidae (Part 2). Publ Seto Mar Biol Lab 10: 35–90

Tanaka O (1963) The pelagic copepods of the Izu Region, middle Japan. Systematic account IX. Families Centropagidae, Pseudodiaptomidae, Temoridae, Metridinidae and Lucicutiidae. Publ Seto Mar Biol Lab 11: 7–55

Tanaka O (1964a) The pelagic copepods of the Izu Region, middle Japan. Systematic account X. Family Heterorhabdidae. Publ Seto Mar Biol Lab 12: 1–37

Tanaka O (1964b) Pelagic copepods of the Izu Region, middle Japan. Systematic account XI. Family Augaptilidae. Publ Seto Mar Biol Lab 12: 39–91

Tanaka O, Omori M (1971) Additional report on calanoid copepods from the Izu Region. Part 4. *Haloptilus*, *Augaptilus*, *Centraugaptilus*, *Pseudaugaptilus* and *Pachyptilus*. Publ Seto Mar Biol Lab 19: 249–268

Thomsen HA, Buck KR, Marino D, Sarno D, Hansen LE, Østergaard JB, Krupp J (1993) *Lennoxia faveolata* gen. et sp. nov. (Diatomophyceae) from South America, California, West Greenland and Denmark. Phycologia 32: 278–283

Unterüberbacher HK (1964) Zooplankton studies in the waters off Walvis Bay with special reference to the Copepoda. Invest Rep Mar Res Lab SW Afr 11: 1–42

Uribe JC (1988a) Antecedente sobre un tercer brote de veneno paralizante de moluscos (VPM), en la region de Magallanes. An Inst Patagonia 18: 97–101

Uribe JC (1988b) Observaciones sobre algunos fenomenos recurrentes en el fitoplancton de seno Union y bahia Bell (region de Magallanes) y su relacion con la estabilidad de la columna de agua. An Inst Patagonia 18: 103–111

Uribe JC (1991) Net-phytoplankton distribution in the Strait of Magellan. Boll Oceanol Teor Appl 9 (2–3 spec issue): 145–150

van Breemen PJ (1908) Copepoden. In: Nordisches Plankton, vol 4, pt 8. Von Lipsius & Tischer, Kiel, pp 1–264

Vervoort W (1946) The bathypelagic Copepoda Calanoida of the Snellius Expedition I. Families Calanidae, Eucalanidae, Paracalanidae and Pseudocalanidae. Biological Results of the Snellius Expedition, 15 Temminckia 8: 1–181

Vervoort W (1949) Some new and rare Copepoda Calanoida from East Indian Seas. Zool Verh (Leiden) 5: 1–53

Vervoort W (1951) Plankton copepods from the Atlantic sector of the Antarctic. K Ned Akad Wet Verh Afd Nat (Tweede Sectie) 47: 1–156

Vervoort W (1957) Copepods from Antarctic and sub-Antarctic plankton samples. Rep BANZ Antarct Res Exped (Ser B) 3: 1–160

Vervoort W (1962) Report on some Copepoda collected during the Melanesia Expedition of the Osaka Museum of Natural History. Publ Seto Mar Biol Lab 10: 393–470

Vervoort W (1963) Pelagic Copepoda. Part I. Copepoda Calanoida of the families Calanidae up to and including Euchaetidae. Atlantide Rep 7: 77–194

Vervoort W (1965a) Notes on the biogeography and ecology of free-living, marine Copepoda. In: Biogeography and ecology in Antarctica. Monogr Biol 15: 381–400

Vervoort W (1965b) Pelagic Copepoda. Part II. Copepoda Calanoida of the families Phaennidae up to and including Acartiidae, containing the description of a new species of Aetideidae. Atlantide Rep 8: 9–216

Vidal JW (1968) Copepodos calanoideos epipelagicos de la Expedicion Marchile II. Gayana Zool 15: 1–98

Ward P, Wood AG (1988) The distribution of the Euchaetidae (Copepoda: Calanoida) around South Georgia. Polar Biol 9: 45–52

Wheeler W (1899) The free-swimming copepods of the Woods Hole Region. Bull US Fish Comm 19: 157–192

Wilson CB (1932) The copepods of the Woods Hole Region Massachusetts. Bull US Natl Mus 158: 1–635

Wilson CB (1942) The copepods of the plankton gathered during the last cruise of the Carnegie. Carnegie Inst Wash Publ 536: 1–237

With C (1915) Copepoda, I. Calanoida Amphascandria. Dan Ingolf-Exped 3: 1–260

Wolfenden RN (1902) The plankton of the Faröe Channel and Shetlands. Preliminary notes on some Radiolaria and Copepoda. J Mar Biol Assoc UK 7: 344–372, 4 pls

Wolfenden RN (1904) Notes on the Copepoda of the North Atlantic Sea and the Faeröe Channel. J Mar Biol Assoc UK 7: 110–146

Wolfenden RN (1905) Plankton studies. Preliminary notes upon new or interesting species, Part I: Copepoda: 1–24

Wolfenden RN (1908) Crustacea VIII. Copepoda. In: National Antarctic Expedition 1901–1904, vol 4. Zoology. British Museum, London, pp 1–46, 7 pls

Appendix 1 Alphabetical list of Identified Species

Appendix 2 Stations and Sample Data for R/V Cariboo Cruise in the Straits of Magellan

Augaptilus glacialis

Station number	Depth (m)	Ind·m^{-3}	Station number	Depth (m)	Ind·m^{-3}	Station number	Depth (m)	Ind·m^{-3}
5	160–140	0.00	13	400–300	0.00	19	100–75	0.00
	115–100	0.00		250–200	0.00		75–50	0.00
	80–60	0.00		150–100	0.00		50–25	0.00
	40–20	0.00		50–25	0.00		25–0	0.00
6	455–400	0.05	14	400–300	0.00	20	80–75	0.00
	300–250	0.00		250–200	0.00		60–40	0.00
	200–150	0.00		150–100	0.00		40–20	0.00
	100–50	0.00		90–50	0.00		20–0	0.00
7	500–400	1.59	15	300–200	0.00	21	40–20	0.00
	300–250	0.00		150–100	0.00		20–0	0.00
	200–150	0.00		80–60	0.00			
	100–50	0.00		40–20	0.00			
9	700–600	0.03	16	80–60	0.00	22	30–20	0.00
	400–300	0.00		60–40	0.00		20–10	0.00
	200–140	0.00		40–20	0.00		10–0	0.00
	100–50	0.00		20–0	0.00			
10	500–400	0.33	17	120–100	0.00	23	30–0	0.00
	300–250	0.00		80–60	0.00			
	200–150	0.00		40–30	0.00			
	100–50	0.00		20–10	0.00			
11	400–300	0.00	18	160–120	0.00	26	30–20	0.00
	250–200	0.03		100–80	0.00		20–0	0.00
	150–100	0.01		60–40	0.00			
	50–25	0.00		20–10	0.00			
12	130–100	0.00						
	100–50	0.00						
	50–25	0.00						
	25–0	0.00						

Haloptilus acutifrons

Station number	Depth (m)	Ind·m^{-3}	Station number	Depth (m)	Ind·m^{-3}	Station number	Depth (m)	Ind·m^{-3}
5	160–140	0.00	13	400–300	0.00	19	100–75	0.00
	115–100	0.00		250–200	0.00		75–50	0.00
	80–60	0.00		150–100	0.00		50–25	0.00
	40–20	0.00		50–25	0.00		25–0	0.00
6	455–400	0.00	14	400–300	0.00	20	80–75	0.00
	300–250	0.00		250–200	0.00		60–40	0.00
	200–150	0.00		150–100	0.00		40–20	0.00
	100–50	0.00		90–50	0.00		20–0	0.00
7	500–400	0.00	15	300–200	0.00	21	40–20	0.00
	300–250	0.00		150–100	0.00		20–0	0.00
	200–150	0.02		80–60	0.00			
	100–50	0.00		40–20	0.00			
9	700–600	0.00	16	80–60	0.00	22	30–20	0.00
	400–300	0.00		60–40	0.00		20–10	0.00
	200–140	0.00		40–20	0.00		10–0	0.00
	100–50	0.00		20–0	0.00			
10	500–400	0.00	17	120–100	0.00	23	30–0	0.00
	300–250	0.00		80–60	0.00			
	200–150	0.00		40–30	0.00			
	100–50	0.00		20–10	0.00			
11	400–300	0.11	18	160–120	0.00	26	30–20	0.00
	250–200	0.01		100–80	0.00		20–0	0.00
	150–100	0.00		60–40	0.00			
	50–25	0.00		20–10	0.00			
12	130–100	0.00						
	100–50	0.00						
	50–25	0.00						
	25–0	0.00						

Heterorhabdus spinifrons

Station number	Depth (m)	Ind·m^{-3}	Station number	Depth (m)	Ind·m^{-3}	Station number	Depth (m)	Ind·m^{-3}
5	160–140	0.00	13	400–300	0.00	19	100–75	0.00
	115–100	0.00		250–200	0.00		75–50	0.00
	80–60	0.00		150–100	0.00		50–25	0.00
	40–20	0.00		50–25	0.00		25–0	0.00
6	455–400	0.00	14	400–300	0.00	20	80–75	0.00
	300–250	0.00		250–200	0.00		60–40	0.00
	200–150	0.00		150–100	0.00		40–20	0.00
	100–50	0.00		90–50	0.00		20–0	0.00
7	500–400	0.00	15	300–200	0.00	21	40–20	0.00
	300–250	0.00		150–100	0.00		20–0	0.00
	200–150	0.00		80–60	0.00			
	100–50	0.00		40–20	0.00			
9	700–600	0.00	16	80–60	0.00	22	30–20	0.00
	400–300	0.00		60–40	0.00		20–10	0.00
	200–140	0.05		40–20	0.00		10–0	0.00
	100–50	0.00		20–0	0.00			
10	500–400	0.02	17	120–100	0.00	23	30–0	0.00
	300–250	0.03		80–60	0.00			
	200–150	0.00		40–30	0.00			
	100–50	0.00		20–10	0.00			
11	400–300	0.00	18	160–120	0.00	26	30–20	0.00
	250–200	0.04		100–80	0.00		20–0	0.00
	150–100	0.20		60–40	0.00			
	50–25	0.00		20–10	0.00			
12	130–100	0.00						
	100–50	0.00						
	50–25	0.06						
	25–0	0.00						

Lucicutia clausi

Station number	Depth (m)	Ind·m^{-3}	Station number	Depth (m)	Ind·m^{-3}	Station number	Depth (m)	Ind·m^{-3}
5	160–140	0.17	13	400–300	0.00	19	100–75	0.00
	115–100	0.03		250–200	0.00		75–50	0.00
	80–60	0.00		150–100	0.00		50–25	0.00
	40–20	0.00		50–25	0.00		25–0	0.00
6	455–400	0.06	14	400–300	0.00	20	80–75	0.00
	300–250	1.00		250–200	0.00		60–40	0.00
	200–150	0.00		150–100	0.00		40–20	0.00
	100–50	0.00		90–50	0.00		20–0	0.00
7	500–400	4.23	15	300–200	0.00	21	40–20	0.00
	300–250	0.31		150–100	0.00		20–0	0.00
	200–150	0.35		80–60	0.00			
	100–50	0.00		40–20	0.00			
9	700–600	0.00	16	80–60	0.00	22	30–20	0.00
	400–300	0.90		60–40	0.00		20–10	0.00
	200–140	0.00		40–20	0.00		10–0	0.00
	100–50	0.05		20–0	0.00			
10	500–400	0.00	17	120–100	0.00	23	30–0	0.00
	300–250	0.28		80–60	0.00			
	200–150	0.06		40–30	0.00			
	100–50	0.00		20–10	0.00			
11	400–300	0.23	18	160–120	0.00	26	30–20	0.00
	250–200	0.06		100–80	0.00		20–0	0.00
	150–100	0.00		60–40	0.00			
	50–25	0.00		20–10	0.00			
12	130–100	0.00						
	100–50	0.00						
	50–25	0.00						
	25–0	0.00						

Metridia lucens

Station number	Depth (m)	Ind·m^{-3}	Station number	Depth (m)	Ind·m^{-3}	Station number	Depth (m)	Ind·m^{-3}
5	160–140	0.00	13	400–300	0.23	19	100–75	1.95
	115–100	3.45		250–200	0.16		75–50	0.00
	80–60	0.00		150–100	7.21		50–25	0.19
	40–20	0.70		50–25	0.61		25–0	15.20
6	455–400	0.00	14	400–300	2.16	20	80–75	0.54
	300–250	17.00		250–200	0.00		60–40	0.00
	200–150	318.62		150–100	0.96		40–20	0.11
	100–50	1.14		90–50	0.39		20–0	0.00
7	500–400	1.59	15	300–200	0.00	21	40–20	3.41
	300–250	3.71		150–100	1.56		20–0	0.00
	200–150	101.36		80–60	0.30			
	100–50	1.55		40–20	0.95			
9	700–600	0.43	16	80–60	0.08	22	30–20	0.00
	400–300	0.72		60–40	0.00		20–10	0.00
	200–140	70.31		40–20	0.00		10–0	0.00
	100–50	0.50		20–0	0.00			
10	500–400	0.59	17	120–100	1.90	23	30–0	0.00
	300–250	9.11		80–60	0.00			
	200–150	12.33		40–30	0.00			
	100–50	42.80		20–10	0.06			
11	400–300	0.94	18	160–120	3.92	26	30–20	0.00
	250–200	2.93		100–80	0.00		20–0	0.00
	150–100	6.89		60–40	0.03			
	50–25	5.49		20–10	0.08			
12	130–100	2.16						
	100–50	0.14						
	50–25	0.15						
	25–0	0.35						

Pleuromamma robusta

Station number	Depth (m)	Ind·m^{-3}	Station number	Depth (m)	Ind·m^{-3}	Station number	Depth (m)	Ind·m^{-3}
5	160–140	1.06	13	400–300	0.00	19	100–75	0.00
	115–100	0.00		250–200	0.00		75–50	0.00
	80–60	0.00		150–100	0.00		50–25	0.00
	40–20	0.00		50–25	0.00		25–0	0.00
6	455–400	0.00	14	400–300	0.00	20	80–75	0.00
	300–250	2.00		250–200	0.00		60–40	0.00
	200–150	0.00		150–100	0.00		40–20	0.00
	100–50	0.00		90–50	0.00		20–0	0.00
7	500–400	0.00	15	300–200	0.00	21	40–20	0.00
	300–250	0.04		150–100	0.00		20–0	0.00
	200–150	0.04		80–60	0.00			
	100–50	0.00		40–20	0.00			
9	700–600	0.16	16	80–60	0.00	22	30–20	0.00
	400–300	0.02		60–40	0.00		20–10	0.00
	200–140	0.00		40–20	0.00		10–0	0.00
	100–50	0.00		20–0	0.00			
10	500–400	0.01	17	120–100	0.00	23	30–0	0.00
	300–250	0.00		80–60	0.00			
	200–150	0.00		40–30	0.00			
	100–50	0.00		20–10	0.00			
11	400–300	0.00	18	160–120	0.00	26	30–20	0.00
	250–200	0.02		100–80	0.00		20–0	0.00
	150–100	0.00		60–40	0.00			
	50–25	0.00		20–10	0.00			
12	130–100	0.00						
	100–50	0.00						
	50–25	0.00						
	25–0	0.00						

Acartia tonsa

Station number	Depth (m)	Ind·m^{-3}	Station number	Depth (m)	Ind·m^{-3}	Station number	Depth (m)	Ind·m^{-3}
5	160–140	0.00	13	400–300	0.00	19	100–75	0.00
	115–100	0.00		250–200	0.00		75–50	0.00
	80–60	0.00		150–100	0.00		50–25	0.00
	40–20	0.00		50–25	5.24		25–0	0.00
6	455–400	0.00	14	400–300	0.00	20	80–75	0.00
	300–250	0.00		250–200	0.00		60–40	0.00
	200–150	0.00		150–100	0.00		40–20	0.66
	100–50	0.00		90–50	0.39		20–0	0.00
7	500–400	0.00	15	300–200	0.00	21	40–20	4.03
	300–250	0.00		150–100	0.00		20–0	8.61
	200–150	0.00		80–60	0.00			
	100–50	0.00		40–20	0.00			
9	700–600	0.00	16	80–60	0.00	22	30–20	28.42
	400–300	0.00		60–40	0.00		20–10	37.78
	200–140	0.00		40–20	0.00		10–0	14.05
	100–50	0.00		20–0	0.12			
10	500–400	0.00	17	120–100	0.00	23	30–0	36.29
	300–250	0.00		80–60	0.00			
	200–150	0.00		40–30	0.28			
	100–50	0.00		20–10	0.60			
11	400–300	0.00	18	160–120	0.00	26	30–20	0.00
	250–200	0.00		100–80	0.00		20–0	0.04
	150–100	0.00		60–40	0.00			
	50–25	0.00		20–10	0.00			
12	130–100	1.08						
	100–50	0.05						
	50–25	0.96						
	25–0	17.58						

Candacia cheirura

Station number	Depth (m)	Ind·m^{-3}	Station number	Depth (m)	Ind·m^{-3}	Station number	Depth (m)	Ind·m^{-3}
5	160–140	0.00	13	400–300	0.00	19	100–75	0.00
	115–100	0.00		250–200	0.07		75–50	0.00
	80–60	0.00		150–100	0.00		50–25	0.00
	40–20	0.00		50–25	0.00		25–0	0.00
6	455–400	0.24	14	400–300	0.00	20	80–75	0.00
	300–250	0.00		250–200	0.00		60–40	0.00
	200–150	0.00		150–100	0.00		40–20	0.00
	100–50	0.00		90–50	0.00		20–0	0.00
7	500–400	0.00	15	300–200	0.00	21	40–20	0.00
	300–250	0.14		150–100	0.03		20–0	0.00
	200–150	0.18		80–60	0.00			
	100–50	0.00		40–20	0.00			
9	700–600	0.00	16	80–60	0.00	22	30–20	0.00
	400–300	0.00		60–40	0.02		20–10	0.00
	200–140	0.81		40–20	0.00		10–0	0.00
	100–50	0.25		20–0	0.00			
10	500–400	0.00	17	120–100	0.00	23	30–0	0.00
	300–250	0.03		80–60	0.00			
	200–150	0.77		40–30	0.00			
	100–50	0.00		20–10	0.00			
11	400–300	0.00	18	160–120	0.00	26	30–20	0.00
	250–200	0.03		100–80	0.00		20–0	0.00
	150–100	0.04		60–40	0.00			
	50–25	0.00		20–10	0.00			
12	130–100	0.00						
	100–50	0.00						
	50–25	0.00						
	25–0	0.00						

Candacia norvegica

Station number	Depth (m)	Ind·m^{-3}	Station number	Depth (m)	Ind·m^{-3}	Station number	Depth (m)	Ind·m^{-3}
5	160–140	0.00	13	400–300	0.00	19	100–75	0.00
	115–100	0.00		250–200	0.00		75–50	0.00
	80–60	0.00		150–100	0.00		50–25	0.00
	40–20	0.00		50–25	0.00		25–0	0.00
6	455–400	0.00	14	400–300	0.00	20	80–75	0.00
	300–250	0.00		250–200	0.00		60–40	0.00
	200–150	0.00		150–100	0.00		40–20	0.00
	100–50	0.00		90–50	0.00		20–0	0.00
7	500–400	0.35	15	300–200	0.00	21	40–20	0.00
	300–250	0.00		150–100	0.00		20–0	0.00
	200–150	0.00		80–60	0.00			
	100–50	0.00		40–20	0.00			
9	700–600	0.21	16	80–60	0.00	22	30–20	0.00
	400–300	0.36		60–40	0.00		20–10	0.00
	200–140	0.00		40–20	0.00		10–0	0.00
	100–50	0.00		20–0	0.00			
10	500–400	0.02	17	120–100	0.00	23	30–0	0.00
	300–250	0.20		80–60	0.00			
	200–150	0.03		40–30	0.00			
	100–50	0.00		20–10	0.00			
11	400–300	0.27	18	160–120	0.00	26	30–20	0.00
	250–200	0.05		100–80	0.00		20–0	0.00
	150–100	0.04		60–40	0.00			
	50–25	0.00		20–10	0.00			
12	130–100	0.00						
	100–50	0.00						
	50–25	0.00						
	25–0	0.00						

Centropages brachiatus

Station number	Depth (m)	Ind·m^{-3}	Station number	Depth (m)	Ind·m^{-3}	Station number	Depth (m)	Ind·m^{-3}
5	160–140	0.00	13	400–300	0.00	19	100–75	0.00
	115–100	0.00		250–200	0.00		75–50	0.00
	80–60	0.00		150–100	0.00		50–25	0.00
	40–20	0.03		50–25	0.15		25–0	0.00
6	455–400	0.00	14	400–300	0.00	20	80–75	0.00
	300–250	0.00		250–200	0.00		60–40	0.00
	200–150	0.00		150–100	0.00		40–20	0.17
	100–50	0.00		90–50	0.00		20–0	0.05
7	500–400	0.00	15	300–200	0.00	21	40–20	0.00
	300–250	0.00		150–100	0.00		20–0	0.29
	200–150	0.00		80–60	0.00			
	100–50	0.00		40–20	0.00			
9	700–600	0.00	16	80–60	0.00	22	30–20	0.05
	400–300	0.00		60–40	0.00		20–10	0.09
	200–140	0.00		40–20	0.00		10–0	0.13
	100–50	0.00		20–0	0.00			
10	500–400	0.00	17	120–100	0.00	23	30–0	0.66
	300–250	0.00		80–60	0.00			
	200–150	0.00		40–30	0.00			
	100–50	0.00		20–10	0.00			
11	400–300	0.00	18	160–120	0.00	26	30–20	0.00
	250–200	0.00		100–80	0.00		20–0	0.00
	150–100	0.00		60–40	0.00			
	50–25	0.34		20–10	0.00			
12	130–100	0.00						
	100–50	0.59						
	50–25	0.00						
	25–0	0.00						

Centropages bradyi

Station number	Depth (m)	Ind·m^{-3}	Station number	Depth (m)	Ind·m^{-3}	Station number	Depth (m)	Ind·m^{-3}
5	160–140	0.02	13	400–300	0.00	19	100–75	0.00
	115–100	0.00		250–200	0.00		75–50	0.00
	80–60	0.02		150–100	0.00		50–25	0.00
	40–20	0.03		50–25	0.00		25–0	0.00
6	455–400	0.00	14	400–300	0.00	20	80–75	0.00
	300–250	0.00		250–200	0.00		60–40	0.00
	200–150	0.00		150–100	0.00		40–20	0.00
	100–50	0.06		90–50	0.00		20–0	0.00
7	500–400	0.00	15	300–200	0.00	21	40–20	0.00
	300–250	0.00		150–100	0.00		20–0	0.00
	200–150	0.00		80–60	0.00			
	100–50	0.00		40–20	0.00			
9	700–600	0.00	16	80–60	0.00	22	30–20	0.00
	400–300	0.00		60–40	0.00		20–10	0.00
	200–140	0.00		40–20	0.00		10–0	0.00
	100–50	0.00		20–0	0.00			
10	500–400	0.00	17	120–100	0.00	23	30–0	0.00
	300–250	0.00		80–60	0.00			
	200–150	0.00		40–30	0.00			
	100–50	0.00		20–10	0.06			
11	400–300	0.00	18	160–120	0.00	26	30–20	0.00
	250–200	0.00		100–80	0.00		20–0	0.00
	150–100	0.00		60–40	0.00			
	50–25	0.00		20–10	0.00			
12	130–100	0.00						
	100–50	0.00						
	50–25	0.00						
	25–0	0.00						

Calanoides patagoniensis

Station number	Depth (m)	Ind·m^{-3}	Station number	Depth (m)	Ind·m^{-3}	Station number	Depth (m)	Ind·m^{-3}
5	160–140	0.00	13	400–300	0.00	19	100–75	0.00
	115–100	0.00		250–200	0.00		75–50	0.00
	80–60	0.00		150–100	0.00		50–25	0.00
	40–20	0.00		50–25	1.17		25–0	0.00
6	455–400	0.00	14	400–300	0.00	20	80–75	0.54
	300–250	0.50		250–200	0.00		60–40	0.03
	200–150	1.23		150–100	0.00		40–20	0.34
	100–50	0.00		90–50	0.00		20–0	0.25
7	500–400	0.00	15	300–200	0.00	21	40–20	0.20
	300–250	0.00		150–100	0.00		20–0	0.06
	200–150	0.00		80–60	0.00			
	100–50	0.00		40–20	0.00			
9	700–600	0.00	16	80–60	0.00	22	30–20	0.27
	400–300	0.00		60–40	0.00		20–10	0.00
	200–140	0.00		40–20	0.10		10–0	0.00
	100–50	0.25		20–0	0.00			
10	500–400	0.01	17	120–100	1.90	23	30–0	0.00
	300–250	0.28		80–60	0.00			
	200–150	0.77		40–30	0.00			
	100–50	24.46		20–10	0.00			
11	400–300	0.00	18	160–120	0.00	26	30–20	0.00
	250–200	0.00		100–80	0.00		20–0	0.00
	150–100	0.06		60–40	0.00			
	50–25	3.43		20–10	0.00			
12	130–100	0.00						
	100–50	0.02						
	50–25	0.05						
	25–0	2.93						

Calanus australis

Station number	Depth (m)	Ind·m^{-3}	Station number	Depth (m)	Ind·m^{-3}	Station number	Depth (m)	Ind·m^{-3}
5	160–140	0.38	13	400–300	0.00	19	100–75	0.00
	115–100	0.24		250–200	0.00		75–50	0.00
	80–60	0.50		150–100	0.33		50–25	2.67
	40–20	1.40		50–25	0.58		25–0	0.00
6	455–400	0.00	14	400–300	0.00	20	80–75	0.00
	300–250	0.37		250–200	0.00		60–40	0.03
	200–150	3.68		150–100	2.39		40–20	0.00
	100–50	0.23		90–50	0.12		20–0	0.10
7	500–400	0.00	15	300–200	0.00	21	40–20	0.00
	300–250	0.04		150–100	0.00		20–0	0.09
	200–150	2.22		80–60	0.00			
	100–50	1.35		40–20	0.00			
9	700–600	0.01	16	80–60	0.00	22	30–20	0.00
	400–300	0.02		60–40	0.00		20–10	0.05
	200–140	5.70		40–20	0.00		10–0	0.00
	100–50	2.78		20–0	0.10			
10	500–400	0.00	17	120–100	1.90	23	30–0	0.05
	300–250	0.28		80–60	0.00			
	200–150	0.28		40–30	0.00			
	100–50	3.06		20–10	0.00			
11	400–300	0.00	18	160–120	1.22	26	30–20	23.00
	250–200	0.03		100–80	0.00		20–0	1.01
	150–100	0.10		60–40	0.00			
	50–25	1.37		20–10	0.00			
12	130–100	0.00						
	100–50	0.00						
	50–25	0.00						
	25–0	0.00						

Calanus simillimus

Station number	Depth (m)	Ind·m^{-3}	Station number	Depth (m)	Ind·m^{-3}	Station number	Depth (m)	Ind·m^{-3}
5	160–140	0.00	13	400–300	0.45	19	100–75	1.95
	115–100	0.03		250–200	0.62		75–50	0.90
	80–60	0.00		150–100	0.66		50–25	55.98
	40–20	0.03		50–25	1.17		25–0	53.21
6	455–400	0.05	14	400–300	11.52	20	80–75	2.70
	300–250	0.25		250–200	5.47		60–40	3.17
	200–150	1.23		150–100	7.14		40–20	1.11
	100–50	0.00		90–50	0.05		20–0	0.44
7	500–400	0.18	15	300–200	6.15	21	40–20	5.43
	300–250	0.08		150–100	3.16		20–0	17.22
	200–150	0.02		80–60	1.55			
	100–50	0.02		40–20	10.78			
9	700–600	0.01	16	80–60	1.07	22	30–20	0.03
	400–300	0.00		60–40	0.16		20–10	0.00
	200–140	1.90		40–20	6.50		10–0	0.00
	100–50	0.08		20–0	0.00			
10	500–400	0.02	17	120–100	3.81	23	30–0	0.05
	300–250	0.06		80–60	0.36			
	200–150	0.16		40–30	1.13			
	100–50	0.00		20–10	1.80			
11	400–300	0.04	18	160–120	16.16	26	30–20	0.66
	250–200	0.05		100–80	0.08		20–0	0.45
	150–100	0.11		60–40	0.32			
	50–25	0.69		20–10	0.23			
12	130–100	0.00						
	100–50	0.45						
	50–25	0.29						
	25–0	1.64						

Neocalanus tonsus

Station number	Depth (m)	Ind·m^{-3}	Station number	Depth (m)	Ind·m^{-3}	Station number	Depth (m)	Ind·m^{-3}
5	160–140	0.00	13	400–300	0.00	19	100–75	0.00
	115–100	0.00		250–200	0.00		75–50	0.00
	80–60	0.00		150–100	0.00		50–25	0.00
	40–20	0.00		50–25	0.00		25–0	0.00
6	455–400	0.00	14	400–300	2.88	20	80–75	0.00
	300–250	0.00		250–200	0.14		60–40	0.00
	200–150	0.00		150–100	0.00		40–20	0.00
	100–50	0.00		90–50	0.00		20–0	0.00
7	500–400	0.00	15	300–200	0.00	21	40–20	0.00
	300–250	0.00		150–100	0.00		20–0	0.00
	200–150	0.02		80–60	0.05			
	100–50	0.00		40–20	0.05			
9	700–600	0.00	16	80–60	0.00	22	30–20	0.00
	400–300	0.00		60–40	0.00		20–10	0.00
	200–140	0.00		40–20	0.00		10–0	0.00
	100–50	0.00		20–0	0.00			
10	500–400	0.00	17	120–100	0.00	23	30–0	0.00
	300–250	0.00		80–60	0.00			
	200–150	0.00		40–30	0.00			
	100–50	0.00		20–10	0.00			
11	400–300	0.00	18	160–120	0.00	26	30–20	0.00
	250–200	0.00		100–80	0.00		20–0	0.00
	150–100	0.00		60–40	0.00			
	50–25	0.00		20–10	0.00			
12	130–100	0.00						
	100–50	0.00						
	50–25	0.00						
	25–0	0.00						

Calocalanus pavoninus

Station number	Depth (m)	$Ind \cdot m^{-3}$	Station number	Depth (m)	$Ind \cdot m^{-3}$	Station number	Depth (m)	$Ind \cdot m^{-3}$
5	160–140	0.00	13	400–300	0.00	19	100–75	0.00
	115–100	0.00		250–200	0.00		75–50	0.00
	80–60	0.00		150–100	0.00		50–25	0.00
	40–20	0.03		50–25	0.58		25–0	0.00
6	455–400	0.00	14	400–300	0.00	20	80–75	0.00
	300–250	0.00		250–200	0.00		60–40	0.00
	200–150	0.00		150–100	0.00		40–20	0.00
	100–50	0.00		90–50	0.00		20–0	0.00
7	500–400	0.00	15	300–200	0.00	21	40–20	0.00
	300–250	0.00		150–100	0.00		20–0	0.00
	200–150	0.00		80–60	0.00			
	100–50	0.02		40–20	0.00			
9	700–600	0.00	16	80–60	0.00	22	30–20	0.00
	400–300	0.00		60–40	0.00		20–10	0.00
	200–140	0.00		40–20	0.00		10–0	0.00
	100–50	0.00		20–0	0.00			
10	500–400	0.00	17	120–100	0.00	23	30–0	0.00
	300–250	0.00		80–60	0.00			
	200–150	0.00		40–30	0.00			
	100–50	0.00		20–10	0.00			
11	400–300	0.00	18	160–120	0.00	26	30–20	0.00
	250–200	0.00		100–80	0.00		20–0	0.00
	150–100	0.00		60–40	0.00			
	50–25	0.00		20–10	0.00			
12	130–100	0.00						
	100–50	0.00						
	50–25	0.00						
	25–0	0.00						

Paracalanus parvus

Station number	Depth (m)	$Ind \cdot m^{-3}$	Station number	Depth (m)	$Ind \cdot m^{-3}$	Station number	Depth (m)	$Ind \cdot m^{-3}$
5	160–140	0.07	13	400–300	0.00	19	100–75	0.00
	115–100	0.00		250–200	0.00		75–50	8.53
	80–60	0.00		150–100	0.02		50–25	0.00
	40–20	0.03		50–25	15.73		25–0	38.00
6	455–400	0.00	14	400–300	0.00	20	80–75	0.00
	300–250	8.51		250–200	0.00		60–40	0.00
	200–150	0.00		150–100	0.00		40–20	5.55
	100–50	0.00		90–50	0.00		20–0	0.00
7	500–400	0.00	15	300–200	0.00	21	40–20	0.00
	300–250	0.00		150–100	3.13		20–0	0.00
	200–150	0.00		80–60	17.30			
	100–50	0.00		40–20	10.78			
9	700–600	0.00	16	80–60	0.00	22	30–20	0.00
	400–300	0.00		60–40	0.00		20–10	0.00
	200–140	0.00		40–20	0.00		10–0	0.07
	100–50	0.50		20–0	0.00			
10	500–400	0.00	17	120–100	0.00	23	30–0	0.00
	300–250	0.00		80–60	0.00			
	200–150	0.00		40–30	0.00			
	100–50	0.00		20–10	0.18			
11	400–300	0.00	18	160–120	0.00	26	30–20	0.00
	250–200	0.00		100–80	0.00		20–0	0.00
	150–100	0.24		60–40	0.00			
	50–25	34.32		20–10	0.08			
12	130–100	6.49						
	100–50	0.00						
	50–25	3.59						
	25–0	5.86						

Eucalanus elongatus hyalinus

Station number	Depth (m)	Ind·m^{-3}	Station number	Depth (m)	Ind·m^{-3}	Station number	Depth (m)	Ind·m^{-3}
5	160–140	0.00	13	400–300	0.00	19	100–75	0.00
	115–100	0.00		250–200	0.00		75–50	0.00
	80–60	0.00		150–100	0.00		50–25	0.00
	40–20	0.00		50–25	0.00		25–0	0.00
6	455–400	0.00	14	400–300	0.00	20	80–75	0.00
	300–250	0.00		250–200	0.00		60–40	0.00
	200–150	0.00		150–100	0.00		40–20	0.00
	100–50	0.00		90–50	0.00		20–0	0.00
7	500–400	0.01	15	300–200	0.00	21	40–20	0.00
	300–250	0.02		150–100	0.00		20–0	0.00
	200–150	0.02		80–60	0.00			
	100–50	0.00		40–20	0.00			
9	700–600	0.00	16	80–60	0.00	22	30–20	0.00
	400–300	0.00		60–40	0.00		20–10	0.00
	200–140	0.00		40–20	0.00		10–0	0.00
	100–50	0.00		20–0	0.00			
10	500–400	0.00	17	120–100	0.00	23	30–0	0.00
	300–250	0.00		80–60	0.00			
	200–150	0.00		40–30	0.00			
	100–50	0.00		20–10	0.00			
11	400–300	0.00	18	160–120	0.00	26	30–20	0.00
	250–200	0.01		100–80	0.00		20–0	0.00
	150–100	0.00		60–40	0.00			
	50–25	0.00		20–10	0.00			
12	130–100	0.00						
	100–50	0.00						
	50–25	0.00						
	25–0	0.00						

Eucalanus longiceps

Station number	Depth (m)	Ind·m^{-3}	Station number	Depth (m)	Ind·m^{-3}	Station number	Depth (m)	Ind·m^{-3}
5	160–140	0.00	13	400–300	0.23	19	100–75	0.00
	115–100	0.00		250–200	0.02		75–50	0.00
	80–60	0.00		150–100	0.03		50–25	0.00
	40–20	0.00		50–25	0.00		25–0	0.00
6	455–400	1.06	14	400–300	0.00	20	80–75	0.00
	300–250	1.50		250–200	0.00		60–40	0.00
	200–150	0.00		150–100	0.00		40–20	0.00
	100–50	0.11		90–50	0.00		20–0	0.00
7	500–400	0.04	15	300–200	0.00	21	40–20	0.00
	300–250	0.00		150–100	0.00		20–0	0.00
	200–150	0.09		80–60	0.00			
	100–50	0.00		40–20	0.00			
9	700–600	0.00	16	80–60	0.00	22	30–20	0.00
	400–300	0.00		60–40	0.00		20–10	0.00
	200–140	0.05		40–20	0.00		10–0	0.00
	100–50	0.03		20–0	0.00			
10	500–400	0.29	17	120–100	0.00	23	30–0	0.00
	300–250	1.43		80–60	0.00			
	200–150	0.00		40–30	0.00			
	100–50	0.00		20–10	0.00			
11	400–300	0.04	18	160–120	0.00	26	30–20	0.00
	250–200	0.01		100–80	0.00		20–0	0.00
	150–100	0.00		60–40	0.00			
	50–25	0.00		20–10	0.00			
12	130–100	0.00						
	100–50	0.00						
	50–25	0.00						
	25–0	0.00						

Rhincalanus nasutus

Station number	Depth (m)	Ind·m^{-3}	Station number	Depth (m)	Ind·m^{-3}	Station number	Depth (m)	Ind·m^{-3}
5	160–140	0.26	13	400–300	0.00	19	100–75	0.00
	115–100	0.03		250–200	0.00		75–50	0.00
	80–60	0.50		150–100	0.02		50–25	0.00
	40–20	0.70		50–25	0.00		25–0	0.00
6	455–400	1.99	14	400–300	0.00	20	80–75	0.00
	300–250	7.00		250–200	0.14		60–40	0.03
	200–150	1.23		150–100	0.00		40–20	0.00
	100–50	0.23		90–50	0.00		20–0	0.00
7	500–400	0.35	15	300–200	0.00	21	40–20	0.00
	300–250	0.50		150–100	0.07		20–0	0.00
	200–150	0.07		80–60	0.00			
	100–50	0.02		40–20	0.00			
9	700–600	0.16	16	80–60	0.00	22	30–20	0.00
	400–300	0.02		60–40	0.00		20–10	0.00
	200–140	0.10		40–20	0.00		10–0	0.00
	100–50	0.03		20–0	0.03			
10	500–400	0.59	17	120–100	0.00	23	30–0	0.00
	300–250	1.43		80–60	0.00			
	200–150	0.16		40–30	0.00			
	100–50	0.30		20–10	0.00			
11	400–300	0.28	18	160–120	0.00	26	30–20	0.00
	250–200	0.06		100–80	0.00		20–0	0.00
	150–100	0.07		60–40	0.00			
	50–25	0.00		20–10	0.00			
12	130–100	0.00						
	100–50	0.00						
	50–25	0.00						
	25–0	0.00						

Aetideus arcuatus

Station number	Depth (m)	Ind·m^{-3}	Station number	Depth (m)	Ind·m^{-3}	Station number	Depth (m)	Ind·m^{-3}
5	160–140	0.00	13	400–300	0.00	19	100–75	0.00
	115–100	0.00		250–200	0.00		75–50	0.00
	80–60	0.00		150–100	0.00		50–25	0.00
	40–20	0.00		50–25	0.00		25–0	0.00
6	455–400	0.05	14	400–300	0.00	20	80–75	0.00
	300–250	0.00		250–200	0.00		60–40	0.00
	200–150	0.00		150–100	0.00		40–20	0.00
	100–50	0.00		90–50	0.00		20–0	0.00
7	500–400	1.23	15	300–200	0.00	21	40–20	0.00
	300–250	0.02		150–100	0.00		20–0	0.00
	200–150	0.00		80–60	0.00			
	100–50	0.00		40–20	0.00			
9	700–600	0.00	16	80–60	0.00	22	30–20	0.00
	400–300	0.18		60–40	0.00		20–10	0.00
	200–140	0.00		40–20	0.00		10–0	0.00
	100–50	0.00		20–0	0.00			
10	500–400	0.00	17	120–100	0.00	23	30–0	0.00
	300–250	0.00		80–60	0.00			
	200–150	0.00		40–30	0.00			
	100–50	0.00		20–10	0.00			
11	400–300	0.00	18	160–120	0.00	26	30–20	0.00
	250–200	0.00		100–80	0.00		20–0	0.00
	150–100	0.00		60–40	0.00			
	50–25	0.00		20–10	0.00			
12	130–100	0.00						
	100–50	0.00						
	50–25	0.00						
	25–0	0.00						

Aetideus australis

Station number	Depth (m)	Ind·m^{-3}	Station number	Depth (m)	Ind·m^{-3}	Station number	Depth (m)	Ind·m^{-3}
5	160–140	0.02	13	400–300	0.00	19	100–75	0.00
	115–100	0.00		250–200	0.02		75–50	0.00
	80–60	0.00		150–100	0.00		50–25	0.00
	40–20	0.00		50–25	0.00		25–0	0.00
6	455–400	0.02	14	400–300	0.00	20	80–75	0.00
	300–250	2.50		250–200	0.00		60–40	0.00
	200–150	0.00		150–100	0.00		40–20	0.00
	100–50	0.00		90–50	0.00		20–0	0.00
7	500–400	0.00	15	300–200	0.00	21	40–20	0.00
	300–250	0.26		150–100	0.00		20–0	0.00
	200–150	0.35		80–60	0.00			
	100–50	0.00		40–20	0.00			
9	700–600	0.14	16	80–60	0.00	22	30–20	0.00
	400–300	0.00		60–40	0.00		20–10	0.00
	200–140	1.95		40–20	0.00		10–0	0.00
	100–50	0.00		20–0	0.00			
10	500–400	0.01	17	120–100	0.00	23	30–0	0.00
	300–250	0.32		80–60	0.00			
	200–150	1.54		40–30	0.00			
	100–50	0.00		20–10	0.00			
11	400–300	0.04	18	160–120	0.00	26	30–20	0.00
	250–200	0.06		100–80	0.00		20–0	0.00
	150–100	0.95		60–40	0.00			
	50–25	0.34		20–10	0.00			
12	130–100	0.00						
	100–50	0.00						
	50–25	0.00						
	25–0	0.00						

Chiridius gracilis and subgracilis

Station number	Depth (m)	Ind·m^{-3}	Station number	Depth (m)	Ind·m^{-3}	Station number	Depth (m)	Ind·m^{-3}
5	160–140	0.07	13	400–300	0.00	19	100–75	0.00
	115–100	0.00		250–200	0.58		75–50	0.00
	80–60	0.00		150–100	0.00		50–25	0.00
	40–20	0.00		50–25	0.03		25–0	0.00
6	455–400	0.06	14	400–300	0.00	20	80–75	0.00
	300–250	4.50		250–200	0.00		60–40	0.00
	200–150	3.68		150–100	0.00		40–20	0.00
	100–50	0.23		90–50	0.00		20–0	0.00
7	500–400	0.07	15	300–200	0.00	21	40–20	0.00
	300–250	0.57		150–100	0.00		20–0	0.00
	200–150	4.24		80–60	0.00			
	100–50	0.06		40–20	0.00			
9	700–600	0.14	16	80–60	0.00	22	30–20	0.00
	400–300	0.04		60–40	0.00		20–10	0.00
	200–140	6.65		40–20	0.00		10–0	0.00
	100–50	0.00		20–0	0.00			
10	500–400	0.90	17	120–100	0.00	23	30–0	0.00
	300–250	0.86		80–60	0.00			
	200–150	0.18		40–30	0.00			
	100–50	12.23		20–10	0.00			
11	400–300	0.00	18	160–120	0.00	26	30–20	0.00
	250–200	0.00		100–80	0.00		20–0	0.00
	150–100	0.05		60–40	0.00			
	50–25	0.00		20–10	0.00			
12	130–100	0.00						
	100–50	0.00						
	50–25	0.00						
	25–0	0.00						

Clausocalanus brevipes

Station number	Depth (m)	Ind·m^{-3}	Station number	Depth (m)	Ind·m^{-3}	Station number	Depth (m)	Ind·m^{-3}
5	160–140	1.06	13	400–300	0.00	19	100–75	46.79
	115–100	1.72		250–200	2.30		75–50	17.04
	80–60	0.50		150–100	22.93		50–25	119.97
	40–20	0.70		50–25	39.04		25–0	266.05
6	455–400	0.00	14	400–300	1.44	20	80–75	25.91
	300–250	0.50		250–200	4.37		60–40	15.84
	200–150	74.76		150–100	47.64		40–20	3.33
	100–50	3.18		90–50	9.89		20–0	0.05
7	500–400	0.00	15	300–200	0.00	21	40–20	7.45
	300–250	0.00		150–100	59.38		20–0	17.22
	200–150	15.89		80–60	84.38			
	100–50	0.35		40–20	221.99			
9	700–600	0.00	16	80–60	47.79	22	30–20	0.00
	400–300	0.00		60–40	9.09		20–10	0.00
	200–140	9.50		40–20	5.04		10–0	0.00
	100–50	12.36		20–0	1.87			
10	500–400	0.00	17	120–100	15.26	23	30–0	0.00
	300–250	0.57		80–60	71.92			
	200–150	7.71		40–30	13.50			
	100–50	30.57		20–10	2.41			
11	400–300	0.00	18	160–120	34.28	26	30–20	0.00
	250–200	2.59		100–80	0.78		20–0	0.00
	150–100	31.62		60–40	0.63			
	50–25	108.11		20–10	0.70			
12	130–100	1.08						
	100–50	8.18						
	50–25	1.19						
	25–0	8.79						

Clausocalanus ingens

Station number	Depth (m)	Ind·m^{-3}	Station number	Depth (m)	Ind·m^{-3}	Station number	Depth (m)	Ind·m^{-3}
5	160–140	0.00	13	400–300	0.00	19	100–75	0.00
	115–100	0.00		250–200	0.00		75–50	0.00
	80–60	0.00		150–100	0.00		50–25	0.00
	40–20	0.00		50–25	1.17		25–0	0.00
6	455–400	0.00	14	400–300	0.00	20	80–75	0.00
	300–250	0.00		250–200	0.00		60–40	0.00
	200–150	0.00		150–100	0.00		40–20	0.00
	100–50	0.00		90–50	0.00		20–0	0.00
7	500–400	0.00	15	300–200	0.00	21	40–20	0.00
	300–250	0.00		150–100	0.00		20–0	0.00
	200–150	0.00		80–60	0.00			
	100–50	0.00		40–20	0.00			
9	700–600	0.00	16	80–60	0.00	22	30–20	0.00
	400–300	0.00		60–40	0.00		20–10	0.00
	200–140	0.00		40–20	0.00		10–0	0.00
	100–50	0.00		20–0	0.00			
10	500–400	0.00	17	120–100	0.00	23	30–0	0.00
	300–250	0.00		80–60	0.00			
	200–150	0.03		40–30	0.00			
	100–50	0.00		20–10	0.00			
11	400–300	0.00	18	160–120	0.00	26	30–20	0.00
	250–200	0.00		100–80	0.00		20–0	0.00
	150–100	0.00		60–40	0.00			
	50–25	0.00		20–10	0.00			
12	130–100	0.00						
	100–50	0.00						
	50–25	2.40						
	25–0	0.00						

Clausocalanus laticeps

Station number	Depth (m)	Ind·m^{-3}	Station number	Depth (m)	Ind·m^{-3}	Station number	Depth (m)	Ind·m^{-3}
5	160–140	0.00	13	400–300	0.00	19	100–75	0.00
	115–100	0.00		250–200	0.00		75–50	0.00
	80–60	0.00		150–100	0.33		50–25	0.00
	40–20	0.00		50–25	0.00		25–0	0.00
6	455–400	0.00	14	400–300	0.00	20	80–75	0.00
	300–250	0.00		250–200	0.00		60–40	0.00
	200–150	0.00		150–100	0.00		40–20	0.00
	100–50	0.00		90–50	0.00		20–0	0.00
7	500–400	0.00	15	300–200	0.00	21	40–20	0.00
	300–250	0.00		150–100	0.00		20–0	0.00
	200–150	0.00		80–60	0.00			
	100–50	0.00		40–20	0.00			
9	700–600	0.00	16	80–60	0.00	22	30–20	0.00
	400–300	0.00		60–40	0.00		20–10	0.00
	200–140	0.00		40–20	0.00		10–0	0.00
	100–50	0.00		20–0	0.00			
10	500–400	0.00	17	120–100	0.00	23	30–0	0.00
	300–250	0.00		80–60	0.00			
	200–150	0.00		40–30	0.00			
	100–50	0.00		20–10	0.00			
11	400–300	0.00	18	160–120	0.00	26	30–20	0.00
	250–200	0.00		100–80	0.00		20–0	0.00
	150–100	0.00		60–40	0.00			
	50–25	0.00		20–10	0.00			
12	130–100	0.00						
	100–50	0.00						
	50–25	0.00						
	25–0	0.00						

Ctenocalanus citer

Station number	Depth (m)	Ind·m^{-3}	Station number	Depth (m)	Ind·m^{-3}	Station number	Depth (m)	Ind·m^{-3}
5	160–140	3.18	13	400–300	13.29	19	100–75	89.69
	115–100	0.00		250–200	33.98		75–50	15.83
	80–60	1.00		150–100	78.63		50–25	37.33
	40–20	1.40		50–25	209.21		25–0	182.43
6	455–400	0.00	14	400–300	0.00	20	80–75	19.43
	300–250	0.00		250–200	2.18		60–40	3.17
	200–150	0.00		150–100	97.67		40–20	13.30
	100–50	2.05		90–50	0.39		20–0	0.00
7	500–400	0.00	15	300–200	1.23	21	40–20	24.36
	300–250	0.25		150–100	14.06		20–0	14.36
	200–150	13.77		80–60	17.30			
	100–50	2.18		40–20	56.04			
9	700–600	0.14	16	80–60	154.41	22	30–20	2.09
	400–300	0.36		60–40	0.00		20–10	1.87
	200–140	12.35		40–20	12.61		10–0	0.07
	100–50	29.77		20–0	0.00			
10	500–400	0.01	17	120–100	47.67	23	30–0	0.00
	300–250	0.57		80–60	89.90			
	200–150	2.31		40–30	6.75			
	100–50	6.11		20–10	6.62			
11	400–300	0.00	18	160–120	205.66	26	30–20	0.00
	250–200	1.29		100–80	18.68		20–0	0.19
	150–100	15.21		60–40	2.07			
	50–25	39.47		20–10	3.56			
12	130–100	15.14						
	100–50	11.36						
	50–25	4.79						
	25–0	14.65						

Drepanopus forcipatus

Station number	Depth (m)	Ind·m^{-3}	Station number	Depth (m)	Ind·m^{-3}	Station number	Depth (m)	Ind·m^{-3}
5	160–140	6.37	13	400–300	0.00	19	100–75	15.59
	115–100	6.21		250–200	0.00		75–50	63.32
	80–60	8.48		150–100	0.98		50–25	58.65
	40–20	5.59		50–25	379.95		25–0	851.36
6	455–400	0.00	14	400–300	0.00	20	80–75	613.25
	300–250	0.50		250–200	0.00		60–40	136.27
	200–150	11.03		150–100	114.36		40–20	56.53
	100–50	6.58		90–50	37.56		20–0	0.00
7	500–400	0.00	15	300–200	0.00	21	40–20	273.08
	300–250	0.00		150–100	0.25		20–0	169.32
	200–150	1.06		80–60	0.14			
	100–50	0.79		40–20	45.26			
9	700–600	0.00	16	80–60	845.58	22	30–20	26.36
	400–300	0.18		60–40	3.03		20–10	12.59
	200–140	0.05		40–20	55.50		10–0	1.63
	100–50	3.78		20–0	0.62			
10	500–400	0.01	17	120–100	476.70	23	30–0	3.65
	300–250	3.13		80–60	957.41			
	200–150	7.71		40–30	13.50			
	100–50	1528.54		20–10	18.65			
11	400–300	0.04	18	160–120	48.97	26	30–20	4406.30
	250–200	1.47		100–80	3.63		20–0	81.55
	150–100	3.57		60–40	4.25			
	50–25	1590.72		20–10	0.93			
12	130–100	20.55						
	100–50	2.72						
	50–25	80.19						
	25–0	896.49						

Microcalanus pygmaeus

Station number	Depth (m)	Ind·m^{-3}	Station number	Depth (m)	Ind·m^{-3}	Station number	Depth (m)	Ind·m^{-3}
5	160–140	0.00	13	400–300	2.93	19	100–75	0.00
	115–100	0.00		250–200	8.06		75–50	0.00
	80–60	0.00		150–100	3.28		50–25	0.00
	40–20	0.00		50–25	1.17		25–0	0.00
6	455–400	0.15	14	400–300	0.00	20	80–75	0.00
	300–250	0.50		250–200	0.00		60–40	0.00
	200–150	0.00		150–100	0.00		40–20	0.00
	100–50	0.00		90–50	0.00		20–0	0.00
7	500–400	0.53	15	300–200	0.00	21	40–20	0.00
	300–250	0.74		150–100	0.00		20–0	0.00
	200–150	0.35		80–60	0.00			
	100–50	0.00		40–20	0.00			
9	700–600	3.42	16	80–60	0.00	22	30–20	0.00
	400–300	0.36		60–40	0.00		20–10	0.00
	200–140	1.90		40–20	0.00		10–0	0.00
	100–50	0.76		20–0	0.00			
10	500–400	2.35	17	120–100	0.00	23	30–0	0.00
	300–250	23.36		80–60	0.00			
	200–150	9.25		40–30	0.00			
	100–50	36.69		20–10	0.00			
11	400–300	0.63	18	160–120	0.61	26	30–20	0.00
	250–200	4.74		100–80	0.78		20–0	0.00
	150–100	6.42		60–40	0.00			
	50–25	1.72		20–10	0.00			
12	130–100	0.00						
	100–50	0.00						
	50–25	0.00						
	25–0	0.00						

Paraeuchaeta antarctica

Station number	Depth (m)	Ind·m^{-3}	Station number	Depth (m)	Ind·m^{-3}	Station number	Depth (m)	Ind·m^{-3}
5	160–140	2.13	13	400–300	2.03	19	100–75	0.21
	115–100	0.00		250–200	2.30		75–50	1.22
	80–60	0.00		150–100	0.34		50–25	0.00
	40–20	0.00		50–25	0.00		25–0	0.00
6	455–400	0.08	14	400–300	0.00	20	80–75	0.21
	300–250	2.00		250–200	13.11		60–40	0.07
	200–150	0.00		150–100	0.00		40–20	0.00
	100–50	0.00		90–50	0.08		20–0	0.10
7	500–400	1.06	15	300–200	7.39	21	40–20	0.08
	300–250	2.00		150–100	1.56		20–0	0.00
	200–150	0.04		80–60	0.21			
	100–50	0.02		40–20	4.31			
9	700–600	0.16	16	80–60	0.08	22	30–20	0.00
	400–300	0.72		60–40	0.00		20–10	0.00
	200–140	0.14		40–20	0.00		10–0	0.07
	100–50	0.05		20–0	0.12			
10	500–400	0.03	17	120–100	5.72	23	30–0	0.00
	300–250	0.23		80–60	0.00			
	200–150	0.74		40–30	0.06			
	100–50	0.00		20–10	0.60			
11	400–300	0.30	18	160–120	17.63	26	30–20	0.00
	250–200	0.23		100–80	0.42		20–0	0.00
	150–100	0.48		60–40	0.98			
	50–25	2.75		20–10	0.93			
12	130–100	0.11						
	100–50	0.07						
	50–25	0.00						
	25–0	0.00						

Paraeuchaeta biloba

Station number	Depth (m)	Ind·m^{-3}	Station number	Depth (m)	Ind·m^{-3}	Station number	Depth (m)	Ind·m^{-3}
5	160–140	0.00	13	400–300	0.00	19	100–75	0.00
	115–100	0.00		250–200	0.00		75–50	0.00
	80–60	0.00		150–100	0.00		50–25	0.00
	40–20	0.00		50–25	0.00		25–0	0.00
6	455–400	0.00	14	400–300	0.00	20	80–75	0.00
	300–250	0.00		250–200	0.00		60–40	0.00
	200–150	0.00		150–100	0.00		40–20	0.00
	100–50	0.00		90–50	0.00		20–0	0.00
7	500–400	0.00	15	300–200	0.00	21	40–20	0.00
	300–250	0.02		150–100	0.00		20–0	0.00
	200–150	0.00		80–60	0.00			
	100–50	0.00		40–20	0.00			
9	700–600	0.00	16	80–60	0.00	22	30–20	0.00
	400–300	0.00		60–40	0.00		20–10	0.00
	200–140	0.00		40–20	0.00		10–0	0.00
	100–50	0.00		20–0	0.00			
10	500–400	0.00	17	120–100	0.00	23	30–0	0.00
	300–250	0.00		80–60	0.00			
	200–150	0.00		40–30	0.00			
	100–50	0.00		20–10	0.00			
11	400–300	0.04	18	160–120	0.00	26	30–20	0.00
	250–200	0.01		100–80	0.00		20–0	0.00
	150–100	0.00		60–40	0.00			
	50–25	0.00		20–10	0.00			
12	130–100	0.00						
	100–50	0.00						
	50–25	0.00						
	25–0	0.00						

Phaenna spinifera

Station number	Depth (m)	Ind·m^{-3}	Station number	Depth (m)	Ind·m^{-3}	Station number	Depth (m)	Ind·m^{-3}
5	160–140	0.00	13	400–300	0.00	19	100–75	0.00
	115–100	0.00		250–200	0.00		75–50	0.00
	80–60	0.00		150–100	0.00		50–25	0.00
	40–20	0.00		50–25	0.00		25–0	0.00
6	455–400	0.00	14	400–300	0.00	20	80–75	0.00
	300–250	0.00		250–200	0.00		60–40	0.00
	200–150	0.00		150–100	0.00		40–20	0.00
	100–50	0.00		90–50	0.00		20–0	0.00
7	500–400	0.00	15	300–200	0.00	21	40–20	0.00
	300–250	0.00		150–100	0.00		20–0	0.00
	200–150	0.00		80–60	0.00			
	100–50	0.00		40–20	0.00			
9	700–600	0.00	16	80–60	0.00	22	30–20	0.00
	400–300	0.00		60–40	0.00		20–10	0.00
	200–140	0.00		40–20	0.00		10–0	0.00
	100–50	0.00		20–0	0.00			
10	500–400	0.00	17	120–100	0.00	23	30–0	0.00
	300–250	0.00		80–60	0.00			
	200–150	0.01		40–30	0.00			
	100–50	0.00		20–10	0.00			
11	400–300	0.00	18	160–120	0.00	26	30–20	0.00
	250–200	0.01		100–80	0.00		20–0	0.00
	150–100	0.02		60–40	0.00			
	50–25	0.00		20–10	0.00			
12	130–100	0.00						
	100–50	0.00						
	50–25	0.00						
	25–0	0.00						

Scaphocalanus curtus

Station number	Depth (m)	Ind·m^{-3}	Station number	Depth (m)	Ind·m^{-3}	Station number	Depth (m)	Ind·m^{-3}
5	160–140	0.00	13	400–300	0.00	19	100–75	0.00
	115–100	0.00		250–200	0.00		75–50	0.00
	80–60	0.00		150–100	0.00		50–25	0.00
	40–20	0.00		50–25	0.00		25–0	0.00
6	455–400	0.00	14	400–300	0.00	20	80–75	0.00
	300–250	0.00		250–200	0.00		60–40	0.00
	200–150	0.00		150–100	0.00		40–20	0.00
	100–50	0.00		90–50	0.00		20–0	0.00
7	500–400	0.00	15	300–200	0.00	21	40–20	0.00
	300–250	0.00		150–100	0.00		20–0	0.00
	200–150	0.00		80–60	0.00			
	100–50	0.00		40–20	0.00			
9	700–600	0.00	16	80–60	0.00	22	30–20	0.00
	400–300	0.00		60–40	0.00		20–10	0.00
	200–140	0.00		40–20	0.00		10–0	0.00
	100–50	0.00		20–0	0.00			
10	500–400	0.01	17	120–100	0.00	23	30–0	0.00
	300–250	0.05		80–60	0.00			
	200–150	0.00		40–30	0.00			
	100–50	0.00		20–10	0.00			
11	400–300	0.04	18	160–120	0.00	26	30–20	0.00
	250–200	0.09		100–80	0.00		20–0	0.00
	150–100	0.24		60–40	0.00			
	50–25	0.00		20–10	0.00			
12	130–100	0.00						
	100–50	0.00						
	50–25	0.00						
	25–0	0.00						

Scaphocalanus farrani

Station number	Depth (m)	Ind·m^{-3}	Station number	Depth (m)	Ind·m^{-3}	Station number	Depth (m)	Ind·m^{-3}
5	160–140	0.00	13	400–300	0.00	19	100–75	0.00
	115–100	0.00		250–200	0.00		75–50	0.00
	80–60	0.00		150–100	0.00		50–25	0.00
	40–20	0.00		50–25	0.00		25–0	0.00
6	455–400	0.00	14	400–300	0.00	20	80–75	0.00
	300–250	0.00		250–200	0.00		60–40	0.00
	200–150	0.00		150–100	0.00		40–20	0.00
	100–50	0.00		90–50	0.39		20–0	0.00
7	500–400	0.42	15	300–200	0.00	21	40–20	0.00
	300–250	0.00		150–100	0.00		20–0	0.00
	200–150	0.00		80–60	0.00			
	100–50	0.00		40–20	0.00			
9	700–600	0.00	16	80–60	0.00	22	30–20	0.00
	400–300	0.00		60–40	0.00		20–10	0.00
	200–140	0.00		40–20	0.00		10–0	0.00
	100–50	0.00		20–0	0.00			
10	500–400	0.02	17	120–100	0.00	23	30–0	0.00
	300–250	0.57		80–60	0.00			
	200–150	0.00		40–30	0.00			
	100–50	0.00		20–10	0.00			
11	400–300	0.13	18	160–120	0.00	26	30–20	0.00
	250–200	0.00		100–80	0.00		20–0	0.00
	150–100	0.00		60–40	0.00			
	50–25	0.00		20–10	0.00			
12	130–100	0.00						
	100–50	0.00						
	50–25	0.00						
	25–0	0.00						

Scolecithricella dentata

Station number	Depth (m)	Ind·m^{-3}	Station number	Depth (m)	Ind·m^{-3}	Station number	Depth (m)	Ind·m^{-3}
5	160–140	0.00	13	400–300	0.00	19	100–75	0.00
	115–100	0.00		250–200	0.05		75–50	0.00
	80–60	0.00		150–100	0.00		50–25	0.00
	40–20	0.00		50–25	0.00		25–0	0.00
6	455–400	0.00	14	400–300	0.00	20	80–75	0.00
	300–250	0.00		250–200	0.00		60–40	0.00
	200–150	0.00		150–100	0.00		40–20	0.00
	100–50	0.00		90–50	0.00		20–0	0.00
7	500–400	0.00	15	300–200	0.00	21	40–20	0.00
	300–250	0.00		150–100	0.00		20–0	0.00
	200–150	0.00		80–60	0.00			
	100–50	0.00		40–20	0.00			
9	700–600	0.00	16	80–60	0.00	22	30–20	0.00
	400–300	0.02		60–40	0.00		20–10	0.00
	200–140	0.00		40–20	0.00		10–0	0.00
	100–50	0.00		20–0	0.00			
10	500–400	0.00	17	120–100	0.00	23	30–0	0.00
	300–250	0.28		80–60	0.00			
	200–150	0.00		40–30	0.00			
	100–50	0.00		20–10	0.00			
11	400–300	0.07	18	160–120	0.00	26	30–20	0.00
	250–200	0.19		100–80	0.00		20–0	0.00
	150–100	0.00		60–40	0.00			
	50–25	0.00		20–10	0.00			
12	130–100	0.00						
	100–50	0.00						
	50–25	0.00						
	25–0	0.00						

Scolecithricella minor

Station number	Depth (m)	Ind·m^{-3}	Station number	Depth (m)	Ind·m^{-3}	Station number	Depth (m)	Ind·m^{-3}
5	160–140	0.00	13	400–300	3.60	19	100–75	0.00
	115–100	0.07		250–200	2.30		75–50	0.00
	80–60	0.10		150–100	1.64		50–25	0.00
	40–20	0.00		50–25	0.12		25–0	7.61
6	455–400	0.00	14	400–300	0.72	20	80–75	0.00
	300–250	0.00		250–200	0.27		60–40	0.00
	200–150	6.13		150–100	0.00		40–20	0.00
	100–50	0.69		90–50	0.00		20–0	0.00
7	500–400	0.18	15	300–200	0.00	21	40–20	0.00
	300–250	0.99		150–100	0.00		20–0	0.00
	200–150	1.41		80–60	0.09			
	100–50	0.89		40–20	0.00			
9	700–600	0.00	16	80–60	0.00	22	30–20	0.00
	400–300	0.00		60–40	0.00		20–10	0.00
	200–140	0.95		10–20	0.00		10–0	0.00
	100–50	1.51		20–0	0.00			
10	500–400	0.00	17	120–100	0.00	23	30–0	0.00
	300–250	0.09		80–60	0.00			
	200–150	1.54		40–30	0.00			
	100–50	0.00		20–10	0.00			
11	400–300	0.04	18	160–120	0.00	26	30–20	0.00
	250–200	0.35		100–80	0.00		20–0	0.00
	150–100	0.71		60–40	0.00			
	50–25	0.34		20–10	0.00			
12	130–100	2.16						
	100–50	2.72						
	50–25	2.40						
	25–0	0.00						

Spinocalanus brevicaudatus

Station number	Depth (m)	Ind·m^{-3}	Station number	Depth (m)	Ind·m^{-3}	Station number	Depth (m)	Ind·m^{-3}
5	160–140	0.00	13	400–300	0.00	19	100–75	0.00
	115–100	0.00		250–200	0.00		75–50	0.00
	80–60	0.00		150–100	0.00		50–25	0.00
	40–20	0.00		50–25	0.00		25–0	0.00
6	455–400	0.03	14	400–300	0.00	20	80–75	0.00
	300–250	1.00		250–200	0.00		60–40	0.00
	200–150	0.00		150–100	0.00		40–20	0.00
	100–50	0.00		90–50	0.00		20–0	0.00
7	500–400	4.23	15	300–200	0.00	21	40–20	0.00
	300–250	0.99		150–100	0.00		20–0	0.00
	200–150	0.35		80–60	0.00			
	100–50	0.04		40–20	0.00			
9	700–600	0.28	16	80–60	0.00	22	30–20	0.00
	400–300	0.54		60–40	0.00		20–10	0.00
	200–140	0.05		40–20	0.00		10–0	0.00
	100–50	0.03		20–0	0.00			
10	500–400	3.24	17	120–100	0.00	23	30–0	0.00
	300–250	1.71		80–60	0.00			
	200–150	0.83		40–30	0.00			
	100–50	0.00		20–10	0.00			
11	400–300	0.91	18	160–120	0.00	26	30–20	0.00
	250–200	0.09		100–80	0.00		20–0	0.00
	150–100	0.00		60–40	0.00			
	50–25	0.00		20–10	0.00			
12	130–100	0.00						
	100–50	0.00						
	50–25	0.00						
	25–0	0.00						

Oithona atlantica

Station number	Depth (m)	Ind·m^{-3}	Station number	Depth (m)	Ind·m^{-3}	Station number	Depth (m)	Ind·m^{-3}
5	160–140	4.24	13	400–300	0.23	19	100–75	0.06
	115–100	9.66		250–200	0.58		75–50	0.10
	80–60	11.97		150–100	0.73		50–25	0.00
	40–20	0.70		50–25	0.00		25–0	0.00
6	455–400	0.20	14	400–300	0.00	20	80–75	0.00
	300–250	11.00		250–200	0.00		60–40	0.00
	200–150	14.70		150–100	0.24		40–20	0.00
	100–50	19.07		90–50	1.18		20–0	0.00
7	500–400	1.41	15	300–200	0.00	21	40–20	0.00
	300–250	10.40		150–100	3.13		20–0	0.00
	200–150	19.07		80–60	0.09			
	100–50	22.62		40–20	0.00			
9	700–600	0.14	16	80–60	0.08	22	30–20	0.00
	400–300	0.18		60–40	0.00		20–10	0.00
	200–140	30.41		40–20	0.00		10–0	0.00
	100–50	54.24		20–0	0.00			
10	500–400	0.59	17	120–100	0.00	23	30–0	0.00
	300–250	9.97		80–60	0.00			
	200–150	8.47		40–30	0.06			
	100–50	24.46		20–10	0.00			
11	400–300	0.11	18	160–120	0.00	26	30–20	0.00
	250–200	2.33		100–80	0.13		20–0	0.00
	150–100	5.71		60–40	0.17			
	50–25	5.21		20–10	0.08			
12	130–100	3.24						
	100–50	4.10						
	50–25	0.00						
	25–0	0.00						

Oithona similis

Station number	Depth (m)	Ind·m^{-3}	Station number	Depth (m)	Ind·m^{-3}	Station number	Depth (m)	Ind·m^{-3}
5	160–140	0.00	13	400–300	1.13	19	100–75	29.25
	115–100	0.00		250–200	16.70		75–50	2.43
	80–60	0.00		150–100	2.88		50–25	58.65
	40–20	0.70		50–25	169.00		25–0	91.22
6	455–400	0.00	14	400–300	1.44	20	80–75	0.00
	300–250	0.50		250–200	6.56		60–40	12.68
	200–150	0.00		150–100	11.91		40–20	129.68
	100–50	0.00		90–50	4.75		20–0	0.00
7	500–400	0.00	15	300–200	0.00	21	40–20	15.52
	300–250	0.50		150–100	4.69		20–0	0.00
	200–150	0.35		80–60	2.16			
	100–50	0.19		40–20	103.44			
9	700–600	0.01	16	80–60	29.41	22	30–20	0.22
	400–300	0.00		60–40	0.00		20–10	2.80
	200–140	0.00		40–20	90.81		10–0	0.07
	100–50	1.01		20–0	0.62			
10	500–400	0.01	17	120–100	1.90	23	30–0	1.86
	300–250	0.28		80–60	58.43			
	200–150	1.54		40–30	50.63			
	100–50	12.23		20–10	15.04			
11	400–300	0.04	18	160–120	3.67	26	30–20	13.05
	250–200	0.69		100–80	24.91		20–0	10.55
	150–100	0.24		60–40	4.93			
	50–25	144.09		20–10	2.17			
12	130–100	0.00						
	100–50	0.00						
	50–25	2.40						
	25–0	5.86						

Lubbockia aculeata

Station number	Depth (m)	Ind·m^{-3}	Station number	Depth (m)	Ind·m^{-3}	Station number	Depth (m)	Ind·m^{-3}
5	160–140	0.00	13	400–300	0.00	19	100–75	0.00
	115–100	0.00		250–200	0.00		75–50	0.00
	80–60	0.00		150–100	0.00		50–25	0.00
	40–20	0.00		50–25	0.00		25–0	0.00
6	455–400	0.00	14	400–300	0.00	20	80–75	0.00
	300–250	0.00		250–200	0.00		60–40	0.00
	200–150	0.00		150–100	0.00		40–20	0.00
	100–50	0.00		90–50	0.00		20–0	0.00
7	500–400	0.01	15	300–200	0.00	21	40–20	0.00
	300–250	0.00		150–100	0.00		20–0	0.00
	200–150	0.00		80–60	0.00			
	100–50	0.00		40–20	0.00			
9	700–600	0.00	16	80–60	0.00	22	30–20	0.00
	400–300	0.00		60–40	0.00		20–10	0.00
	200–140	0.00		40–20	0.00		10–0	0.00
	100–50	0.00		20–0	0.00			
10	500–400	0.00	17	120–100	0.00	23	30–0	0.00
	300–250	0.00		80–60	0.00			
	200–150	0.00		40–30	0.00			
	100–50	0.00		20–10	0.00			
11	400–300	0.00	18	160–120	0.00	26	30–20	0.00
	250–200	0.09		100–80	0.00		30–20	0.00
	150–100	0.01		60–40	0.00			
	50–25	0.00		20–10	0.00			
12	130–100	0.00						
	100–50	0.00						
	50–25	0.00						
	25–0	0.00						

Lubbockia minuta

Station number	Depth (m)	Ind·m^{-3}	Station number	Depth (m)	Ind·m^{-3}	Station number	Depth (m)	Ind·m^{-3}
5	160–140	0.00	13	400–300	0.00	19	100–75	0.00
	115–100	0.00		250–200	0.00		75–50	0.00
	80–60	0.00		150–100	0.00		50–25	0.00
	40–20	0.00		50–25	0.00		25–0	0.00
6	455–400	0.00	14	400–300	0.00	20	80–75	0.00
	300–250	0.00		250–200	0.00		60–40	0.00
	200–150	0.00		150–100	0.00		40–20	0.00
	100–50	0.00		90–50	0.00		20–0	0.00
7	500–400	0.00	15	300–200	0.00	21	40–20	0.00
	300–250	0.00		150–100	0.00		20–0	0.00
	200–150	0.00		80–60	0.00			
	100–50	0.00		40–20	0.00			
9	700–600	0.00	16	80–60	0.00	22	30–20	0.00
	400–300	0.00		60–40	0.00		20–10	0.00
	200–140	0.00		40–20	0.00		10–0	0.00
	100–50	0.00		20–0	0.00			
10	500–400	0.02	17	120–100	0.02	23	30–0	0.00
	300–250	0.00		80–60	0.00			
	200–150	0.00		40–30	0.00			
	100–50	0.00		20–10	0.00			
11	400–300	0.00	18	160–120	0.00	26	30–20	0.00
	250–200	0.00		100–80	0.00		20–0	0.00
	150–100	0.00		60–40	0.00			
	50–25	0.00		20–10	0.00			
12	130–100	0.00						
	100–50	0.00						
	50–25	0.00						
	25–0	0.00						

Oncaea conifera

Station number	Depth (m)	Ind·m^{-3}	Station number	Depth (m)	Ind·m^{-3}	Station number	Depth (m)	Ind·m^{-3}
5	160–140	1.06	13	400–300	0.23	19	100–75	0.00
	115–100	0.00		250–200	0.02		75–50	0.00
	80–60	0.00		150–100	0.34		50–25	0.00
	40–20	0.00		50–25	0.00		25–0	0.00
6	455–400	2.14	14	400–300	0.00	20	80–75	6.48
	300–250	49.01		250–200	0.14		60–40	0.00
	200–150	171.57		150–100	0.00		40–20	0.00
	100–50	0.91		90–50	0.00		20–0	0.00
7	500–400	39.64	15	300–200	0.00	21	40–20	0.00
	300–250	71.79		150–100	0.00		20–0	0.00
	200–150	173.05		80–60	0.00			
	100–50	6.19		40–20	0.00			
9	700–600	3.03	16	80–60	0.00	22	30–20	0.00
	400–300	9.03		60–40	0.00		20–10	0.00
	200–140	107.37		40–20	0.00		10–0	0.00
	100–50	6.56		20–0	0.00			
10	500–400	0.59	17	120–100	0.00	23	30–0	0.00
	300–250	11.10		80–60	0.00			
	200–150	23.12		40–30	0.00			
	100–50	12.23		20–10	0.00			
11	400–300	3.56	18	160–120	0.00	26	30–20	0.00
	250–200	6.73		100–80	0.00		20–0	0.00
	150–100	21.16		60–40	0.00			
	50–25	6.86		20–10	0.00			
12	130–100	2.16						
	100–50	0.45						
	50–25	0.00						
	25–0	0.00						

Oncaea curvata

Station number	Depth (m)	Ind·m^{-3}	Station number	Depth (m)	Ind·m^{-3}	Station number	Depth (m)	Ind·m^{-3}
5	160–140	0.00	13	400–300	0.00	19	100–75	0.00
	115–100	0.00		250–200	0.58		75–50	0.00
	80–60	0.00		150–100	0.00		50–25	0.00
	40–20	0.00		50–25	0.00		25–0	0.00
6	455–400	0.00	14	400–300	0.00	20	80–75	0.00
	300–250	0.00		250–200	0.00		60–40	0.00
	200–150	0.00		150–100	0.00		40–20	0.00
	100–50	0.00		90–50	0.00		20–0	0.00
7	500–400	0.00	15	300–200	0.00	21	40–20	0.00
	300–250	0.00		150–100	0.00		20–0	0.00
	200–150	0.00		80–60	0.00			
	100–50	0.00		40–20	0.00			
9	700–600	0.00	16	80–60	0.00	22	30–20	0.00
	400–300	0.00		60–40	0.00		20–10	0.00
	200–140	0.00		40–20	0.00		10–0	0.00
	100–50	0.00		20–0	0.00			
10	500–400	0.00	17	120–100	0.00	23	30–0	0.00
	300–250	0.00		80–60	0.00			
	200–150	0.00		40–30	0.00			
	100–50	0.00		20–10	0.00			
11	400–300	0.04	18	160–120	0.24	26	30–20	0.00
	250–200	0.00		100–80	0.00		20–0	0.00
	150–100	0.00		60–40	0.00			
	50–25	0.00		20–10	0.00			
12	130–100	0.00						
	100–50	0.00						
	50–25	0.00						
	25–0	0.00						

Oncaea englishi

Station number	Depth (m)	$Ind \cdot m^{-3}$	Station number	Depth (m)	$Ind \cdot m^{-3}$	Station number	Depth (m)	$Ind \cdot m^{-3}$
5	160–140	0.00	13	400–300	0.23	19	100–75	0.00
	115–100	0.00		250–200	0.00		75–50	0.00
	80–60	0.02		150–100	0.00		50–25	0.00
	40–20	0.03		50–25	0.00		25–0	0.00
6	455–400	2.61	14	400–300	0.00	20	80–75	0.00
	300–250	0.00		250–200	0.00		60–40	0.00
	200–150	0.00		150–100	0.00		40–20	0.00
	100–50	0.00		90–50	0.00		20–0	0.00
7	500–400	60.25	15	300–200	0.00	21	40–20	0.00
	300–250	0.50		150–100	0.00		20–0	0.00
	200–150	0.35		80–60	0.00			
	100–50	0.02		40–20	0.00			
9	700–600	5.90	16	80–60	0.00	22	30–20	0.00
	400–300	20.78		60–40	0.00		20–10	0.00
	200–140	0.95		40–20	0.00		10–0	0.00
	100–50	0.50		20–0	0.00			
10	500–400	8.83	17	120–100	0.00	23	30–0	0.00
	300–250	0.00		80–60	0.00			
	200–150	0.00		40–30	0.00			
	100–50	0.00		20–10	0.00			
11	400–300	3.83	18	160–120	0.00	26	30–20	0.00
	250–200	0.43		100–80	0.00		20–0	0.00
	150–100	1.66		60–40	0.00			
	50–25	0.00		20–10	0.00			
12	130–100	0.00						
	100–50	0.00						
	50–25	0.00						
	25–0	0.00						

MIX
Papier aus verantwortungsvollen Quellen
Paper from responsible sources
FSC® C105338

If you have any concerns about our products,
you can contact us on
ProductSafety@springernature.com

In case Publisher is established outside the EU,
the EU authorized representative is:
Springer Nature Customer Service Center GmbH
Europaplatz 3, 69115 Heidelberg, Germany

Printed by Libri Plureos GmbH
in Hamburg, Germany